Practical Cryostat Design and Construction

Practical Cryostat Design and Construction: Technologies and Customization introduces the innovative work being done across experiments and applications at cryogenic temperatures. Because custom engineering is heavily experience-based and has limited resources, this book will share the authors' vast experience in this field. The book covers both cryostat design and fabrication techniques, making it of interest to scientists and engineers across the cryogenic community as they work toward designing and developing their cryogenic platforms.

Key Features:

- Emphasizes the design and construction of customized cryogenic systems, bridging fundamental cryogenic principles with practical engineering implementation.
- Covers a broad spectrum of cryogenic research and engineering topics including: helium-3 systems, dilution refrigerators, adiabatic demagnetization refrigerators, superconducting magnet integration, vacuum technology, thermal management, vibration reduction, and cryogenic wiring and cabling.

Designed for readers across academia and industry, particularly early-career students, researchers, and engineers involved in cryogenic technology and low-temperature experimental systems.

Zuyu Zhao received his B.S. degree from Fudan University in 1982, Class 77. He came to the United States in 1983 with the World Bank Scholarship program and graduated from Northwestern University with a Ph.D. degree in physics in 1990. He then spent two and a half years working at Harvard University as a Research Associate and set up a new ultra-low temperature lab pursuing Bose–Einstein condensation on spin polarized hydrogen. Dr. Zhao joined Janis Research Company (later American Cryogenics dba JanisULT) in 1993 and focused on developing custom ultra-low temperature facilities for the research and science community. He served as the Principal Scientist and Executive Vice President at both Janis Research and JanisULT as well as the trustee of the company. Dr. Zhao also served from July 1, 2007 through June 30, 2013 as an elected member of the American Institute of Physics (Physics Today) Advisory Committee. Before he retired in late 2023, he had served as the Chief Cryogenic Scientist for 15 months after the business merged to FormFactor LLC and helped with smooth business and technical transfer.

Zhitao Wang received his B.S. degree from Hebei University in 2001. He then pursued his Ph.D. at the Institute of Physics, Chinese Academy of Sciences, during which he spent two years at Tohoku University as an exchange student. After receiving his Ph.D. in 2006, Dr. Wang worked at Trinity College Dublin, Pacific Northwest National Laboratory, and Tufts University (joint with Harvard University). His research focused on surface science at cryogenic temperatures, particularly low-temperature scanning probe microscopy. In 2017, Dr. Wang joined Janis Research Company (later JanisULT) as a cryogenic engineer, specializing in custom low-temperature systems. After FormFactor acquired JanisULT in 2022, he became a cryogenic scientist, working on the development of customized helium-3 systems and dilution refrigerators. From 2024 to 2026, he worked at Commonwealth Fusion Systems, focusing on supercritical helium cooling systems for high-field high-temperature superconducting magnets. In 2026, Dr. Wang joined IonQ as a cryogenic engineer.

Series in Materials Science and Engineering

The series publishes cutting edge monographs and foundational textbooks for interdisciplinary materials science and engineering. It is aimed at undergraduate and graduate level students, as well as practicing scientists and engineers. Its purpose is to address the connections between properties, structure, synthesis, processing, characterization, and performance of materials.

Handbook of Silicon Carbide Materials and Devices
Zhe Chuan Feng, Ed.

Bioelectronics: Materials, Technologies, and Emerging Applications
Ram K. Gupta and Anuj Kumar, Eds.

Advances in 3D Bioprinting
Roger J. Narayan, Ed.

Hydrogels: Fundamentals to Advanced Energy Applications
Ram K. Gupta and Anuj Kumar, Eds.

Emerging Energy Materials: Applications and Challenges
Govind Nair, H. Nagabhushana, Nirupama Dhoble and Sanjay J. Dhoble, Eds.

Handbook of 2D Materials for a Sustainable Future Vol 1
Mangalaraja Ramalinga Viswanathan, Arulraj Arunachalam, Pandiyarajan Thangaraj, and Ernesto Chicardi Augusto, Eds

Handbook of 2D Materials for a Sustainable Future Vol 2
Mangalaraja Ramalinga Viswanathan, Arulraj Arunachalam, Arun Thirumurugan and P. Sakthivel, Eds

Unravelling the Power of 2D Materials: Revolutionizing Flexible Devices
Ram K. Gupta, Ed.

Flow Batteries: Fundamentals, Technological Advancement and Challenges
Ram K. Gupta, Ed.

Electromagnetic Interference Shielding Technologies: Theory and Applications
Reenu Elizabeth John, Steffy Sara Varghese, Anoop Chandran and N.V. Unnikrishnan

Practical Cryostat Design and Construction: Technologies and Customization
Zuyu Zhao and Zhitao Wang

Series Preface

The series publishes cutting edge monographs and foundational textbooks for interdisciplinary materials science and engineering.

Its purpose is to address the connections between properties, structure, synthesis, processing, characterization, and performance of materials. The subject matter of individual volumes spans fundamental theory, computational modeling, and experimental methods used for design, modeling, and practical applications. The series encompasses thin films, surfaces, and interfaces, and the full spectrum of material types, including biomaterials, energy materials, metals, semiconductors, optoelectronic materials, ceramics, magnetic materials, superconductors, nanomaterials, composites, and polymers.

It is aimed at undergraduate and graduate level students, as well as practicing scientists and engineers.
Proposals for new volumes in the series may be directed to Carolina Antunes, Commissioning Editor at CRC Press, Taylor & Francis Group (Carolina.Antunes@tandf.co.uk).

Practical Cryostat Design and Construction

Technologies and Customization

Zuyu Zhao and Zhitao Wang

CRC Press
Taylor & Francis Group
Boca Raton London New York

CRC Press is an imprint of the
Taylor & Francis Group, an **informa** business

Designed cover image: Shutterstock

First edition published 2027
by CRC Press
4 Park Square, Milton Park, Abingdon, Oxon, OX14 4RN

and by CRC Press
2385 NW Executive Center Drive, Suite 320, Boca Raton FL 33431

British Library Cataloguing-in-Publication Data
A catalogue record for this book is available from the British Library

ISBN: 9781041034537 (hbk)
ISBN: 9781041359531 (pbk)
ISBN: 9781042000050 (ebk)

DOI: 10.1201/9781042000050

Typeset in Times
by Newgen Publishing UK

Contents

Preface

Cryogenics is a challenging and comprehensive field that integrates physics, chemistry, materials science, mechanical engineering, and manufacturing technologies. Cryogenic techniques have continuously evolved for centuries and remain a driving force behind innovation in many disciplines, which not only deepen our understanding of fundamental science but also enable the development of cutting-edge technologies. In recent years, the rapid growth of quantum information technologies has significantly increased the demand for practical knowledge of cryogenic techniques. While the majority of cryogenic-related publications focus on scientific principles, theoretical analysis, and engineering fundamentals, most readers—particularly those in universities and research institutes—primarily seek practical guidance for designing, constructing, and implementing their own experiments using standard cryogenic platforms.

We, the authors, have had the privilege of working on a wide range of customized and innovative cryogenic projects that have contributed to advancing the frontiers of scientific research. Many of these accomplished projects involve pioneering work in the development of novel "wet" and cryogen-free systems. They include, but are not limited to, sub-Kelvin angle-resolved photoemission spectroscopy cryostats (0.6 K), helium-3 cryostats (0.3 K), adiabatic demagnetization refrigerators (50 mK), and dilution refrigerators (below 10 mK).

This book is a compilation of selected work accomplished over the past four decades. Unlike most existing cryogenic engineering books that provide comprehensive coverage from scientific foundations to working principles, this book focuses on practical techniques and the design of customized cryostats for specific applications. It is intended serve as a practical complement to traditional cryogenic engineering texts. Detailed engineering methods and practical design techniques developed during cryostat construction are also presented. We would be greatly rewarded if the knowledge and experience shared in this book help readers who rely on cryogenic technologies in their work.

We would like to express our sincere gratitude to Prof. John B. Ketterson, Prof. William P. Halperin, Prof. Shireen Adenwala, Prof. Issac F. Silvera, Prof. Qi-Kun Xue, Dr. Vladimir Shvarts, Dr. Jeonghoon Ha, and all other colleagues at Janis Research Company and Formfactor for their scientific guidance, generous support, and dedicated collaboration, all of which have been instrumental to the success of these projects. Finally, we express our deepest gratitude to our families for their unwavering support and understanding throughout this long journey.

1 He-3 Cryostat—Basic Concept, Design, and Construction

1.1 INTRODUCTION

As an isotope of helium-4 (He-4), helium-3 (He-3) exhibits unique properties at temperatures below its normal boiling point. Due to its significantly higher saturated vapor pressure, liquid He-3 undergoes efficient evaporative cooling, resulting in low base temperatures. To cool itself from its condensed temperature of 1.5 K to 0.3 K, only approximately 20% of the liquid He-3 is consumed during its own heat of evaporation. In this temperature range, liquid He-3 has a relatively large specific heat, making it an effective cooling reservoir for experimental applications. Therefore, evaporative cooling of liquid He-3 is the most practical method for maintaining an experimental platform around 0.3 K while ensuring reasonably high cooling power. This chapter presents the design, construction, and performance of various He-3 cryostats.

The unique properties of liquid He-3 that enable its use in sub-kelvin refrigeration include the following:

- Liquid He-3 has low base temperatures (below 0.3 K) via evaporative cooling due to its significantly higher saturated vapor pressure.
- Liquid He-3 is typically condensed at much lower temperatures (around 1.5 K) compared to its normal boiling point (3.2 K) and therefore requires only about 20% evaporation to cool from 1.5 K to the particularly interesting regime of 0.3 K. In contrast, liquid He-4 is usually formed at 4.2 K and need to evaporate about 40% of its volume to cool down to the lambda point (2.186 K).
- In the temperature range of 0.3 K to 1.5 K, liquid He-3 has a relatively large specific heat, making it an effective cooling reservoir for experimental applications.
- Within the above interesting temperature range, liquid He-3 does not exhibit superfluidity, which makes it easier to thermally isolate from surrounding higher-temperature regions.

These unique properties make evaporative cooling of liquid He-3 the most practical method to establish experimental platforms at sub-kelvin while maintaining reasonably high cooling power.

1.2 SINGLE-SHOT, WET HE-3 CRYOSTAT

The single-shot, wet He-3 cryostats are widely used due to the multiple advantages:

- Cost-effective
- User-friendly operation

DOI: 10.1201/9781042000050-1

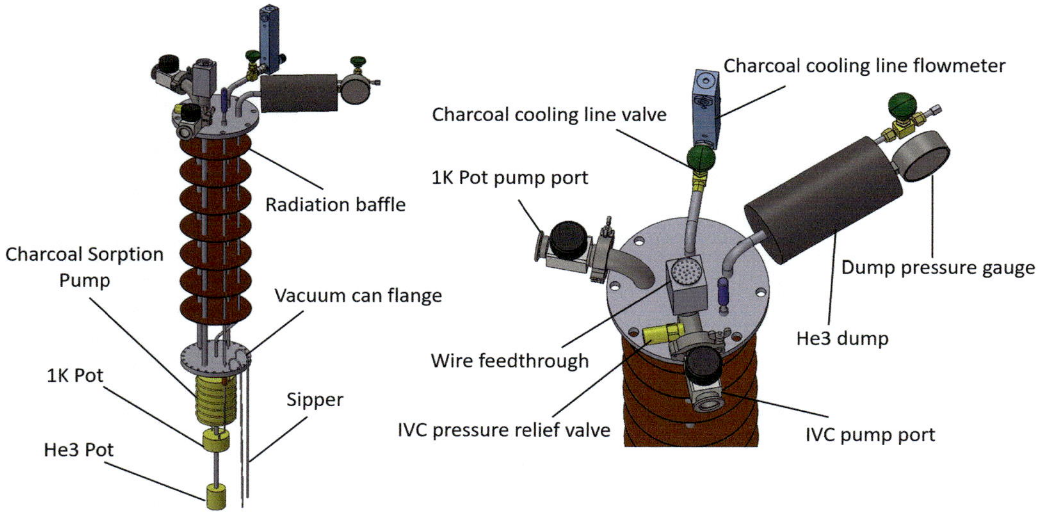

FIGURE 1.1 Single-shot-type wet He-3 insert. (Projects accomplished by authors at Janis Research Company. Model courtesy of FormFactor, Inc.)

- Excellent temperature stability, with fluctuations within a few millikelvin.
- Customizable configurations, including flexible sample space, customized wiring, RF cabling, optical windows, and integration with superconducting magnets, etc.

The core of a single-shot, wet He-3 cryostat includes a charcoal sorption pump, a 1K pot, and a He-3 pot, which are enclosed in an inner vacuum chamber (IVC), as illustrated in Figure 1.1. The charcoal sorption pump is connected to the He-3 pot through a small diameter tube that passes through the 1K pot, providing a He-3 passage. The charcoal sorption pump is typically a cylindrical copper container filled with activated charcoal. It plays a dual role: releasing He-3 gas for condensation and adsorbing He-3 for evaporative cooling. The sorption pump is cooled by flowing cold He-4 gas or liquid through a capillary soldered around it. The 1K pot, which contains liquid He-4, can be cooled to 1.5 K or below through evaporative cooling to condense He-3 gas in the passage tube into liquid. The He-3 pot holds liquid He-3. Liquid He-3 is condensed into He-3 pot when the charcoal sorption pump is heated up to 40–50 K while the 1K pot is maintained at 1.5 K. The charcoal begins to adsorb He-3 gas when its temperature is lower than 20 K, thereby reducing the vapor pressure in He-3 pot and resulting in the evaporative cooling. Its adsorption capacity reaches a maximum at temperatures below 5 K.

Outside IVC, two small diameter capillaries ("sippers") supply liquid helium from the liquid helium reservoir to the 1K pot and charcoal cooling line. On the room temperature top plate, valves and a flowmeter are employed to control the cryogen flows. A He-3 dump, equipped with a pressure gauge and a service valve, is also installed on the top plate to store He-3 gas released from the core components at elevated temperatures.

1.2.1 Key Performance Specifications

1.2.1.1 Base Temperature

The base temperature is the lowest temperature the He-3 pot can reach without any external heating. It is independent of the quantity of He-3 installed in the cryostat and is primarily determined by two factors.

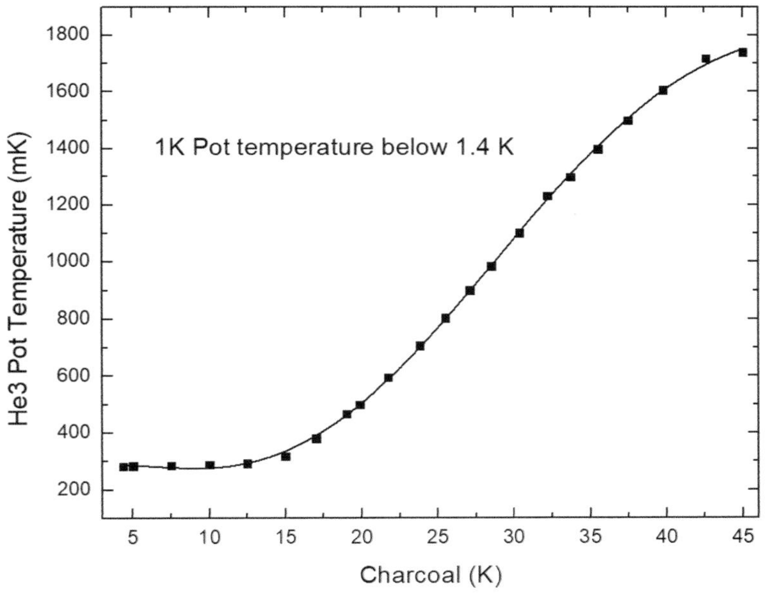

FIGURE 1.2 Typical relationship between He-3 pot temperature versus temperature of the charcoal sorption pump measured from a typical single-shot-type wet He-3 cryostat. (Projects accomplished by authors at Janis Research Company. Image courtesy of FormFactor, Inc.)

- Parasitic heat load from the He-3 cryostat to the He-3 pot.
- The temperature of the charcoal sorption pump. The sorption pump should be maintained below 5 K to maximize its absorption capacity and achieve the lowest temperature on the He-3 pot. On the other hand, adjusting the pump temperatures can alter the vapor pressures and set the He-3 pot to any temperatures between the base temperature and the condensing temperature (Figure 1.2).

1.2.1.2 Holding Time at Base Temperature or Specific Temperatures

The holding time of a He-3 cryostat refers to the duration at a specific temperature, typically the base temperature. This depends on the amount of liquid He-3 in the He-3 pot and the total heat load at the holding temperature. In addition, the temperature of the charcoal sorption pump also impacts the holding time. A higher sorption pump temperature reduces pumping speed, resulting in a higher vapor pressure in He-3 pot. This increases He-3 pot temperature but extends the holding time. For example, when the 1K pot is maintained at 1.5 K and the charcoal sorption pump is set to 45 K, the cryostat is operated at the condensing condition. In this case, He-3 vapor cannot be adsorbed by the sorption pump and is, instead, continuously condensed into 1.5 K liquid by 1K pot, which eventually collects in the He-3 pot. Under these extreme conditions, the temperature of the He-3 pot is about 1.5 K, and the holding time can become effectively unlimited due to the ongoing condensation process.

1.2.1.3 Cooling Power at Certain Temperatures

Cooling power refers to the amount of heat that can be added to He-3 pot while it maintains at a specific temperature. The cooling power depends on He-3 pot temperature, charcoal sorption pump temperature, and the parasitic heat load from the He-3 cryostat to the He-3 pot and becomes zero at the base temperature (Figure 1.3). During cooling power measurements, the charcoal sorption pump is typically set to its lowest temperature to maximize its adsorption capacity.

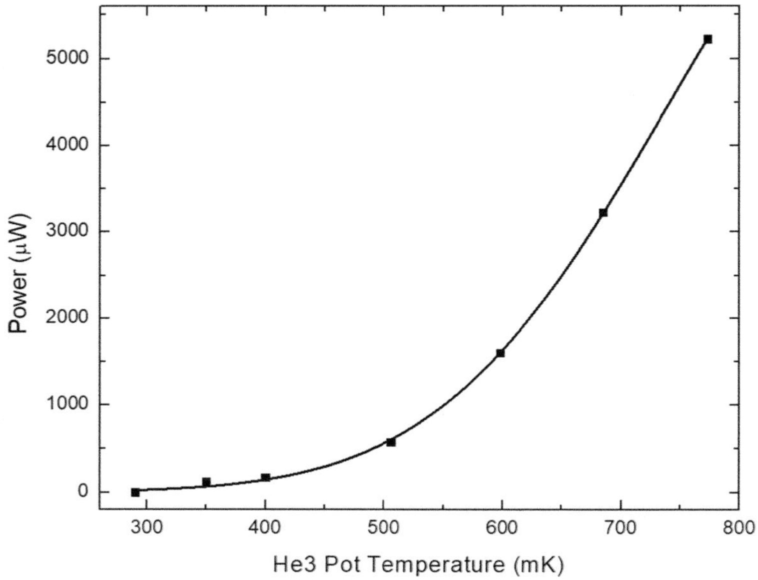

FIGURE 1.3 Typical cooling power of single-shot-type wet He-3 cryostat. (Projects accomplished by authors at Janis Research Company. Image courtesy of FormFactor, Inc.)

1.2.2 Determine the Amount of He-3

The amount of He-3 required for a He-3 cryostat primarily depends on the desired holding time. For example, achieving a 150-hour holding time at a base temperature of 270 mK requires a minimum of six STP liters of He-3 gas for a basic single-shot wet He-3 insert without additional heat load from the user's wiring or experiments. Experimental results have confirmed that, for a single-shot wet He-3 cryostat, the holding time is proportional to the amount of He-3 when the cryostat contains between 3 and 55 STP liters of He-3 gas.

When estimating the total amount of He-3 required for the design, the following considerations should be taken into account:

- Residual gaseous He-3: Some "residual" He-3 always remains in the gas phase during condensation and should be excluded when estimating the amount of liquid He-3 in the He-3 pot. The quantity of residual gas depends on the volume of the gas dump and the He-3 vapor pressure at the condensing temperature, but it is independent of the total He-3 charge. In systems with a large dump volume, this residual gas amount can be significant and must be considered.
- Liquid He-3 required for cooling: A portion of the liquid He-3 is required to cool the He-3 pot and the remaining liquid from the condensation temperature to the desired temperature. As discussed above, after cooling from 1.5 K, only about 80% of the liquid He-3 is available to hold the He-3 pot at 0.3 K.
- Holding time versus base temperature: The total amount of He-3 in the system determines the holding time but does not impact the base temperature.
- Additional parasitic heat loads: These are typically associated with the additional wires, RF cables and devices installed to cryostats for user experiments and must be considered when estimating the required amount of He-3.

1.2.3 Determine the Key Structural Parameters

Once the total amount of He-3 gas is determined, the next step is to define the following four structural parameters [1, 2].

a. The size of He-3 gas dump

As a general guideline, the maximum pressure of the He-3 gas dump should not exceed 10 atmospheric pressures (atm) for safety considerations, i.e., the dump capacity should be at least 10% of the installed total He-3 gas volume in STP liters. American Society of Mechanical Engineers (ASME) code should be followed wherever applicable.

b. Charcoal sorption pump

- Figure 1.4 demonstrates examples of the quantities of He-3 gas and the corresponding charcoal installed in well-performing He-3 cryostats, which can be used to determine the charcoal required for a new cryostat.
- Once the required amount of charcoal is determined, the next step is to design the charcoal sorption pump, including its volume and dimensions. According to Reference [1], about each gram (1 g) of charcoal requires the minimum of 2 cm^3 space, which should be considered as a general guideline for sizing the sorption pump volume.
- Since charcoal has poor thermal conductivity, copper isothermal plates are often installed inside to ensure uniform temperature distribution. These plates occupy space and should be considered in the overall design.

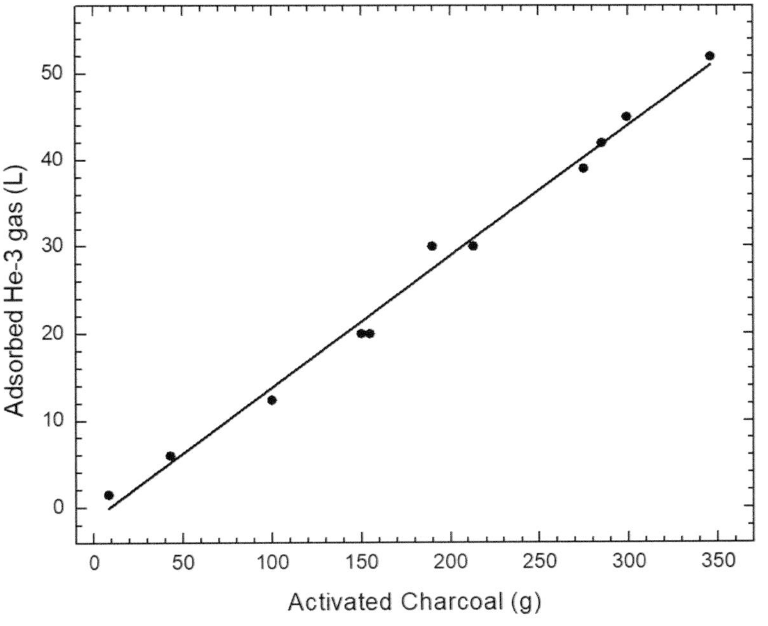

FIGURE 1.4 Quantities of charcoal and He-3 gas installed in several functioning He-3 cryostats. All data were collected with charcoal maintained below 5 K and under the saturated vapor pressure (SVP) of He-3 near 300 mK. (Projects accomplished by authors at Janis Research Company. Image courtesy of FormFactor, Inc.) *Source*: Zuyu Zhao in his projects; Ventura, G. and Risegari, L., *The Art of Low Temperature Experimental Technologies,* Elsevier Oxford UK (2008) [3]: Palumbo, P., Aquilin, E., Cardoni, P., P. de Bernardis, A. De Ninno, Martinis, L., Masi, S., and Scaramuzzi, F., *Balloon-borne 3He cryostat for millimeter bolometric photometry*, Cryogenics Vol. 34, number 12 (1994) [4]; Masi, S., Aquilini, E,. Cardoni, P. de Bernardis, Martinis, L, Scaramuzzi, F., and Sforna, D., *A self-contained* ^{3}*He refrigerator suitable for long duration balloon experiment*. Cryogenics, Volume 38, Issue 3, March 1998, Pages 319–34 [5].

c. The volume of He-3 Pot

To ensure proper He-3 condensation, the He-3 pot must have sufficient volume to accommodate the entire amount of liquid He-3 at the condensing temperature. According to Reference [1] for the thermal properties of He-3, the molar volume at an elevated condensing temperature (e.g., 1.5 K) is greater than at He-3 cryostat base temperatures (e.g., 300 mK). As a practical guideline, the volume of He-3 pot should be approximately 20% greater than the volume of the maximally condensed liquid He-3 at condensing temperatures, in order to ensure optimal cryostat performance.

d. Thermal isolation between the 1K pot and the He-3 pot

The design of the connecting tube between the 1K pot and the He-3 pot plays a critical role in minimizing the parasitic heat load on the He-3 pot, which directly impacts the overall performance of the cryostat. Key parameters—such as tube length, diameter, and wall thickness—must be carefully optimized to balance mechanical stability with thermal isolation.

A commonly used configuration involves stainless steel tubing with an outer diameter of 0.375 "and a wall thickness of 0.010", maintaining an effective length of 5.0" between the two pots.

Increasing the tube length enhances thermal isolation between the 1K pot and the He-3 pot, thereby improving the performance of the He-3 cryostat. Experimental results show that, for identical diameter and wall thickness, a wet He-3 cryostat equipped with a 25-inch-long tube achieves approximately three times the holding time of one with a 5-inch-long tube, while maintaining a similar base temperature.

1.3 SINGLE-SHOT TYPE CRYOGEN-FREE ("DRY") HE-3 CRYOSTAT

Instead of relying costly and operationally inconvenient liquid He-4, a single-shot He-3 cryostat can be implemented with a two-stage cryocooler [1], thereby functioning as a cryogen-free He-3 cryostat (Figure 1.5). Similar to the wet cryostat, a charcoal sorption pump is still used to reduce the

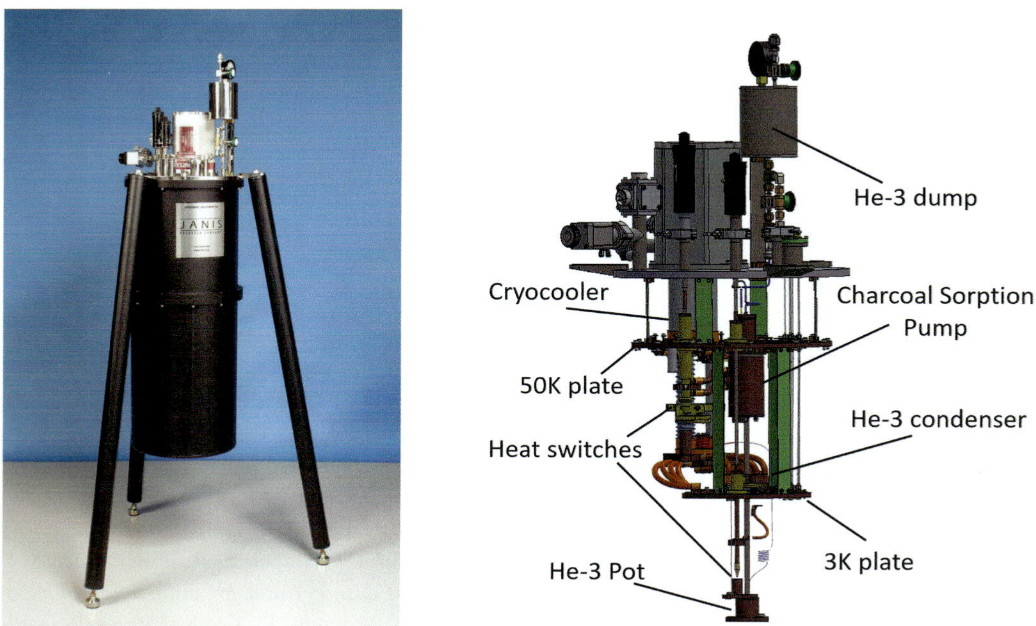

FIGURE 1.5 Typical examples of single-shot-type cryogen-free He-3 cryostat with pulse tube refrigerator (PTR). (Projects accomplished by authors at Janis Research Company. Image courtesy of FormFactor, Inc.)

saturated vapor pressure of liquid He-3 for evaporative cooling. However, a special heat exchanger, described in Chapter 6, is employed in this cryostat to replace the 1K pot. These two types of cryostats have comparable performance, except in the efficiency of utilizing the installed He-3.

1.3.1 The Amount of Residual Gaseous He-3

While He-3 gas typically condenses at around 1.5 K in a wet He-3 cryostat, it condenses at higher temperatures in cryogen-free He-3 cryostat, limited by the temperature of the cryocooler's second stage, typically around 2.4 K. According to Table 1.1, the He-3 vapor pressure at 2.4 K is roughly 5.7 times higher than at 1.5 K. As a result, a larger fraction of He-3 remains in the gaseous phase at the end of condensation in cryogen-free systems.

TABLE 1.1
^{3}He Vapor Pressure (kPa) as a Function of Temperature

T (K)	0	0.01	0.02	0.03	0.04	0.05	0.06	0.07	0.08	0.09
0.2	1.47E-06	2.98E-06	5.69E-06	1.03E-05	1.78E-05	2.96E-05	4.75E-05	7.38E-05	1.11E-04	1.64E-04
0.3	2.35E-04	3.32E-04	4.58E-04	6.22E-04	8.30E-04	1.09E-03	1.42E-03	1.82E-03	2.30E-03	2.89E-03
0.4	3.59E-03	4.41E-03	5.38E-03	6.51E-03	7.82E-03	9.32E-03	0.011	0.013	0.0152	0.0177
0.5	0.0205	0.0236	0.0271	0.0309	0.0352	0.0399	0.045	0.0506	0.0567	0.0633
0.6	0.071	0.079	0.087	0.096	0.105	0.116	0.127	0.139	0.152	0.166
0.7	0.18	0.195	0.211	0.229	0.247	0.267	0.287	0.308	0.33	0.353
0.8	0.378	0.404	0.431	0.459	0.489	0.52	0.552	0.586	0.621	0.657
0.9	0.695	0.734	0.775	0.817	0.861	0.907	0.954	1.003	1.054	1.106
1	1.16	1.216	1.274	1.333	1.395	1.459	1.523	1.59	1.66	1.731
1.1	1.804	1.88	1.957	2.037	2.118	2.202	2.288	2.376	2.466	2.559
1.2	2.654	2.752	2.851	2.954	3.059	3.165	3.275	3.387	3.501	3.618
1.3	3.738	3.86	3.985	4.112	4.242	4.375	4.511	4.649	4.79	4.934
1.4	5.081	5.231	5.383	5.538	5.691	5.858	6.022	6.189	6.36	6.533
1.5	6.709	6.889	7.071	7.257	7.446	7.638	7.834	8.033	8.235	8.44
1.6	8.649	8.861	9.076	9.295	9.511	9.742	9.972	10.2	10.44	10.68
1.7	10.92	11.17	11.42	11.68	11.93	12.19	12.46	12.73	13	13.28
1.8	13.56	13.84	14.13	14.42	14.72	15.02	15.32	15.63	15.94	16.26
1.9	16.58	16.99	17.23	17.56	17.9	18.24	18.58	18.93	19.28	19.64
2	27.26	27.71	20.74	21.11	21.49	21.87	22.26	22.65	23.05	23.45
2.1	23.85	24.26	24.68	25.1	25.52	25.95	26.38	26.82	27.26	27.71
2.2	28.16	28.61	29.08	29.54	30.01	30.49	30.97	31.45	31.94	32.44
2.3	32.94	33.44	33.95	34.47	34.99	35.51	36.04	36.58	37.12	37.67
2.4	38.22	38.77	39.33	39.9	40.47	41.05	41.63	42.22	42.82	43.41
2.5	44.02	44.63	45.24	45.87	46.49	47.12	47.76	48.4	49.05	49.71
2.6	50.37	51.04	51.71	52.38	53.07	53.76	54.45	55.15	55.86	56.58
2.7	57.29	58.02	58.75	59.49	60.23	60.98	61.74	62.5	63.27	64.04
2.8	64.82	65.61	66.41	67.21	68.01	68.83	69.65	70.47	71.3	72.14
2.9	72.99	73.84	74.7	75.57	76.44	77.32	78.21	79.1	80	80.91
3	81.83	82.75	83.68	84.61	85.96	86.51	87.46	88.43	89.4	90.38
3.1	91.37	92.36	93.37	94.38	95.39	96.42	97.45	98.49	99.54	100.6
3.2	101.66	102.73	103.82	104.9	106	107.1	108.22	109.34	110.47	111.61

Source: Temperatures above 0.65 K are given on the International Temperature Scale of 1990 (ITS-90). Temperatures below 0.65 K lie outside the defined range of ITS-90 and are expressed as thermodynamic temperatures derived from the ^{3}He vapor-pressure scale, consistent with ITS-90 within its range of validity.

1.3.2 Holding Time

Because of the higher condensing temperature, a cryogen-free He-3 cryostat requires, compared with a wet cryostat, a significantly larger amount of liquid He-3 to evaporate and cool itself from 2.4 K to 0.3 K, resulting in a shorter holding time. Figure 1.6 illustrates several examples of the typical relationship between the installed quantity of He-3 and the holding time in a bare single-shot cryogen-free He-3 cryostat. These examples highlight the following consideration.

- A large fraction of liquid He-3 evaporates to cool the remaining liquid He-3 to the base temperature.
- The holding time is approximately proportional to the installed amount of He-3.
- The data exhibit a larger standard deviation, which is primarily attributed to variations in the quality of the multi-layer insulation (MLI) assembly, particularly on the 3 K radiation shield. Improper MLI installation can permit excess infrared (IR) radiation to reach the sample space, resulting in a non-negligible impact on the performance of the He-3 cryostat.

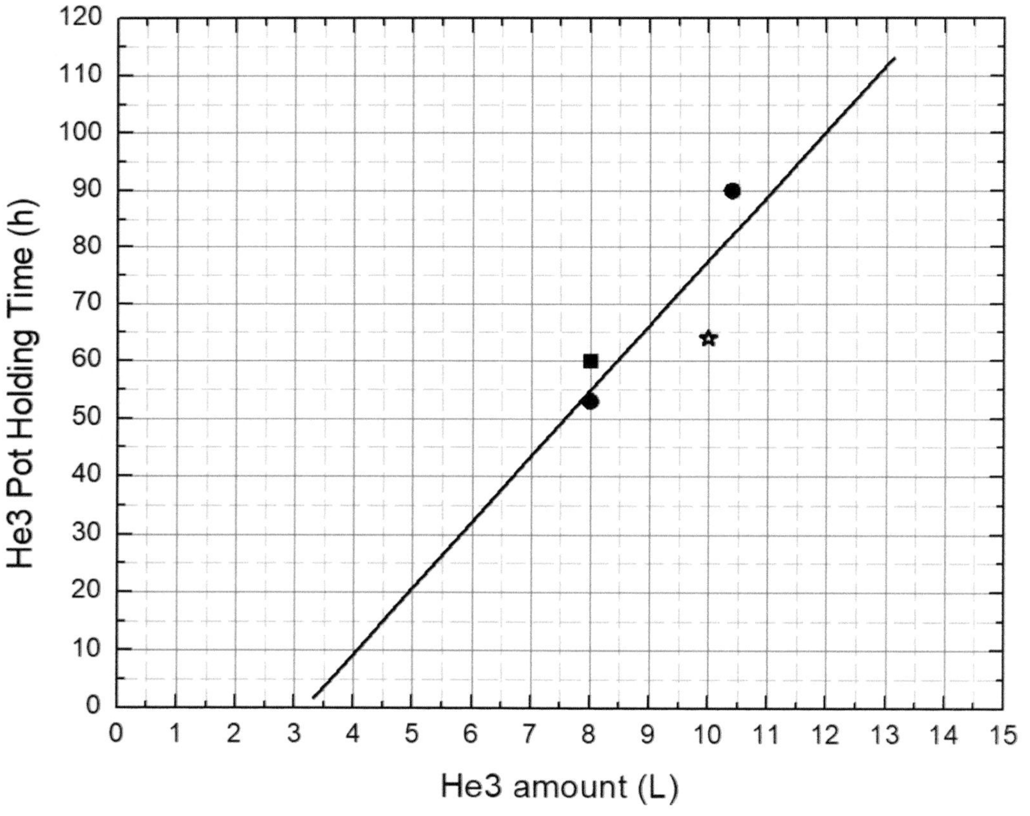

FIGURE 1.6 Typical relationship between quantity of He-3 gas versus the holding time of single-shot type cryogen-free He-3 cryostat. All data were collected with charcoal maintained below 5 K and under the saturated vapor pressure (SVP) of He-3 near 300 mK. (Projects accomplished by authors at Janis Research Company. Image courtesy of FormFactor, Inc.)

1.4 CONTINUOUS-FLOW HE-3 CRYOSTAT

1.4.1 Continuous-Flow Cryogen-Free He-3 Cryostat

The continuous-flow cryogen-free He-3 cryostat operates as a closed-loop recirculating system [1, 6, 7], as illustrated in Figure 1.7. Unlike single-shot cryostats that rely on charcoal sorption pumps, this type of cryostats uses an external hermetic pump to reduce the saturated vapor pressure of liquid He-3, enabling evaporative cooling. The He-3 vapor is subsequently recondensed into the He-3 pot for continuous circulation. Proper operation also requires a versatile gas handling system (GHS) to manage gas flow, pressure, and recovery.

The configuration of a typical continuous-flow, cryogen-free He-3 cryostat is illustrated in Figure 1.8. A Joule-Thomson He-3 condensing unit (HE3CU) is employed to condense He-3 (Figures 1.9 and 1.10). The primary components of the HE3CU include a recuperative heat exchanger and a demountable condensing impedance. The impedance consists of a small CuNi capillary with typically values on the order of 10^9 to 10^{10} cm^{-3}. The details regarding the construction and measurement of the heat exchanger and impedance are provided in Chapter 5 and Appendix A1.

The refrigerant of He-3 can be replaced with He-4, converting the He-3 cryostat into a cost-effective sub-K cryostat. Figures 1.11(a) and (b) show the base temperatures and cooling power of a continuous-flow cryogen-free He-3 cryostat equipped with He-3 refrigerant and He-4 refrigerant, respectively. These results were obtained using a Pfeiffer HiPace 400 turbo pump in the GHS. Lower base temperatures and higher cooling power can be achieved by using turbo pumps with lower ultimate vacuum and higher throughput. Further performance improvements are expected by the direct integration of the turbo pump onto the He-3 cryostat.

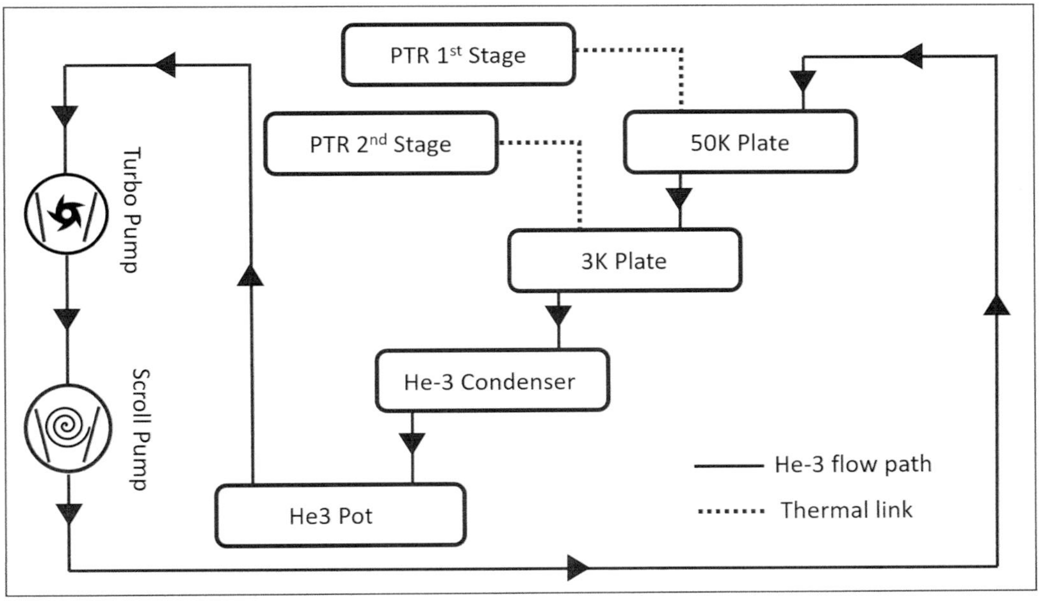

FIGURE 1.7 Close-loop He-3 circulation diagram in the continuous-flow cryogen-free He-3 cryostat.

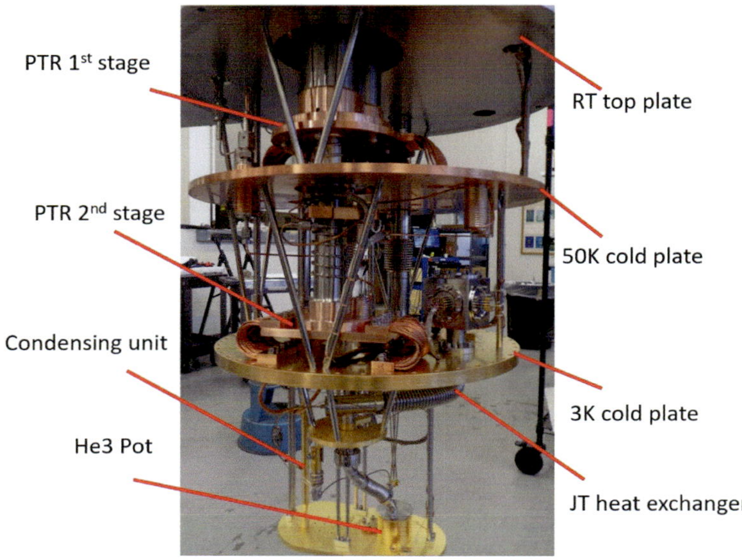

FIGURE 1.8 Typical configuration of continuous-flow cryogen-free He-3 cryostat [6]. (Projects accomplished by authors at Janis Research Company. Image courtesy of FormFactor, Inc.)

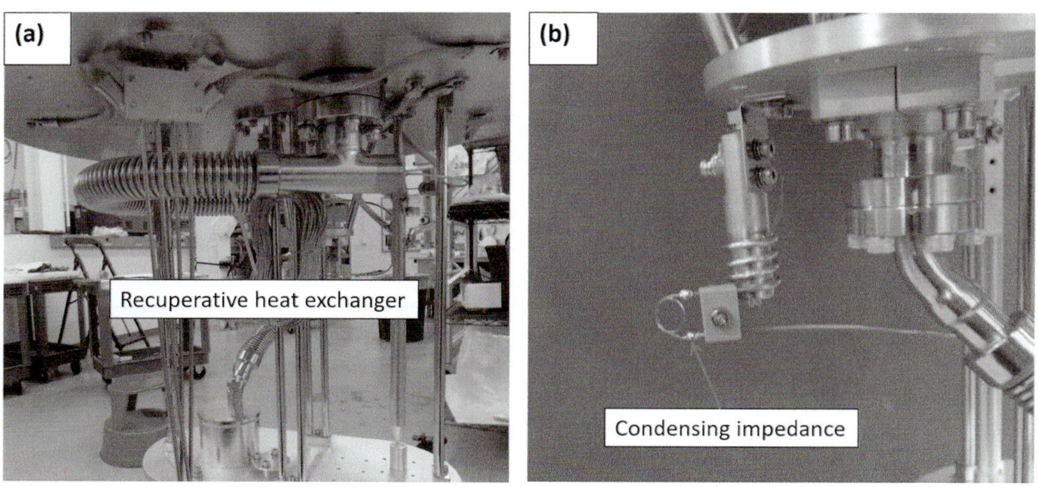

FIGURE 1.9 (a) Recuperative heat exchanger; (b) Demountable condensing impedances. (Projects accomplished by authors at Janis Research Company. Image courtesy of FormFactor, Inc.)

1.4.2 Continuous-Flow "Wet" He-3 Cryostat

A continuous-flow "wet" He-3 cryostat has the similar working principle, operation, and performance as that of a "dry" system described in Section 1.4.1. Both types of systems employ similar impedance with comparable construction and function. The primary differences are as follows:

a. The "wet" He-3 system uses liquid He-4 to reach 4 K.

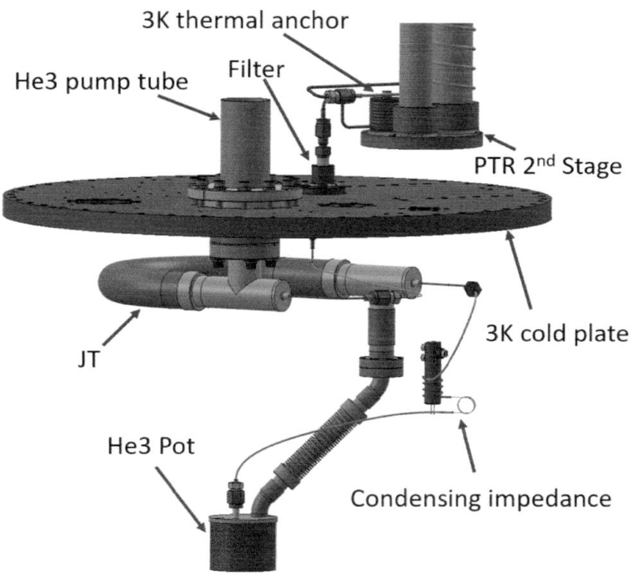

FIGURE 1.10 Complete Joule-Thomson He-3 condensing unit (HE3CU) assembly. (Projects accomplished by authors at Janis Research Company. Image courtesy of FormFactor, Inc.)

b. The "wet" He-3 system often employs 1K pot instead of Joule-Thomson heat exchanger for He-3 condensation. In this configuration, the returning He-3 condenses inside the 1K pot, achieving lower condensing temperatures. Additionally, the 1K pot provides significantly higher cooling power than the Joule-Thomson heat exchanger.

1.5 OPTICAL HE-3 CRYOSTAT

Optical He-3 cryostats are He-3 cryostats equipped with optical windows and/or optical feedthroughs. In these systems, one of the major sources of heat load is incident light on the He-3 pot, particularly infrared radiation. Its magnitude depends on several factors, including the window material, the solid angle between the window and the sample, the laboratory environment, and the sample surface finish. These variables make precise estimation of the heat load challenging. Additionally, the diffusive reflection of radiation within the sample space can significantly impact cryostat performance and remains difficult to model or quantify with accuracy.

1.5.1 Multiple Windows

The blackbody radiative energy density, P, i.e., the power per unit area, is proportional to the fourth power of the temperature T of the blackbody:

$$P = \sigma T^4 \tag{1.1}$$

where σ is Stefan-Boltzmann constant, $5.670373 \times 10^{-8}\ \mathrm{W \cdot m^{-2} \cdot K^{-4}}$.

The equation indicates that the radiative energy emitted by a blackbody significantly decreases at lower temperatures (see Figure 1.12). To reduce the heat load to cryostat, multiple windows can be installed at different temperature zones. Figure 1.13 shows an optical He-3 cryostat, where five

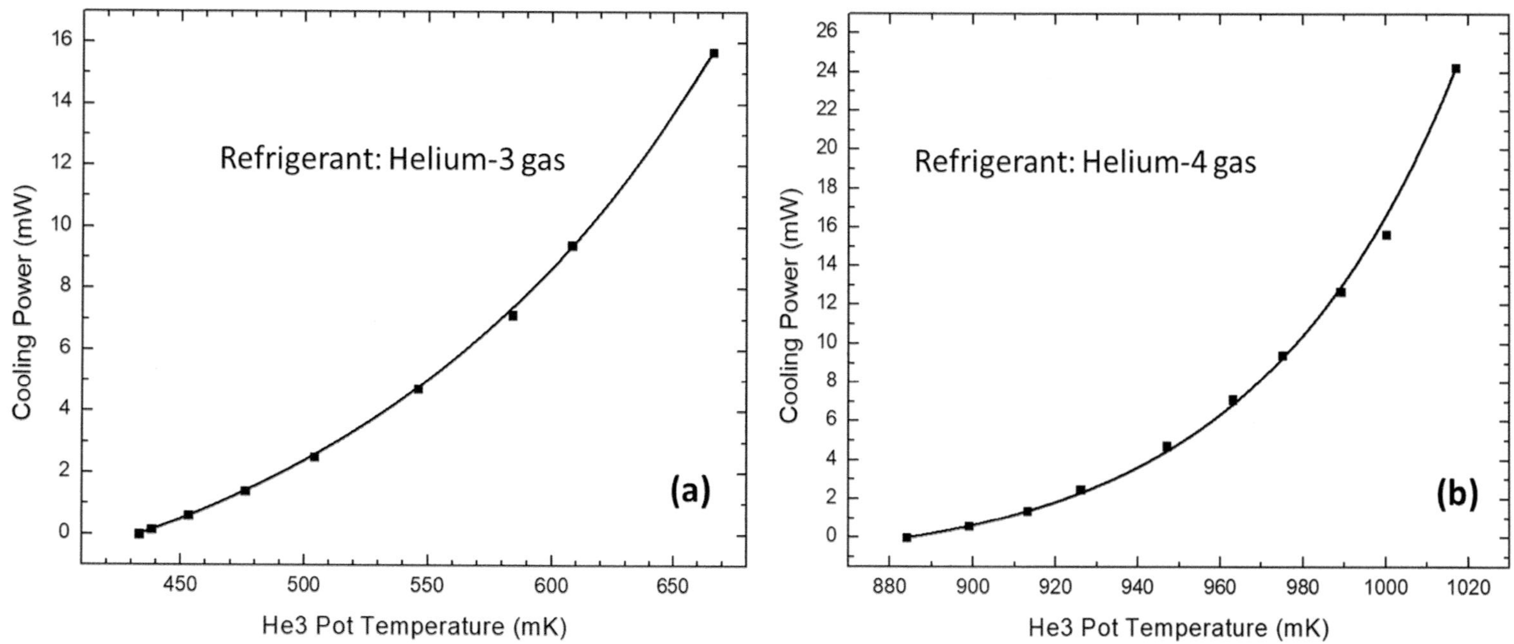

FIGURE 1.11 (a) Typical performance of continuous-flow cryogen-free He-3 cryostat with He-3; (b) Typical performance of continuous-flow cryogen-free He-3 cryostat with He-4. (Projects accomplished by authors at Janis Research Company. Image courtesy of FormFactor, Inc.)

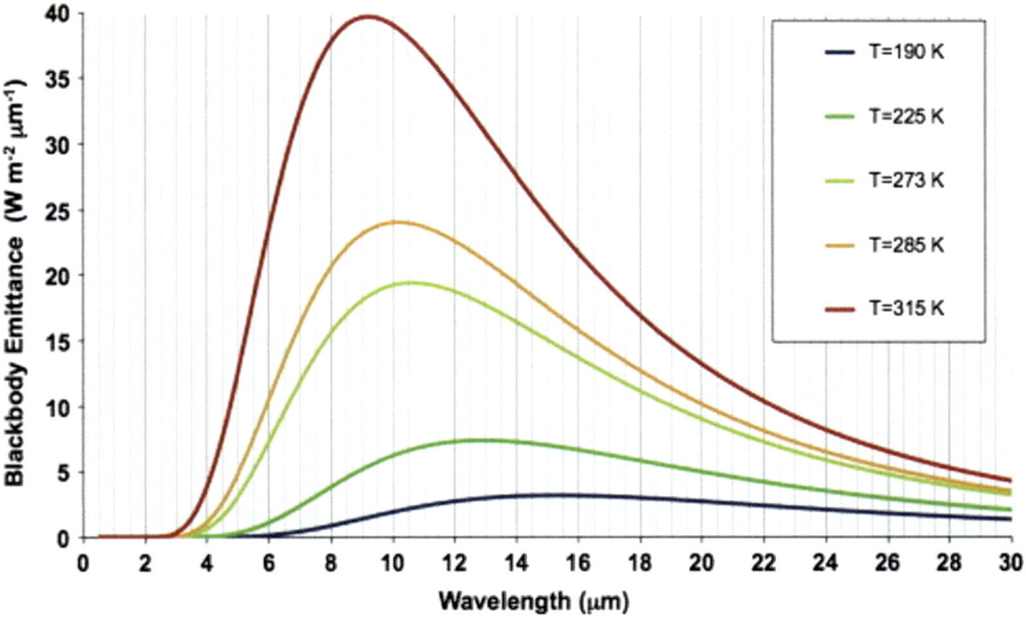

FIGURE 1.12 Black body radiation as a function of temperatures.

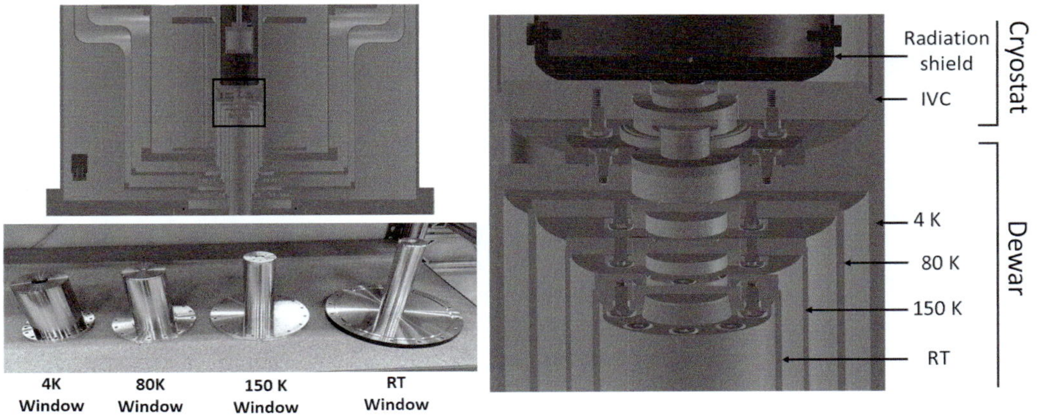

FIGURE 1.13 Multiple windows on the optic He-3 cryostat. (Projects accomplished by authors at Janis Research Company. Image courtesy of FormFactor, Inc.)

windows are installed at room temperature on the outer vacuum shroud, about 150 K on the first radiation shield, about 80 K on the second radiation shield, 4.2 K at the bottom of the liquid helium reservoir, and 4.2 K at the bottom of the inner vacuum can.

- When the 4 K windows were blocked, the optical cryostat functioned as a conventional He-3 cryostat and reached a base temperature of 269 mK.

- When all windows were unblocked, the base temperature of the He-3 pot slightly increased to 277 mK, indicating that the incident room-temperature radiation is effectively suppressed by the multiple-window design.
- Since the optical radiation, particularly from lasers, may cause bubbling in liquid helium, an alternative design is needed if bubbling is unacceptable for user experiments.
- The wet optical He-3 system shown in Figure 1.13 contains two 4K hermetic windows that must remain leak-tight after thermal cycling. These windows are installed on indium-sealed and demountable retainers using Versimid epoxy [8] (Figure 1.14). A tolerance of 0.005"–0.008"

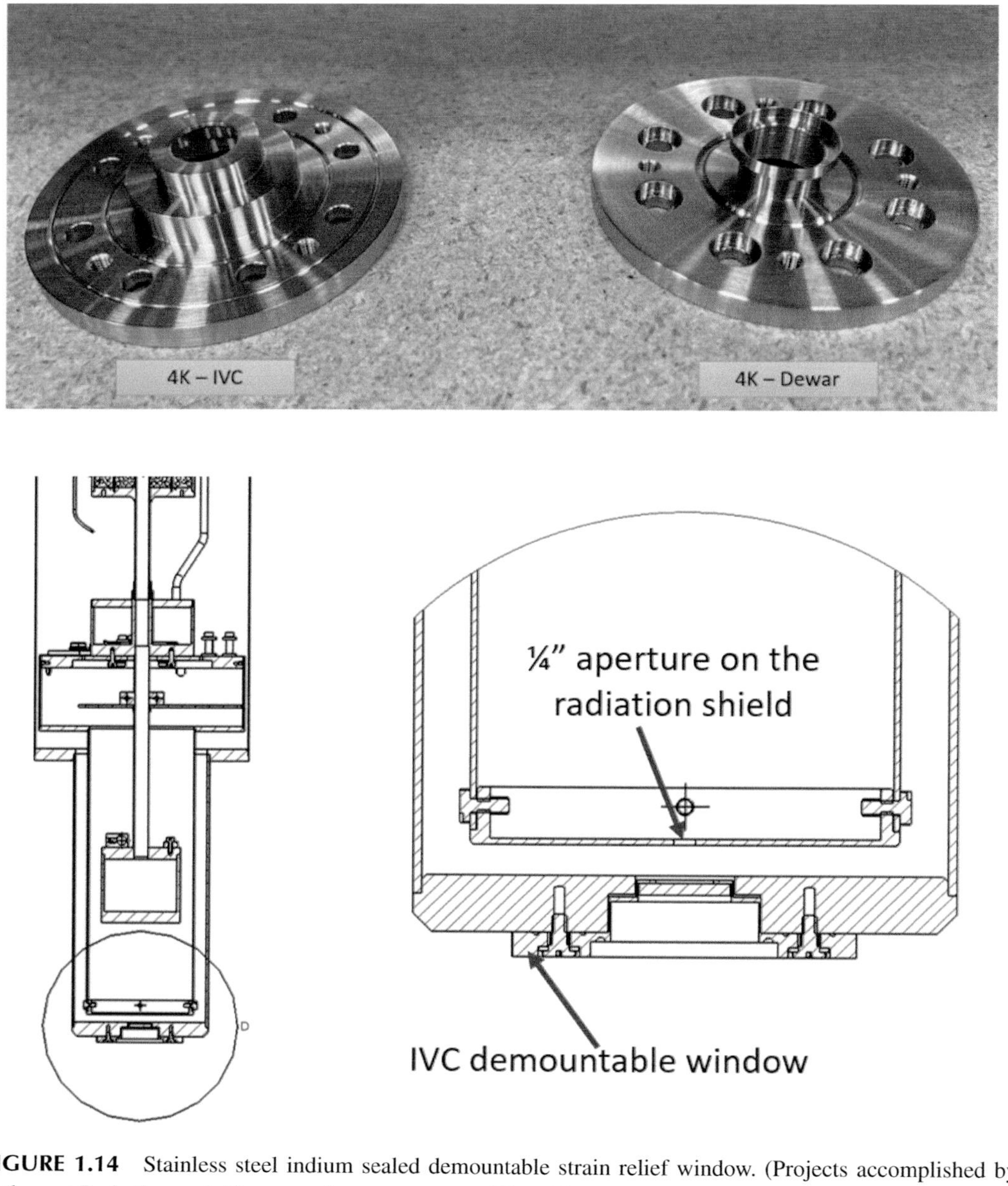

FIGURE 1.14 Stainless steel indium sealed demountable strain relief window. (Projects accomplished by authors at Janis Research Company. Image courtesy of FormFactor, Inc.)

is made between the window and the retainer, allowing the epoxy to adequately "wet" both surfaces. This design helps the strain relief, thereby enhancing the window reliability, and also allows for easy replacement.

1.5.2 Minimize the Solid Angle

In addition to minimizing the window's clear view, the solid angle between the sample and the window should also be minimized, as this effectively reduces the heat load from radiation incident on the sample through the window.

1.5.3 1K Shield (Wet System)

Installing a 1K radiation shield in a wet He-3 cryostat (Figure 1.14) is an effective method for further reducing the heat load on the sample from diffuse reflections of radiation within the cryostat. An aperture, rather than a window, on the 1K shield is typically sufficient, but its diameter should be minimized to limit thermal radiation.

1.5.4 Window Material

Figure 1.15 illustrates the transmission properties of different window materials. Selecting the right window material is an effective way to filter out light with undesired wavelengths.

Close attention should be paid when using plastic windows (e.g., Teflon, polyethylene, polypropylene, Kapton, etc.), as cryostats with plastic windows typically experience elevated infrared (IR) radiation heat loads within specific wavelength ranges. This arises because these materials generally possess high transmissivity across much of the IR spectrum, as illustrated in Figure 1.16.

Bounthavy reported the performance of a wet single-shot He-3 cryostat equipped with two sets of polyethylene windows installed on three different temperature shields: the outer vacuum shroud (300 K), the liquid nitrogen radiation shield (78 K), and the liquid helium radiation shield (4 K),

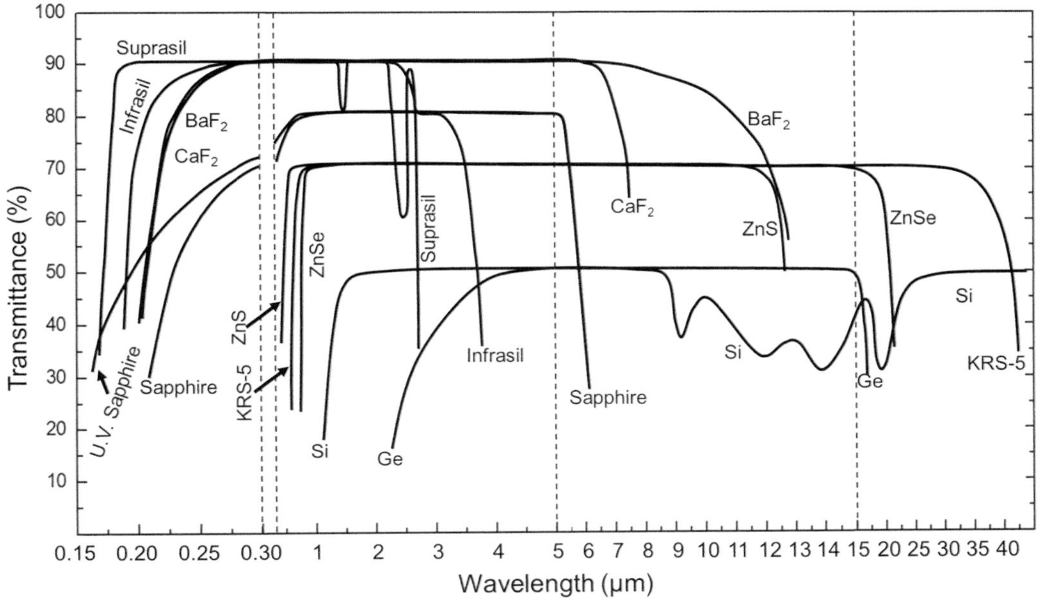

FIGURE 1.15 Transmission of different window material.

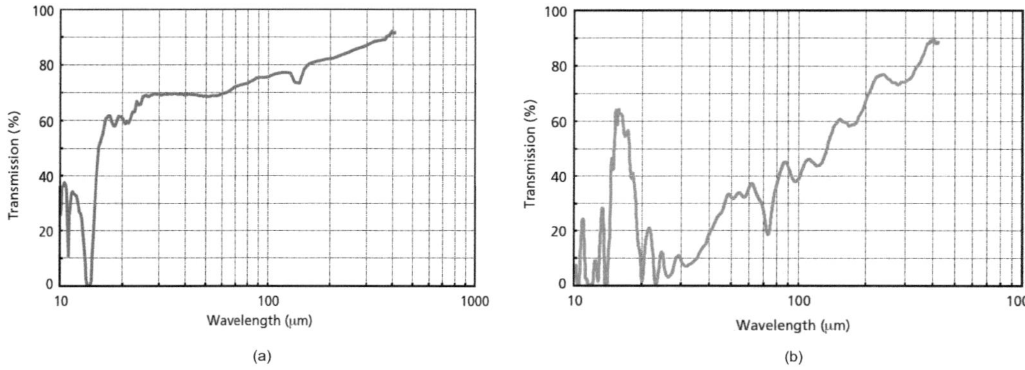

FIGURE 1.16 (a) Transmission of 1.0-mm-thick polythene window; (b) Transmission of 125-μm-thick Mylar window. (Reprint with permission from Oxford Instrument, 300 Baker Ave. #150, Concord, MA, USA, [9].)

respectively [10]. The 4 K polyethylene window, providing a 0.5-inch clear view, was positioned approximately 1.0 inch from the sample. The total heat load was estimated at approximately 7.6 mW, primarily due to infrared radiation. This substantial heat load led to a high boil-off rate of liquid He-3, which subsequently increased the temperature of the charcoal sorption pump. Consequently, the sample could not reach the desired low temperature.

1.5.5 He-3 Cryostat with an Open Slot for Angle-Resolved Photoemission Spectroscopy (ARPES)

ARPES requires a specialized type of cryostat designed for photoemission experiments. Unlike traditional optical cryostats, ARPES cryostat features open slots at multiple temperature stages instead of windows, allowing photoelectrons emitted from the sample to travel through these slots and reach the room temperature analyzer. These slots typically have widths of 2.5-3.0 mm and a length of 8.0 mm. The distance from the room-temperature slot (where the analyzer is located) to the sample ranges from 30 to 38 mm, providing sufficient information to estimate the solid angle subtended by the analyzer from the sample position.

The sample faces the room-temperature analyzer and is exposed to radiation from it, contributing a heat load of approximately 12–15 mW. Despite the large heat load, the sample temperature often needs to reach below 1 K, posing a significant experimental challenge.

As demonstrated in Reference [10], a single-shot He-3 cryostat is generally not sufficient to handle milliwatt-level heat loads from radiation. The solution is to employ high-power, continuously flowing He-3 cryostats equipped with larger hermetic pumps (Figure 1.17), such as the Pfeiffer HiPace 1200 turbo pump or the Pfeiffer OKTA-1000 Roots blower pump. These pumps provide higher He-3 circulation rates, delivering sufficient cooling power below 1 K to compensate for the milliwatt-level radiation heat load. With different system configurations (Figure 1.18), ARPES cryostats achieved base temperatures below 600 mK on the He-3 pot and approximately 840 mK on a rotating sample holder.

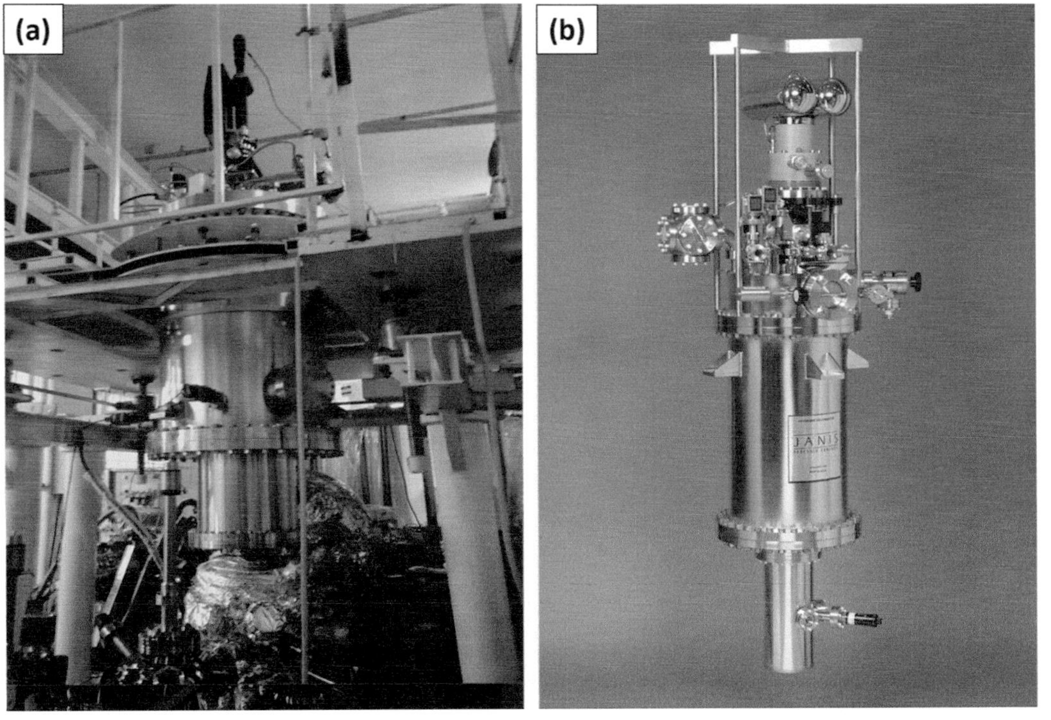

FIGURE 1.17 (a) Wet high power continuous flow type He-3 cryostat for sub-K ARPES; (b) Cryogen-free high power continuous flow type He-3 cryostat for sub-K ARPES cryostat. (Projects accomplished by authors at Janis Research Company. Image courtesy of FormFactor, Inc.)

1.5.6 Typical Plastic Window Materials and Installation

Table 1.2 presents examples of plastic window material employed in our projects for reader's reference (Figure 1.19).

1.5.7 Window Coatings for Optical He-3 Cryostats

Two types of surface coating are commonly used for optical He-3 cryostats: optical filter and anti-reflection coating.

1.5.7.1 Optical Filter

Optical filter (Figure 1.20) selectively transmit light within a specific range of wavelengths while absorbing the remainder, which is commonly used in microscopy, spectroscopy, chemical analysis, and machine vision. In optical He-3 cryostats, these optical filters can be applied to windows to allow only light within the desired wavelength range to pass through. They can be implemented as separate lenses or be coated on window surfaces.

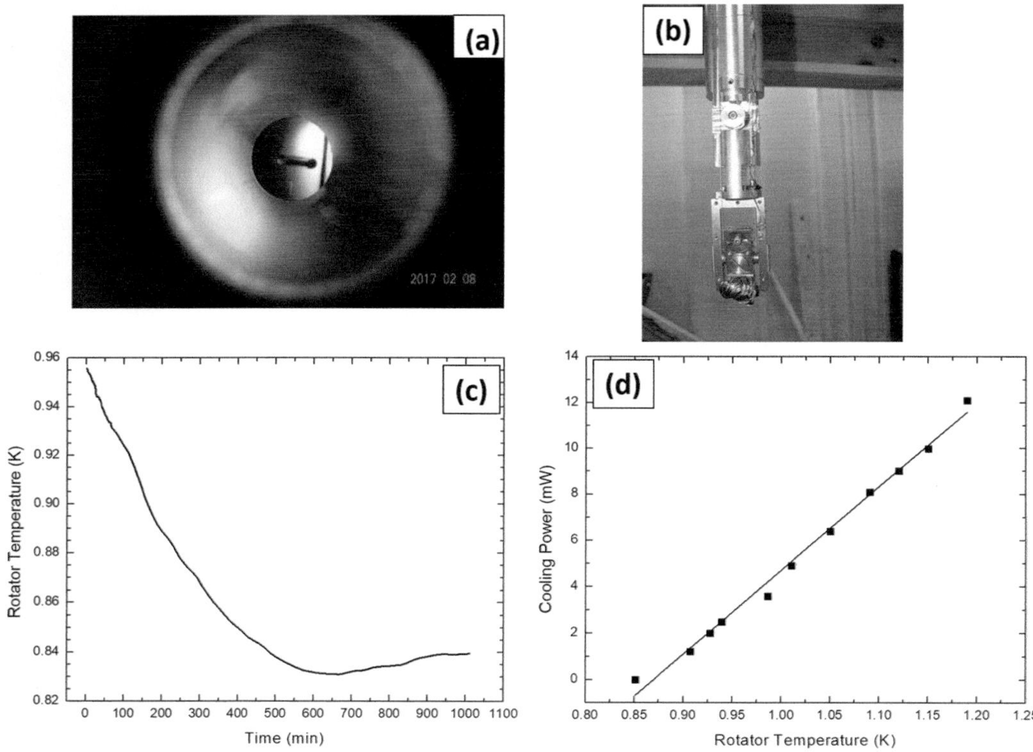

FIGURE 1.18 (a) Slots on ARPES cryostat, located at room temperature, 300 K, 78 K and 4 K, respectively; (b) Rotating sample holder; (c) Rotating sample holder on ARPES cryostat reached base temperature 0.84 K; (d) Cooling power of the rotating sample holder on ARPES cryostat. (Projects accomplished by authors at Janis Research Company. Image courtesy of FormFactor, Inc.)

1.5.7.2 Anti-Reflection Coating and Optical Filter

An antireflection (AR) film or coating is an optical layer applied to an optical window to minimize the reflection of incident light. The AR layer is designed so that light of a selected wavelength, when reflected from the AR layer, has the same or similar amplitude but an opposite phase relative to light to that reflected from the window, as illustrated in Figure 1.21. This results in destructive interference, minimizing the total reflected light at the selected wavelength.

The optical filter and AR coating are often applied to the same window. Typically, the filter is coated on the surface facing the light source, while the AR coating is applied to the opposite surface to enhance transmission within the desired wavelength range (Figure 1.21).

TABLE 1.2
Plastic/Film Window Material

No.	Material	Color	Thickness	Application and Remarks	Max. C.V. Dia.
1	Mylar	Translucent ("White")	0.014"	Room temperature shrouds and window blocks	1.5" for O-ring seal 1.25" for wrapping seal
2	Mylar	Translucent ("White")	0.010"	Cold sample tubes	0.75" for epoxy or indium seal
3	Mylar	Translucent ("White")	0.0055"	Cold sample tubes upon customer's special requests	
4	Mylar	Clear	0.010"		
5	Aluminized mylar	Clear	0.0005"	Radiation shields	
6	Polypropylene (PPN)	Clear	0.002"	Cold windows	0.5"
7	Polypropylene (PPN)	Clear	0.005"	Room temperature windows	1.5"
8	Polyethylene	Clear	0.010"	O-ring sealed windows	
9	Kapton	Light brown; clear	0.0025"	Wrapped on warm and cold shrouds or upon customer's special request	
10	Kapton	Light brown; clear	0.005"	Vacuum shroud window	

Source: Projects Accomplished by Authors at Janis Research Company. Data courtesy of FormFactor, Inc.

Note:
. This information package is for reference only. Material and dimensions may vary with different applications.
. Use Araldite 2011 for all epoxied film windows.
. Use mylar on tubes only but do not install mylar on window block.

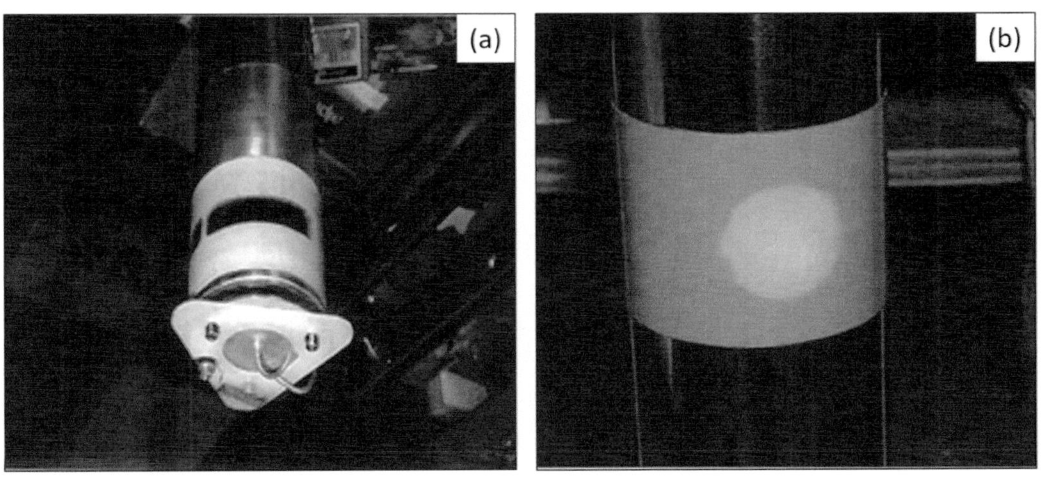

FIGURE 1.19 (a) Kapton window installation on cylindrical cryostat; (b) Mylar window installation on cylindrical cryostat. (Projects accomplished by authors at Janis Research Company. Image courtesy of FormFactor, Inc.)

FIGURE 1.20 IR cut (shortpass) optical filter. *Information source*: Edmund Optics; www.edmundoptics.com/ (Public domain, [11].)

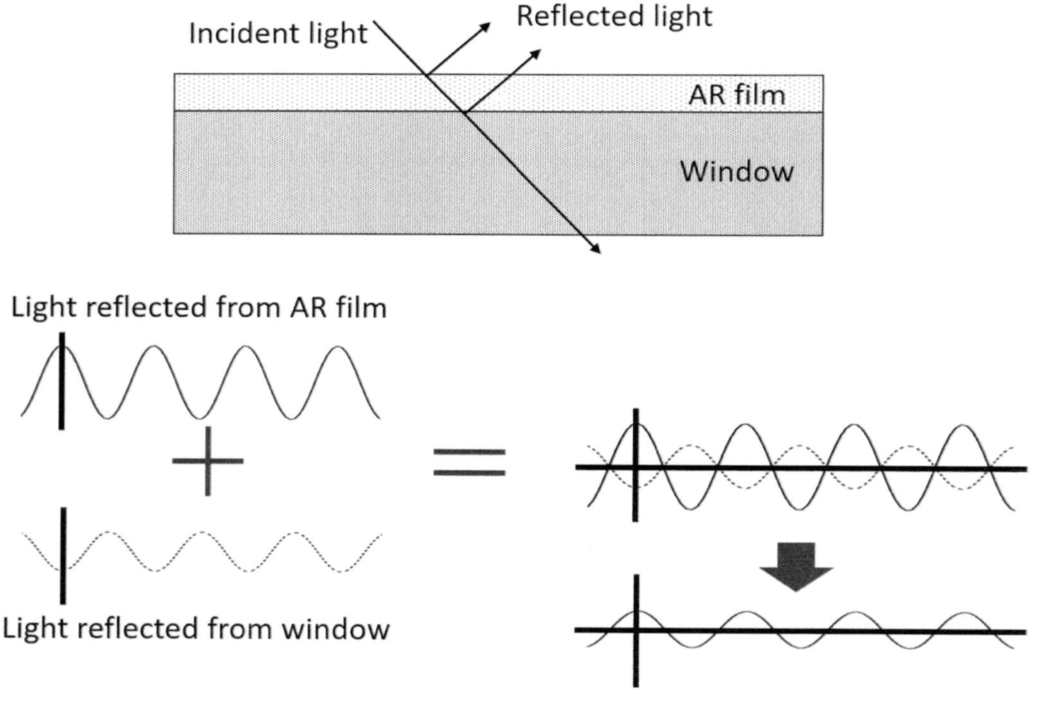

FIGURE 1.21 AR film structure.

REFERENCES

1. Zuyu Zhao and Chao Wang (Eds.), *Cryogenic Engineering and Technologies: Principles and Applications of Cryogen-Free Systems*, CRC Press (Taylor & Francis Group) Boca Raton, USA.(2020).
2. C. F. Mate, R. Harris-Lowe, W. L. Davis, and J. G. Daunt, 3He Cryostat with Adsorption Pumping, *The Review of Scientific Instruments*, 36(3), 369–373, (March 1965).
3. G. Ventura and L. Risegari, *The Art of Low Temperature Experimental Technologies*, Elsevier Science, Amsterdam, Netherlands, (2008).
4. P. Palumbo, E. Aquilin, P. Cardoni, P. de Bernardis, A. De Ninno, L. Martinis, S. Masi, and F. Scaramuzzi, Balloon-Borne 3He Cryostat for Millimeter Bolometric Photometry, *Cryogenics*, 34(12), 1001–1005, (1994).
5. S. Masi, E. Aquilini, P. Cardoni, P. de Bernardis, L. Martinis, F. Scaramuzzi, and D. Sforna, A Self-Contained ^{3}He Refrigerator Suitable for Long Duration Balloon Experiment, *Cryogenics*, 38(3), 319–334, (March 1998).
6. O. Korneta, C. Diep, V. Luppov, and Z. Zhao, Breakthrough Instruments and Products Janis Research Company, LLC Cryogen free sub-Kelvin ARPES System, Cold Facts, *Product Show Case*, (2022).
7. J.C. Burton, E. Van Cleve, P. Taborek, A Continuous 3He Cryostat with Pulse -tube Pre-cooling and Optical Access, *Cryogenics*, 51, 209–213, (2011).
8. Versimid epoxy consists of DER-331 resin and Versamid 140 hardener, supplied pre-measured at an 11:14 weight ratio. The components are mixed at room temperature until uniform and cured at approximately 65°C (150°F). It is commonly used for low-temperature sealing applications, particularly cryogenic windows.
9. Oxford Instruments plc, Tubney Woods, Abingdon, Oxfordshire, UK, Website: www.oxinst.com
10. Y. Bounthavy, *Internship Report: Advanced Light Source – Beamline 1.4*. Lawrence Berkeley National Laboratory, Advanced Light Source, Berkeley, CA, USA, (2005).
11. Edmund Optics 101 East Gloucester Pike, Barrington, NJ, USA Website: www.edmundoptics.com

2 Sample Loading and Sample Holders in Customized Cryostats

To reduce turnaround time between experiments, it is highly desirable for users to change samples without warming up the cryostat. This capability is typically referred to as "top-loading" when the sample is inserted from the top of the cryostat, and "bottom-loading" when it is inserted from the bottom of the cryostat.

Both top-loading and bottom-loading configurations provide unobstructed access paths for loading samples from room temperature and can also be integrated with superconducting magnets, allowing experiments under low-temperature and high-magnetic-field conditions in a single setup. Depending on the experimental requirements, the sample environment may be either vacuum or liquid helium.

When the sample resides in vacuum, this access path shares the vacuum environment with samples and cryostat cores. To minimize radiative heat loads, various types of shutters are commonly installed within the cryostat. In top-loading systems, the sample can be immersed in liquid helium, and the central access typically consists of a hermetically sealed tube.

This chapter focuses on the "wet" He-3 cryostat as a primary example and presents the various sample-loading mechanisms, loader designs, and radiation shutters developed in our previous projects. Although the discussion focuses on He-3 systems, the underlying design principles are equally applicable to dilution refrigerators (DRs).

2.1 TOP-LOADING "WET" HE-3 CRYOSTAT

2.1.1 Top-Loading He-3 Insert with Sample in Liquid

A typical top-loading He-3 insert with sample in liquid helium is illustrated in Figure 2.1. The system consists of the He-3 insert and a sample probe assembly. A hermetically sealed stainless steel sample delivery tube extends from the top of the He-3 insert into the Inner Vacuum Chamber (IVC) can, where it is integrated with the He-3 core: charcoal sorption pump, the 1K pot, and the He-3 pot formed at the tube's lower end. The He-3 core is similar to the wet He-3 system discussed in Chapter 1 but incorporates a larger He-3 passage diameter tube for the sample transfer.

The top-loading sample probe delivers samples that are loaded into the He-3 liquid through the delivery tube and kept fully inserted in the cryostat during experiments. To minimize heat load on the liquid He-3, the probe is thermally anchored to the delivery tube at the 1K region via direct contact. The sample probe assembly is a critical component of this cryostat configuration. As shown in Figure 2.1, the assembly consists of two main elements: the load-lock and the sample rod.

DOI: 10.1201/9781042000050-2

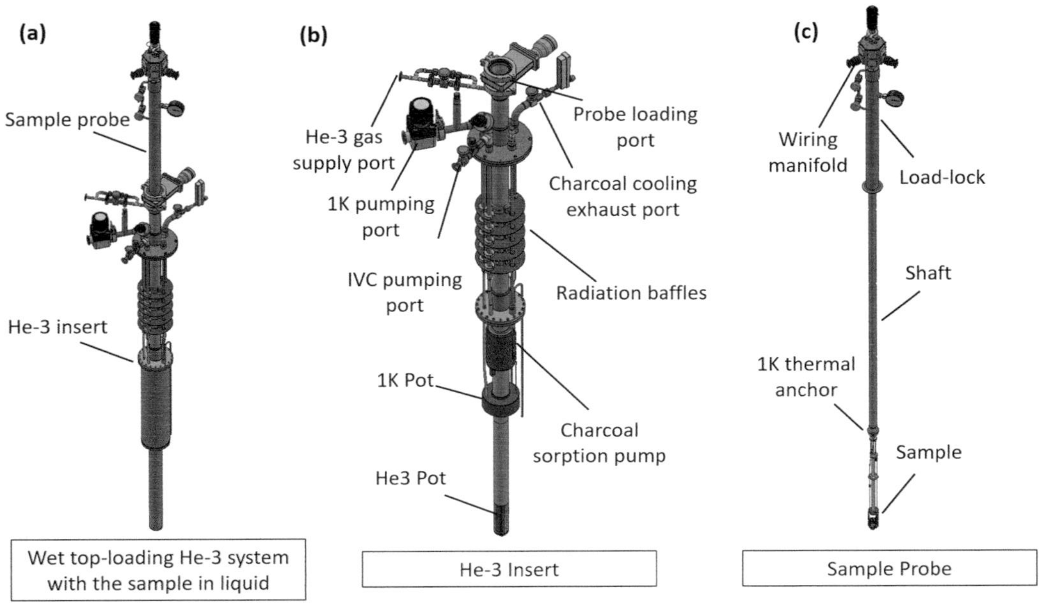

FIGURE 2.1 (a) and (b) Typical top-loading (wet) He-3 insert with sample in (He-3) liquid; (c) Typical top-loading sample probe assembly. (Projects accomplished by authors at Janis Research Company. Image courtesy of FormFactor, Inc.)

2.1.1.1 Load-Lock

With a hermetic gate valve, the load-lock forms a sealed and isolated chamber that encloses the section of the sample rod below the 1K thermal anchor during sample exchange. As shown in Figures 2.1 and 2.2, the load-lock assembly begins with a stainless-steel vacuum tube. Inside the tube, tandem double O-rings provide a hermetic seal around the sample rod both during stationary and in motion. Externally, the vacuum chamber is equipped with two evacuation valves and a compound pressure gauge. The upper valve connects to a tube terminated between the double O-rings for differential pumping, while the lower one controls evacuation of the load-lock chamber. The gauge monitors the load-lock vacuum. A Teflon bushing near the top of the vacuum tube prevents the sample rod from contacting the tube inner surface and being scratched. The load-lock assembly is mounted directly on top of the hermetic gate valve at the cryostat's entry point (Figure 2.1 (a) and Figure 2.2 (b)).

2.1.1.2 Sample Rod

The sample rod assembly begins with the wiring manifold located at the top (Figure 2.3), followed by Ti-6Al-4V titanium tubing that is selected for its high mechanical strength and low thermal conductivity. Typically, a 0.25-inch diameter stainless-steel tube is installed concentrically inside the titanium tube to accommodate auxiliary devices such as fiber optics.

At the top of the titanium tubing is a brass knurled knob with female threads, referred to as the force generator (Figure 2.3). When threaded onto its brass mating component with male threads, located on the top of the load-lock vacuum tube, the force generator pushes the sample rod downward, ensuring good thermal contact at the tapered interface between the 1K stage and the 1K

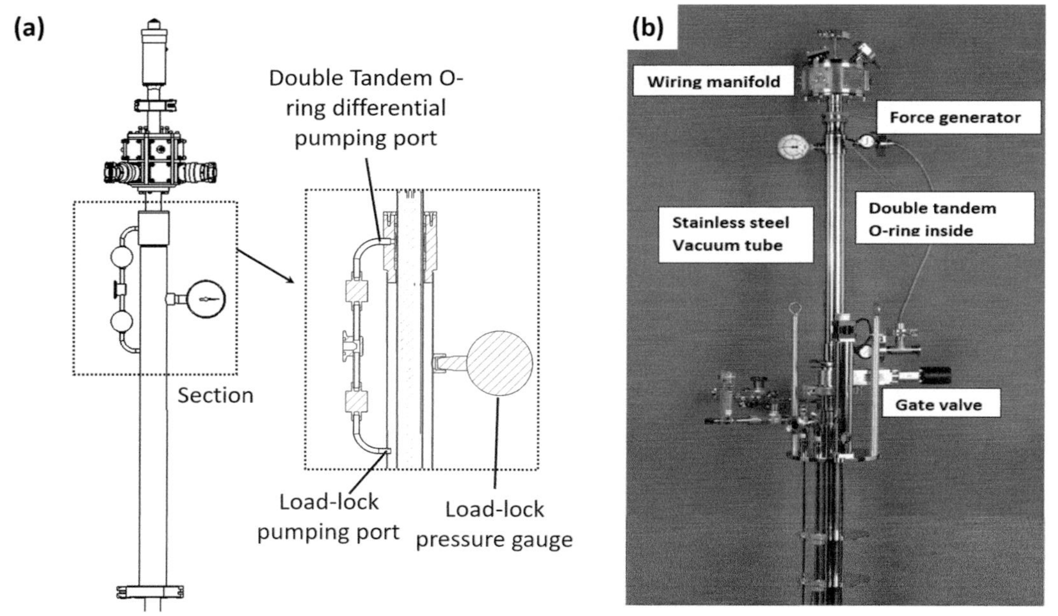

FIGURE 2.2 (a) Load-lock sketch of a top-loading assembly; (b) Load-lock photo. (Projects accomplished by authors at Janis Research Company. Image courtesy of FormFactor, Inc.)

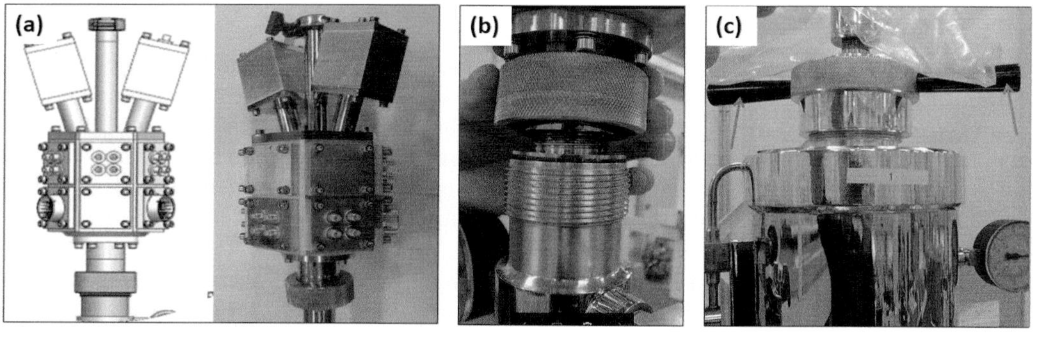

FIGURE 2.3 (a) Manifold of a sample rod assembly; (b), (c) Force generator. (Projects accomplished by authors at Janis Research Company. Image courtesy of FormFactor, Inc.)

thermal anchor on the sample rod. For applications requiring higher compressive force, handles can be attached to the knurled knob to generate greater torque, as shown in Figure 2.3. As the generator with handles can produce significant torques, extra care must be taken during use. The force generator should always be loosened during system warm-up to prevent bending of the sample probe due to differential thermal expansion.

The 1K thermal anchor stage (Figure 2.4 (a)) is mounted at the bottom end of the titanium sample tube and fabricated from gold-plated Oxygen-Free High Conductivity (OFHC) copper to ensure high thermal conductivity. This stage is designed to mate precisely with the tapered interface of the 1K pot in a "wet" cryostat to provide efficient thermal contact. Therefore, proper manufacturing and careful assembly of this stage are critical to guarantee reliable thermal performance. A slot is

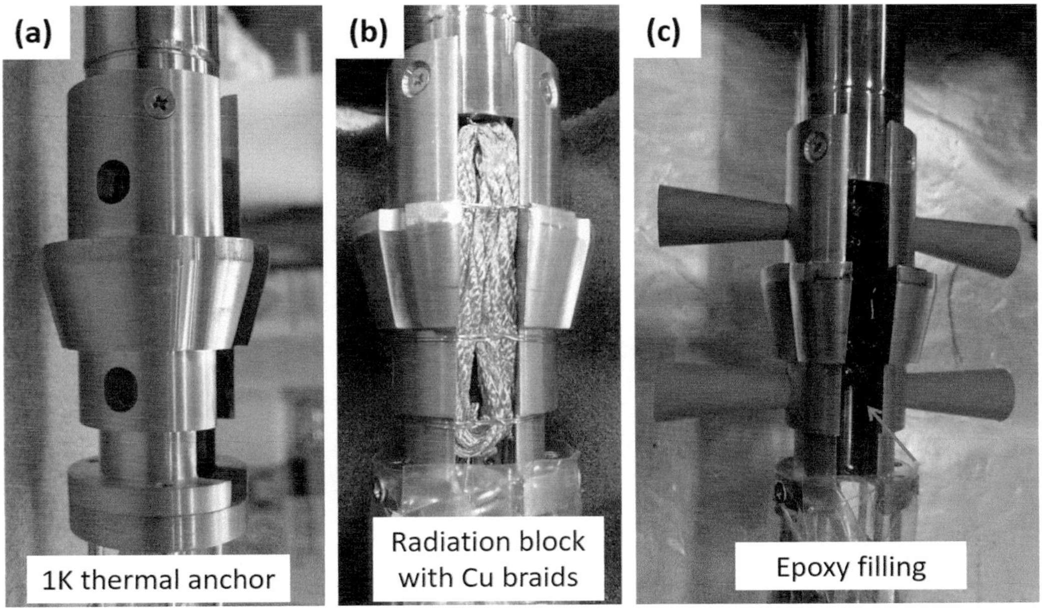

FIGURE 2.4 Tapered 1 K thermal anchor stage. (Projects accomplished by authors at Janis Research Company. Image courtesy of FormFactor, Inc.)

machined into the thermal anchor stage to thermally anchor electrical wires and RF cables, which should be fixed in place using Stycast 2850 epoxy (Figure 2.4 (c)). Alternatively, the wires can also be thermally anchored directly to the stage bottom surface.

The 1K pot stage also serves as a thermal radiation shield, replacing additional radiation baffles or shutters inside the cryostat. To further reduce radiation from room temperature, the remaining open space within the stage can be filled with OFHC copper braids as an infrared radiation shield (Figure 2.4 (b)).

Before opening the gate valve for a sample exchange, the lower evacuation valve on the load-lock must be opened to thoroughly pump the load-lock in order to maintain vacuum integrity and prevent He-3 from contamination. For the same purposes, during sample exchanges, the space between the two O-rings should be continuously pumped via the upper evacuation valve to remove any leaked air.

Due to the weight of the probe assembly and the pressure differential between the inside and outside of the load-lock, manipulating the sample rod can be physically challenging. A clamp (Figure 2.5) is commonly used to secure the sample rod in place and prevent it from unintentionally sliding downward. Special care must be taken when withdrawing the sample rod from a cold cryostat. If the sample rod is not properly handled and warmed, its cold surface could freeze the double O-rings, causing cracking or loss of elasticity. This may compromise the vacuum seal, resulting in air leaks into the He-3 space.

2.1.1.3 Sample Change Procedures on a Wet He-3 Cryostat with Sample in (He-3) Liquid

- Sample loading
  - Secure the sample rod using a clamp.
  - Pump the load-lock chamber and the space between the double O-rings.
  - Ensure that the He-3 pot is "dry" and the charcoal sorption pump is at the lowest possible temperatures (<5 K) to minimize the He-3 vapor pressure inside the sample space.

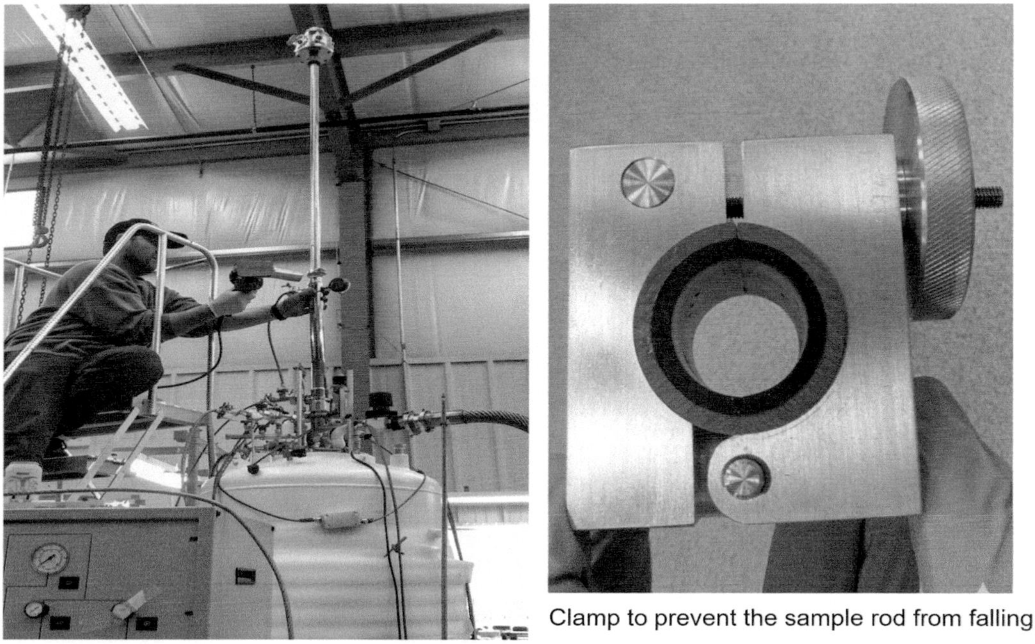

FIGURE 2.5 When pulling up the sample rod for sample change, the user is holding the sample rod and unfreezes the tandem O-rings. (Projects accomplished by authors at Janis Research Company. Image courtesy of FormFactor, Inc.)

  - Open the gate valve, and stop pumping the load-lock chamber. But continue pumping the double O-rings space.
  - Slowly insert the sample rod to the cryostat.
- Sample removal
  - Ensure the He-3 pot is "dry" and the charcoal sorption pump is at the lowest possible temperatures (< 5 K).
  - Pump the space between the double O-rings using a turbo pumping station.
  - Slowly retract the sample rod from the cryostat to prevent the O-ring from freezing.
  - Monitor the turbo pump inlet pressure. If the pressure suddenly increases, stop moving the sample rod immediately and secure it with a clamp. Wait until the O-rings thaw and the pressure falls below 1.0×10^{-5} mbar before resuming the retraction. A heat gun may be used to accelerate thawing (Figure 2.5).
  - Once the sample rod clears the gate valve, close the gate valve and proceed with sample exchange.

2.1.1.4 Discussion

During the sample change, a small amount of He-3 gas is present in the load-lock, and a small quantity loss is inevitable. Because the He-3 pot is "dry" and the charcoal temperature is maintained below 5 K, the He-3 pressure is lower than the saturated vapor pressure at 300 mK (≈0.0018 torr). Under these conditions, the estimated loss is less than 1.0×10^{-7} moles per sample exchange for a 1-liter load-lock, which is negligible and does not affect the cryostat's performance. However, care still must be taken to avoid introducing air into the sample region.

This type of He-3 cryostat does not require a sophisticated liquid helium dewar, and a bucket-type dewar is sufficient for most needs. Because the sample is immersed in liquid helium, a long sample delivery probe is necessary to minimize the heat load from room temperature to the cryogen. Depending on the experiment configurations, this type of cryostats typically reaches a base temperature of 300–320 mK. Additionally, the load-lock further offers rapid experiment turnover. Overall, the top-loading cryostat is considered as a user-friendly cryostat and particularly well-suitable for transport measurements.

2.1.2 Top-Loading He-3 Insert with Sample in Vacuum

Figure 2.6 illustrates an example of a top-loading He-3 insert with the sample in an ultra-high vacuum (UHV) environment (the sample transfer rod is not shown). Similar to the wet He-3 system discussed in Chapter 1, the core components consist of a charcoal sorption pump, a 1K pot, and a He-3 pot. However, instead of simple cylindrical structures, these components are built as concentric cylinders with large open centers, integrated with the sample delivery tube running from the room-temperature top plate to the IVC can, serving as the pathway for sample loading. In addition, all materials used in the vacuum space are UHV-compatible.

To minimize the room temperature radiation through the sample delivery tube, two radiation shutters are installed inside the IVC. One is positioned below the IVC 4K top flange, and another one is located below the 1K pot. They are thermally anchored to the top flange and the 1K pot, respectively, using flexible copper braids. These shutters remain closed during experiments and open only during sample transfer.

Similar to the top-loading sample-in-liquid He-3 cryostat, this system only needs a bucket-type liquid helium dewar. Both cryostats require a long probe with the load-lock mechanism for sample transfer. Unlike the sample-in-liquid cryostat, where the probe remains in place during experiments, the probe in this system is often completely removed after sample exchange. This type of cryostats is commonly used for scanning probe microscopy (SPM) measurements at sub-kelvin temperatures, where the SPM head is thermalized to the He-3 pot via thermally conductive springs/rods or through direct mounting. Depending on the experiment configurations, the cryostat typically reaches a base temperature of 270 mK.

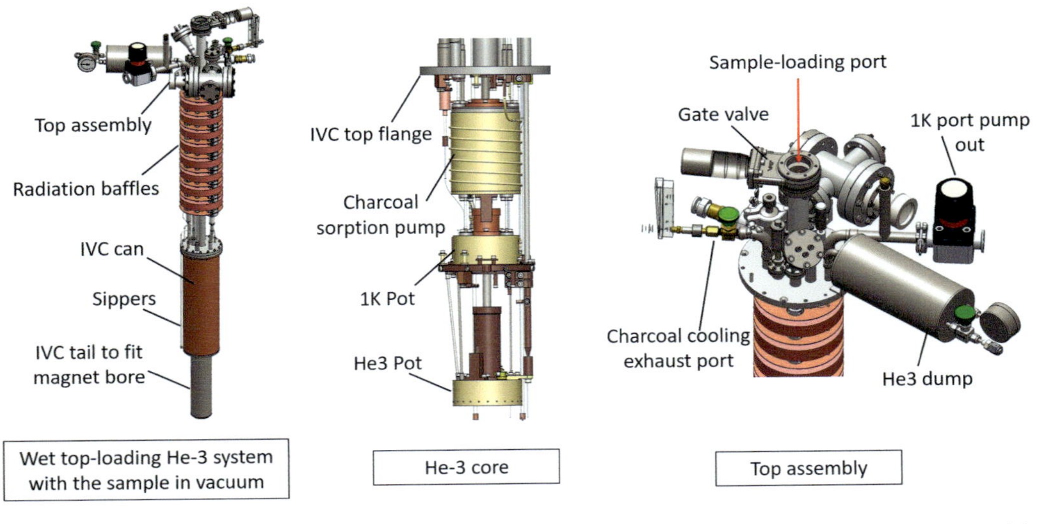

FIGURE 2.6 Typical top-loading (wet) He-3 insert with sample in vacuum. (Projects accomplished by authors at Janis Research Company. Image courtesy of FormFactor, Inc.)

2.2 BOTTOM-LOADING "WET" CRYOSTAT

2.2.1 Characteristic Features of Bottom-Loading "Wet" Cryostat

A bottom-loading cryostat is a system that allows the user to load or remove a sample from the bottom of the cryostat. Compared to the top-loading systems, it has a more complicated design and features a self-contained He-3 insert and an advanced liquid helium dewar.

The self-contained He-3 insert (Figure 2.7) includes not only the complete He-3 core (charcoal sorption pump, 1K pot, and He-3 pot) but also the entire radiation baffles. The insert is installed in a long IVC that is referred to as the sample tube (Figure 2.8). In this type of cryostat, the IVC top flange serves as the room-temperature plate. Additionally, a dedicated internal liquid helium reservoir is installed between the lowest radiation baffle and the charcoal sorption pump, which supplies liquid helium to the 1K pot and charcoal cooling lines. Although this reservoir has limited capacity due to space constraints, the design enables continuous helium transfer from the main dewar reservoir to it (see Figure 2.9).

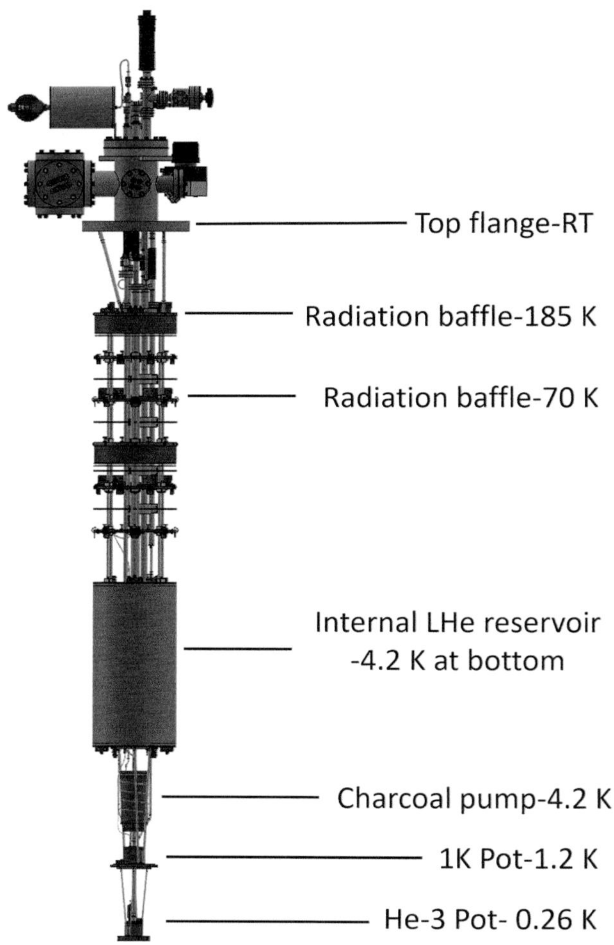

FIGURE 2.7 Self-contained He-3 insert of a bottom-loading cryostat and the temperature distribution. (Projects accomplished by authors at Janis Research Company. Image courtesy of FormFactor, Inc.)

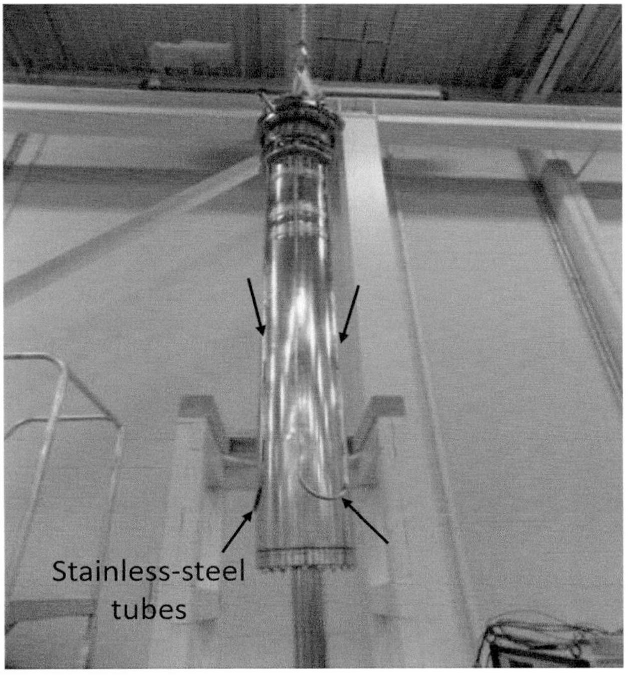

FIGURE 2.8 Stainless-steel tubes soldered on the outer surface of the sample tube. (Projects accomplished by authors at Janis Research Company. Image courtesy of FormFactor, Inc.).

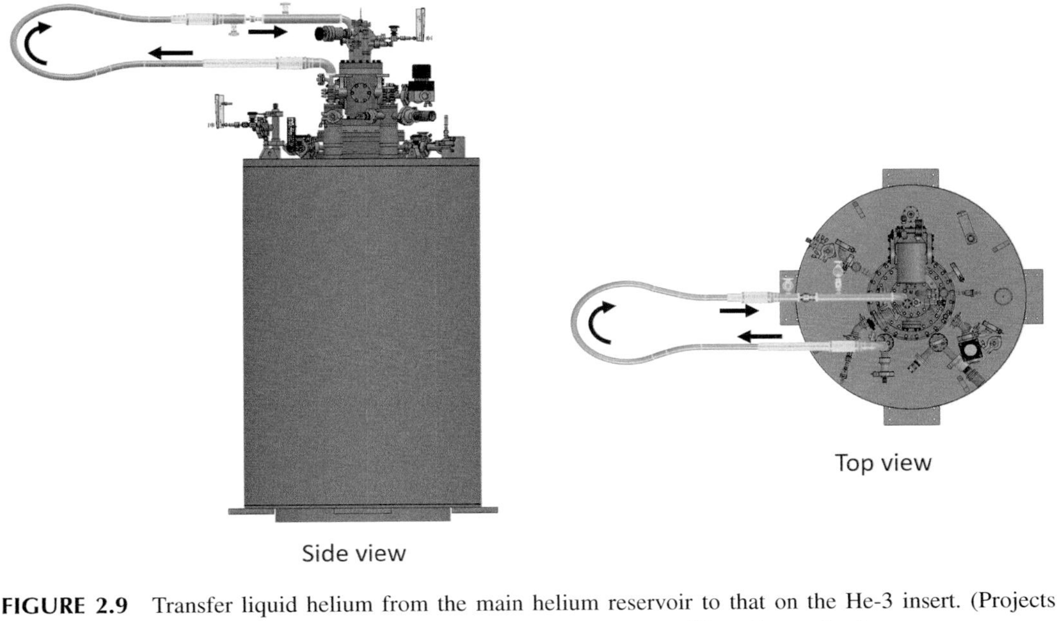

FIGURE 2.9 Transfer liquid helium from the main helium reservoir to that on the He-3 insert. (Projects accomplished by authors at Janis Research Company. Image courtesy of FormFactor, Inc.)

Because samples are loaded from the room temperature zone through the bottom of the dewar, multiple radiation shutters are required. They remain closed during experiments to block thermal radiation and open during sample transfer (Figure 2.10). These shutters are thermally anchored to the radiation bottom plates of the dewar at different temperature stages—typically at 4 K, 30 K, and 80 K, depending on the dewar design. To provide sufficient cooling for the bottom radiation plates, cooling loops made of 0.25-inch diameter copper tubing are soldered onto the plates. One end of the loop connects to the main liquid helium reservoir through a needle valve that controls the helium flow rate. Cold helium vapor flows through the loop and is vented to ambient (or a helium recovery system) from the opposite end of the loop. Furthermore, a liquid nitrogen jacket may be incorporated into the dewar as the 77K radiation to minimize the liquid helium loss rate. This type of cryostats (Figures 2.11 and 2.12) also incorporates the following unique features.

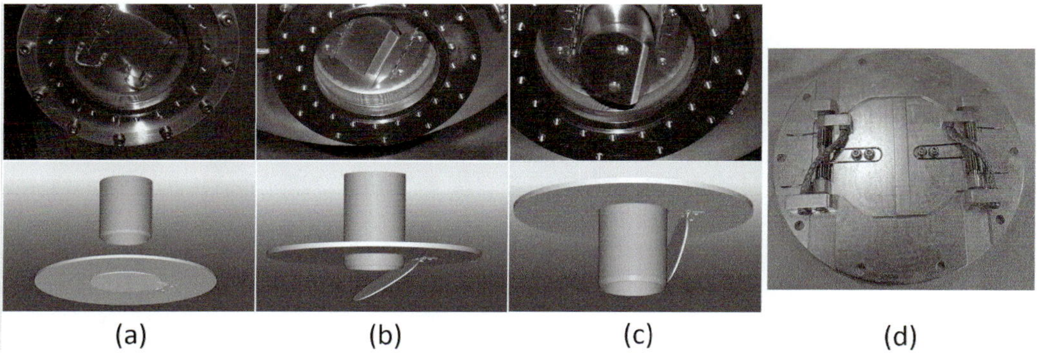

FIGURE 2.10 (a) Shutter closed before being opened by the insert; (b) the insert pushing the shutter open; (c) shutter fully open; and (d) photograph of the shutter installed on the bottom-loading He-3 cryostat. (Projects accomplished by authors at Janis Research Company. Photo courtesy of FormFactor, Inc.)

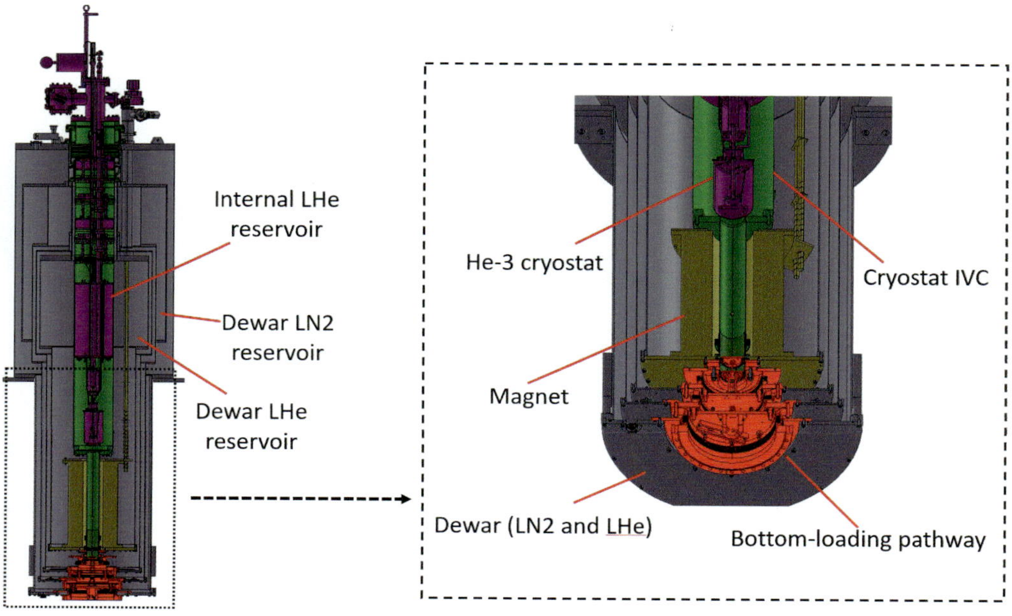

FIGURE 2.11 Custom-designed bottom-loading cryostat with fixed He-3 insert.

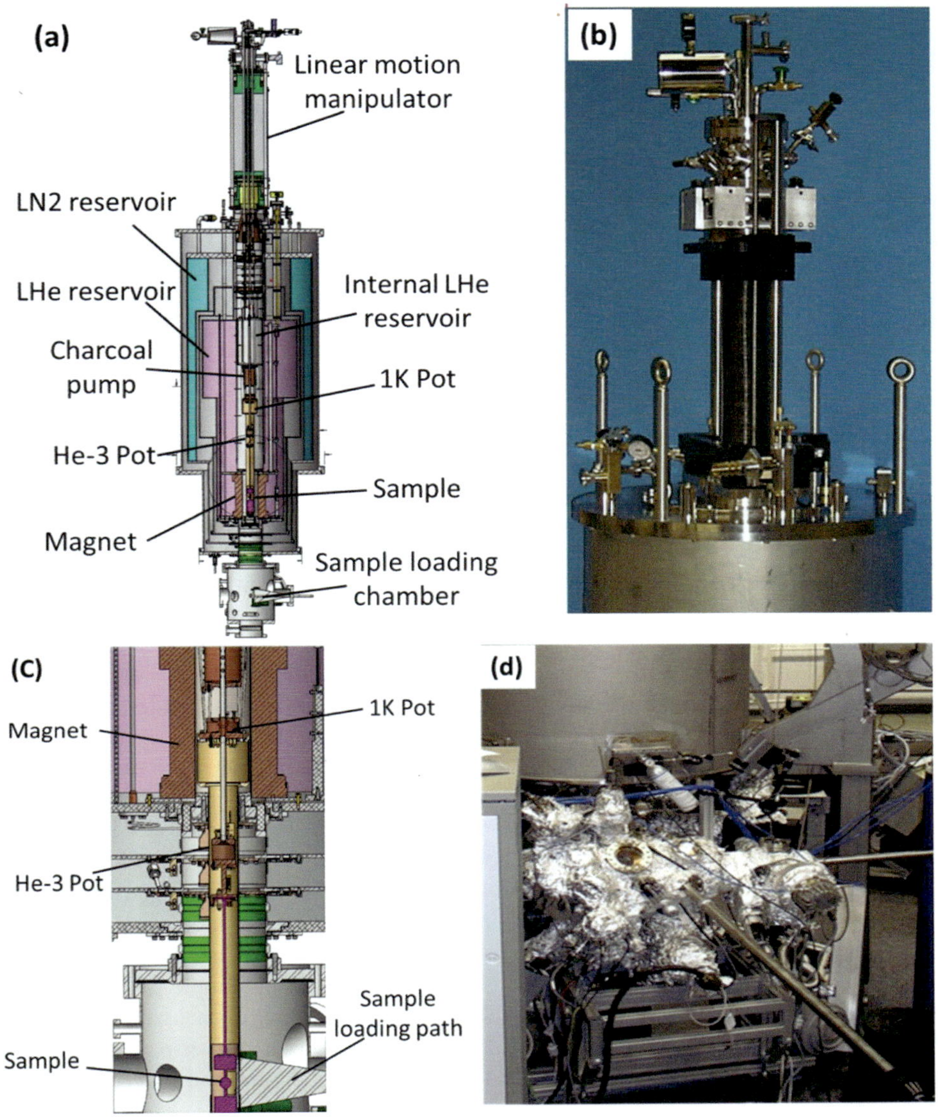

FIGURE 2.12 (a) Custom-designed bottom-loading cryostat with movable He-3 insert. (b) Linear motion manipulator. (c) Sample loading configuration and interfaces. (d) Sample loading chamber image. (Projects accomplished by authors at Janis Research Company. Image courtesy of FormFactor, Inc.)

1. **Bakeable Sample Tube**: A thin-wall stainless steel tube with the diameter of 0.25″ can be soldered onto the outer surface of the sample tube (Figure 2.8). This design makes the whole He-3 insert bakeable by circulating hot air through this tube. With further proper material selection, dedicated processing, and optimized design, the insert can be made UHV-compatible, suitable for surface science experiments such as STM and ARPES.
2. **Finger Stocks for Thermal Contact**: Unlike the vapor-cooled radiation baffles in the top-loading cryostats, the baffles in this system are enclosed in the sample tube under vacuum. Alternatively, Be-Cu finger stocks are installed on the radiation baffles of the He-3 insert,

FIGURE 2.13 Be-Cu finger stocks for thermal contact. (Projects accomplished by authors at Janis Research Company. Image courtesy of FormFactor, Inc.)

as illustrated in Figure 2.13. The finger stocks provide tight contact with the sample tube exposed to cold helium vapor, which effectively thermalizes the radiation baffles and results in a optimal temperature distribution during normal operation, as illustrated in Figure 2.7.

3. **Interfaces between the Inert and the Dewar**: When integrating the sample tube, dewar, and even a sample preparation/loading chamber, multiple spaces and volumes are involved. These include the liquid helium reservoir maintained slightly above ambient pressure, the dewar OVC (vacuum jacket), the ambient environment outside the dewar, and the sample tube. Despite the complexity, only two vacuum seals need to be considered, as the sample tube extends from the helium reservoir through the dewar OVC and connects to a preparation/loading chamber on the ambient side (Figure 2.12 (c)). One is the metal seal between the sample tube and the bottom plate of the main liquid helium reservoir, which separates the sample space, liquid helium and the OVC vacuum environment. Another one is the O-ring seal between the sample tube and the dewar bottom plate at room temperature, which separates the sample space, the OVC vacuum environment and the ambient environment. The sample tube may then be connected to a preparation or loading chamber in the ambient side.
4. **Shorter Sample Probe**: Both the sample stage in the sample tube and the preparation/loading chamber are located near the dewar bottom, providing shorter sample delivery distances in both vertical and horizontal directions. This simplifies sample exchanges and enhances the mechanical stability of the cryostat. This design also offers a shorter path from the sample space to the vacuum pump, which enhances pumping efficiency and makes higher vacuum levels to be reached more easily.
5. **Improved Accessibility:** In the bottom-loading design, most of the cryostat's functional components are still located on the top of the insert. These include the valves and gauges of the charcoal sorption pump, the 1K pot, the He-3 pot and the dewar, cryogen transfer access for both dewar and the insert, and wiring manifolds. The area beneath the cryostat remains open and accessible, allowing easier sample preparation and experimental operation (Figures 2.11 and 2.12).

2.2.2 Two Designs of Bottom-Loading Cryostat

The bottom-loading cryostats can be further categorized into two subtypes: fixed He-3 insert and movable He-3 insert.

In the cryostat with a fixed He-3 insert, all components on the He-3 insert remain at fixed positions and do not move relative to the dewar (Figure 2.11). The sample must be vertically loaded into the cold He-3 stage from the bottom. The radiation shutters are typically installed on linear or rotatable manipulators that are operated from the dewar side. Copper braids are connected between each shutter and a dewar radiation cold plate to ensure effective thermalization.

In the cryostat with a movable He-3 insert, the whole insert is mounted to a linear motion manipulator with a typical travel range of 27 inches (Figure 2.12). This allows the sample to be lowered from the cold liquid helium dewar into a room temperature sample chamber attached to the bottom of the dewar. The sample is typically delivered to the sample stage from the side when the stage is lowered into the sample chamber. In this design, the radiation shutters are mounted on spring-loaded hinges. They are pushed open by the lowering He-3 insert and automatically closes when the insert is retracted. Copper braids are used to thermalize the shutters.

2.2.3 Summary

Bottom-loading cryostats require a dewar with a more sophisticated bottom design, which generally makes them more expensive than top-loading systems. However, they offer several practical advantages: the sample delivery probe has a shorter travel path, and the larger pumping port combined with the short distance between the pump and the sample space makes it easier to achieve UHV conditions. Although the cryostat loses the vapor cool ability for the radiation baffles, the figure-stock design and the internal liquid-helium reservoir still provide a optimal temperature distribution on the insert. Bottom-loading "wet" He-3 cryostats typically achieve a base temperature of 270 mK or lower, depending on the wiring configuration and other installed components. Based on customer feedback, bottom-loading cryostats offer a much higher signal-to-noise ratio, making them particularly well-suited for surface science measurements.

2.3 BOTTOM LOADING "WET" DR

A similar design used for bottom-loading cryostats with a fixed He-3 insert is applied to DRs (Figure 2.14), in which the DR inserts remain stationary during sample changes. This type of DR cryostat can achieve a base temperature of 8 mK or lower and, compared to the top-loading DRs, offers a much higher signal-to-noise ratio and more stable temperatures, making it ideal for SPM experiments.

2.4 ALTERNATIVE SHUTTER DESIGNS

Besides the "push-open" shutter described above, we also developed several different types of shutters. In principle, all of the shutter designs can be applied to both He-3 systems and DRs.

2.4.1 Sliding Shutter on Flat Surface

The sliding shutter is typically mounted on rails fixed to a flat plate, and a linear motion manipulator is used to move it along the rails (Figure 2.15). To prevent galling or seizing, the shutter and rail should be made from different materials. In the design, the shutter material is gold-plated OFHC copper, while the rails are made of PEEK. Copper braids are used as thermal links to connect the shutter to a cold surface. Springs may also be added to provide a restoring force to the shutter. In cryogenics, flat plates are typically installed horizontally, and the shutter therefore moves in horizontal directions. This design significantly reduces the required vertical space and minimizes the overall system height. Sliding shutters are commonly used in bottom-loading systems with fixed inserts.

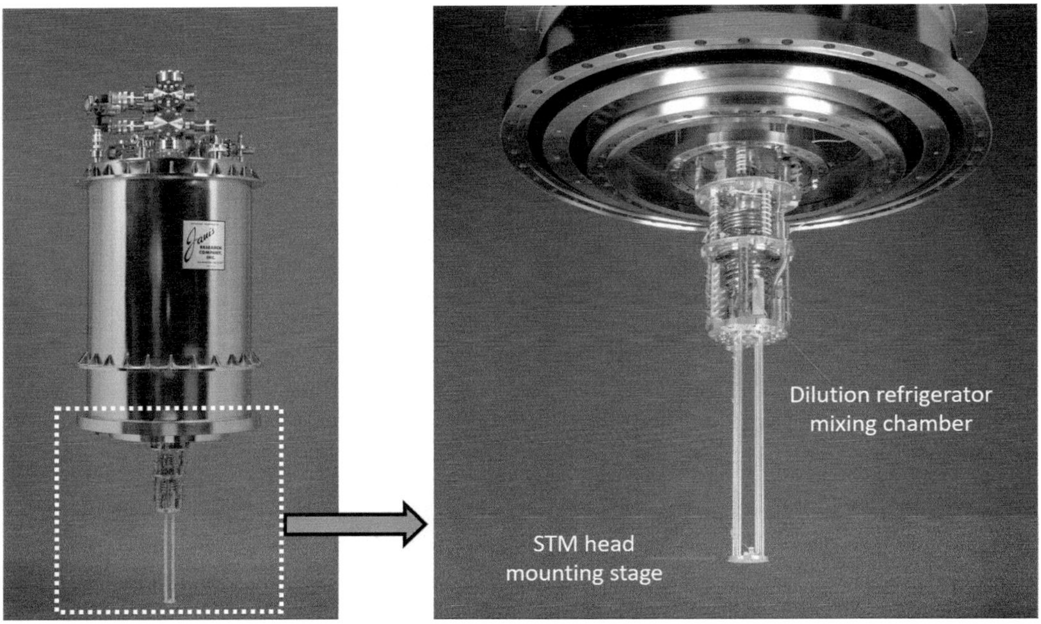

FIGURE 2.14 Bottom loading wet dilution refrigerator. (Projects accomplished by authors at Janis Research Company. Image courtesy of FormFactor, Inc.)

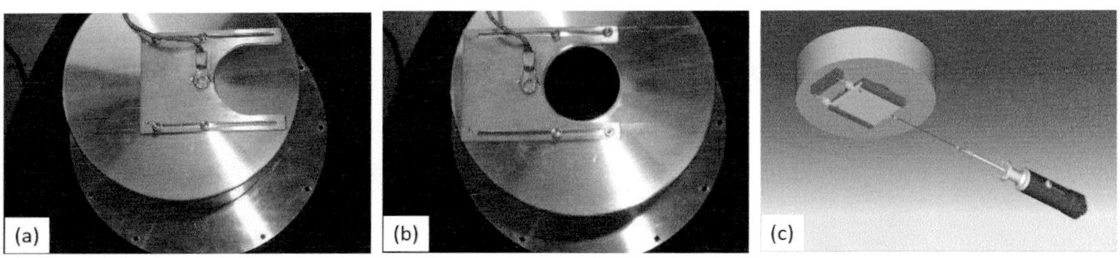

FIGURE 2.15 (a) Shutter closed, (b) shutter open, and (c) linear motion manipulator operates the shutter with the spring for the restoring force. (Projects accomplished by authors at Janis Research Company. Image courtesy of FormFactor, Inc.).

2.4.2 Flap-Open Shutter

The flapping shutters shown in Figure 2.16 require rotating manipulators for operation and are typically used in bottom-loading systems. Each shutter is generally actuated by a dedicated rotary manipulator. Copper braids are used as thermal links to connect the shutter to a cold surface. Because the shutters open in the vertical plane, the actuation mechanism requires additional vertical clearance compared with a sliding shutter.

2.4.3 Rotating Shutter

This design employs a rotating manipulator to open the shutter within the horizontal plane (Figure 2.17). Copper braids are also used as thermal links to connect the shutter to a cold surface. It

FIGURE 2.16 (a) Shutter is open. (b) Operating rotating manipulator. (Projects accomplished by authors at Janis Research Company. Image courtesy of FormFactor, Inc.)

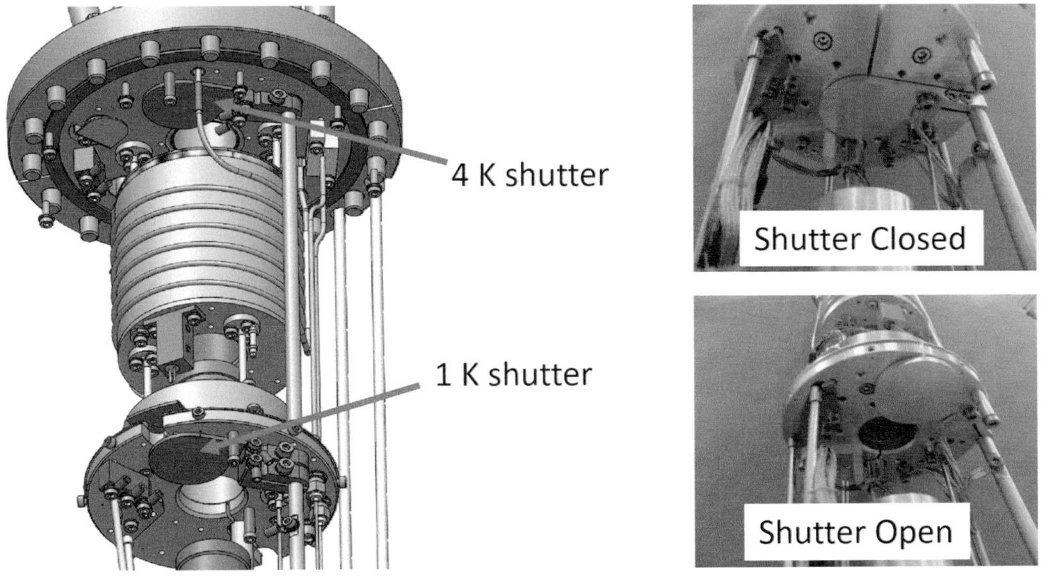

FIGURE 2.17 Rotating shutter. (Projects accomplished by authors at Janis Research Company. Image courtesy of FormFactor, Inc.)

is typically implemented in top-loading systems, where the shutters are integrated with heat switches for the initial cooldown of He-3 pots or the mixing chambers in DRs.

2.4.4 Side-Sliding Shutter

A side-sliding shutter is required when the sample is loaded from the side. The shutters are mounted on a curved side surface (Figure 2.18) and are actuated vertically or horizontally using wobble sticks. This design has been widely used for low-temperature microscopes and, for us, in bottom-loading cryostats with movable He-3 inserts and high-power He-3 cryostats for sub-K ARPES

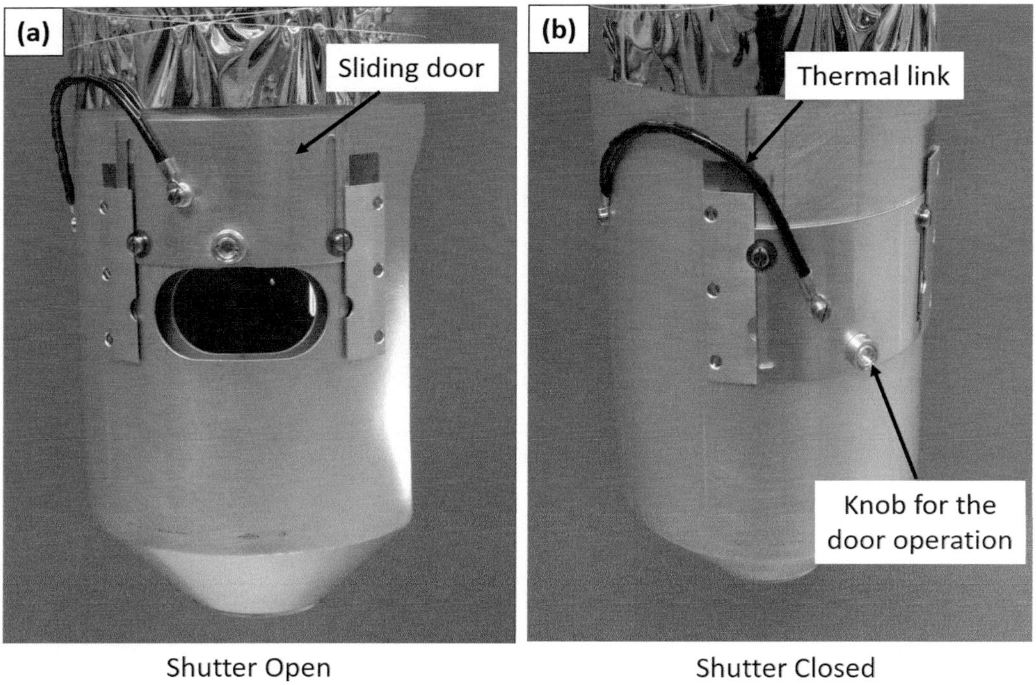

FIGURE 2.18 Sliding shutter. (Projects accomplished by authors at Janis Research Company. Image courtesy of FormFactor, Inc.)

measurements. Multiple side-sliding doors are often required for a single cryostat, which increases both the design complexity and operational challenges.

2.5 SAMPLE STORAGE AND SAMPLE HOLDER DESIGN

2.5.1 Sample Storage Assembly

In a top-loading cryostat, such as He-3 system or DR, it is highly desirable to maintain low temperatures while loading a sample. This requires precooling the sample before it is inserted into the sample stage. Fortunately, both He-3 cryostat and DR have stages that provide substantial large cooling power at the temperatures close to that of the sample stage. These include the 1K pot in a wet He-3 cryostat, the Pulse Tube Refrigerator (PTR) second stage in a cryogen-free He-3 cryostat, and the Still stage in DR. These stages will be collectively referred to as the *cold stages* throughout the discussion. To ensure efficient sample precooling, a sample storage assembly is installed at a cold stage. Figure 2.19 shows an example of such an assembly installed on a cold stage—the 1K pot.

2.5.1.1 Design and Components

The sample storage assembly consists of the following main components:

- Assembly base plate: It is made of OFHC copper and bolted to the cold stage to ensure efficient thermalization of the entire assembly.
- Thermal storage cart with a sample holder dock and a removable sample holder: The removable sample holder with a pre-loaded sample can be placed on the dock for precooling. The

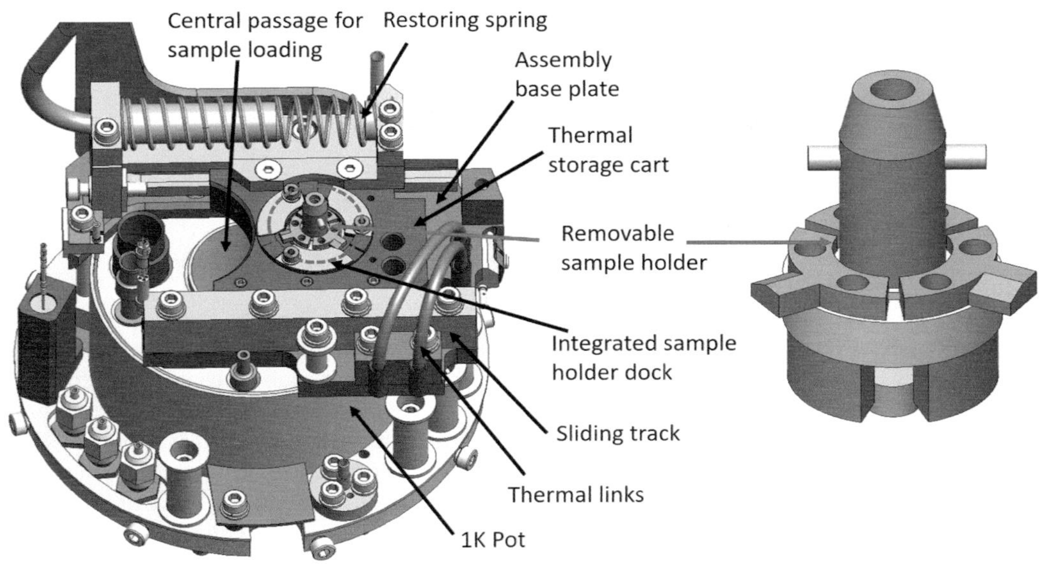

FIGURE 2.19 Sample storage assembly. (Projects accomplished by authors at Janis Research Company. Model courtesy of FormFactor, Inc.)

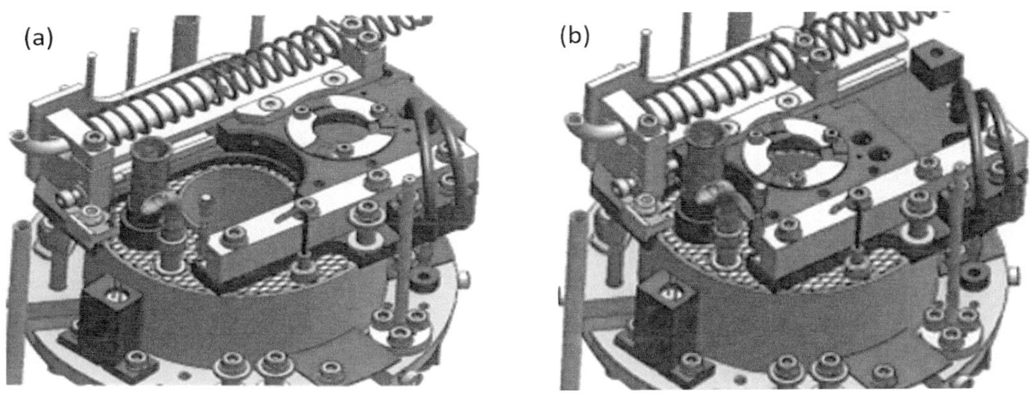

FIGURE 2.20 (a) Storage cart in closed position (sample loading position). (b) Storage cart in open position. (Projects accomplished by authors at Janis Research Company. Model courtesy of FormFactor, Inc.)

cart, which carries the holder dock, slides along the track and also functions as a 1K radiation shutter.

- Thermal links: They are made of OFHC copper braids to thermalize the storage cart to the base plate.
- Sliding track: The cart moves along this track, either blocking the central access (acting as a shutter) or leaving it open to load the sample to the He-3 stage.
- Restoring spring: Connected to the cart via a stainless-steel cable, the restoring spring pulls the cart back to the central "closed" position, blocking the access path, which is similar to the sliding shutter mechanism described earlier.

2.5.1.2 Operation

- Position the cart at the center of the Cold Stage—the "closed" position.
- Load the sample holder onto the dock for pre-cooling.
- Once the sample is precooled, remove the sample holder from the dock and move the cart to the open position, as shown in Figure 2.20 (a).
- Load the sample holder onto the He-3 stage, then return the cart to the closed position, as shown in Figure 2.20 (b).

2.5.2 Rotational Sample Holder

The study of material properties as a function of lattice plane orientation—defined by the crystal's Miller indices relative to the experimental setup—is critically important across condensed matter and materials science. The orientations are associated with many key experimental conditions, including the directions of external magnetic fields, applied pressure, and incident radiation. These orientation-dependent studies are widespread across multiple disciplines, including anisotropic transport measurements in magnetic fields [1–7], ARPES [8, 9], neutron diffraction [10, 11], X-ray diffraction [12], and broader materials science research [13–15]. Because adjusting the orientation of external conditions such as magnetic fields is often limited or even not practical, angular control is typically achieved by rotating the sample installed on a rotating sample holder, usually within the vertical plane around a horizontal axis. The following sections present the rotating sample holder design from us and introduce both commercial and other customized sample holders.

2.5.2.1 Structure

Figure 2.21 illustrates the structure of a low-friction rotator.

Two wheels, integrated with the sample holder, are mounted within a yoke. Each wheel is equipped with a sapphire VEE bearing, supported by Navipoint pivots installed on the yoke, as shown in Figure 2.22. This configuration enables the wheels to rotate freely with minimal friction.

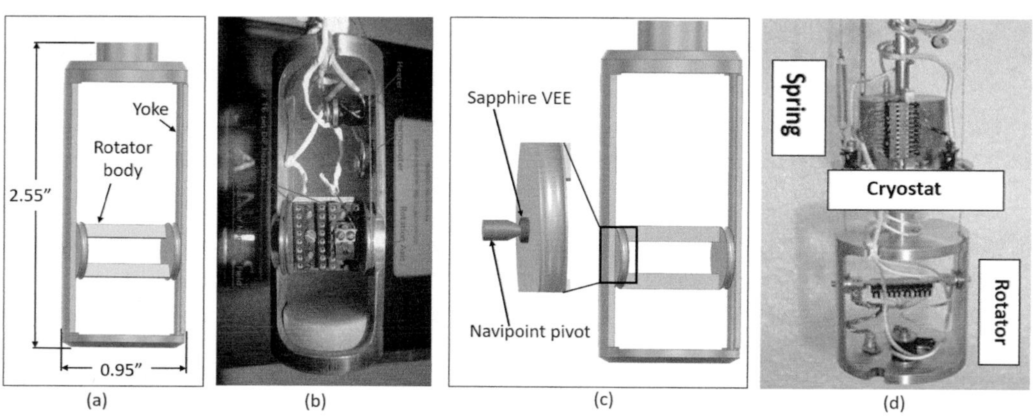

FIGURE 2.21 Low friction rotating sample holder. (Projects accomplished by authors at Janis Research Company. Image courtesy of FormFactor, Inc.)

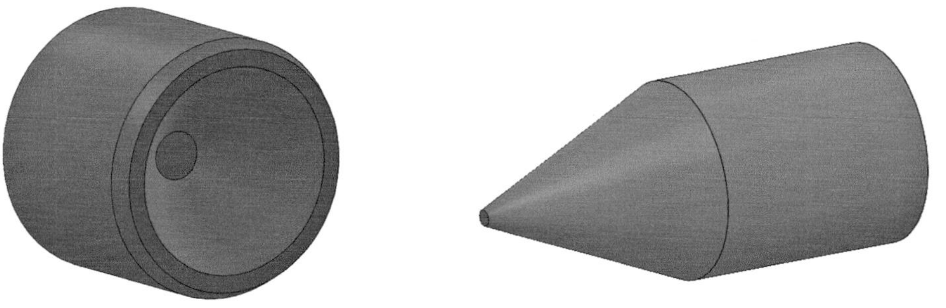

FIGURE 2.22 Sapphire VEE and NaviPoint pivot [16].

The user can pull a thread attached to one of the wheels to rotate the sample holder along the horizontal axis. A restoring extension string is connected to another wheel, providing a counteracting force to balance the pulling force. When the pulling string is released, the rotator returns in the opposite direction under the continuous tension of the restoring spring. Because the wheel-sample holder assembly generates minimal heat during rotation, this rotator is well-suited for applications at ultra-low temperatures.

2.5.2.2 Discussion

a. This sample holder design is ideal for a sample-in-liquid cryostat. When operated in vacuum, however, two thermal links need to be added to the rotator, as illustrated in Figure 2.23 (a). The first link is a solid piece of OFHC copper to thermally anchor the "Yoke" to the cryostat for cooling. The second one is a flexible OFHC copper braid that connects the "Yoke" to the rotating sample holder, ensuring thermal equilibrium between them. The copper braid must be sufficiently long and flexible enough to avoid interference with the motion of the rotator.
   In systems such as ARPES, where the sample holder can generate heat on the order of 10 to 15 mW, it is essential to optimize the flexible thermal link by incorporating additional copper braids to minimize the temperature gradient along it, as illustrated in Figure 2.23 (b).
b. The pulling string can be made from thin, flexible stainless-steel wire, Kevlar threads, or similar materials. A suitable choice is McMaster-Carr's 1×7×0.018" flexible stainless-steel wire. If nonmagnetic properties are required, the wire should be annealed.
c. Because stainless-steel extension springs are typically strongly magnetic, nonmagnetic beryllium copper (Be–Cu) springs are often preferred for the restoring spring. For example, the spring is fabricated from 0.032″-diameter Be–Cu Alloy 25 (CDA 172) wire in the hard, stress-relieved temper. The wire is wound onto a 0.125″-diameter stainless-steel mandrel, unless otherwise specified.
d. To maintain alignment of the sample with the rotation axis during motion, the sample holder should be offset slightly to compensate for the sample's thickness.
e. The rotator is typically enclosed within a cylindrical space. While the width of the sample is limited by the spacing between the two wheels, the length can extend close to the cylinder's inner diameter, as illustrated in Figure 2.24. This space is generally sufficient to accommodate a rectangular multipinchip holder.

FIGURE 2.23 (a) Rotator with flexible thermal link, operated in a vacuum. (b) Rotator with optimized flexible thermal link for sub-K ARPES cryostat. (Projects accomplished by authors at Janis Research Company. Image courtesy of FormFactor, Inc.)

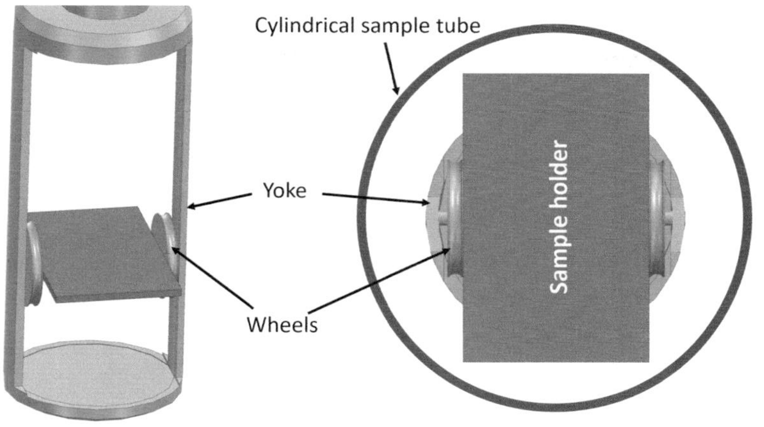

FIGURE 2.24 Allowed sample size on the rotating sample holder.

2.5.2.3 Key Design Parameters

1. Rotating angle: The sample holder is designed to rotate from 0° to 90°, which is sufficient for most angle-dependent measurements. In special cases, the rotation range can be extended up to 180° if needed.
2. The travel distance of the manipulator and rotating angle resolution

   The MDC standard linear motion manipulator is commonly used to operate the rotator. It is constructed from high-grade, vacuum-compatible materials. Linear position is measured along a laser-etched, black-anodized barrel graduated in 0.001" increments. One full revolution of the barrel corresponds to 0.025" of linear travel, as shown in Figure 2.25.

   Assuming the wheel diameter is 0.5", the linear motion manipulator must move a distance equal to the wheel's circumference,

$$\text{Circumference} = \pi \times 0.5 \approx 1.57''$$

   for the sample holder to complete a full 360° rotation.

   Given the 0.001" graduation increments, the angular resolution can be calculated as:

$$\text{Resolution} = (360 \times 0.001) \div 1.57 \approx 0.23° \quad (2.1)$$

   Linear motion is transmitted through the chamber wall via precision fine-pitch lead screws coupled to bellows-sealed, bearing-supported shafts [17]. A snap ring installed inside the linear motion manipulator services as a retention mechanism, securing the linear motion components by fitting into a groove on the shaft to prevent unintended movement or dislodging during operation. However, the snap ring is not designed to withhold high loads and may slip out of the groove, leading to unwanted shaft movement or dislodging. To mitigate this risk, the user can bind the two keyholes together to prevent the snap ring from opening and slipping out, as shown in Figure 2.26 (b). A more robust solution is to replace the snap ring with a spiral retaining ring, secured by a pair of retaining clamps, providing enhanced stability under higher loads.

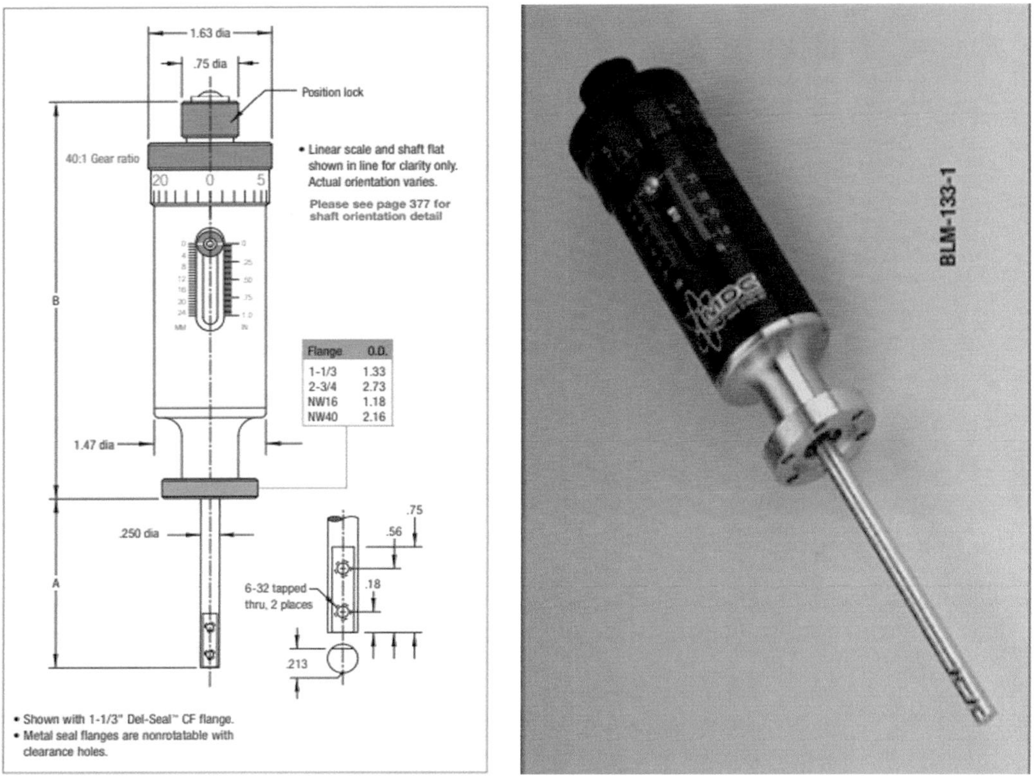

FIGURE 2.25 Linear motion manipulator and rotating resolution. (Reprint with permission from MDC Precision, LLC, www.mdcprecision.com/, [17].)

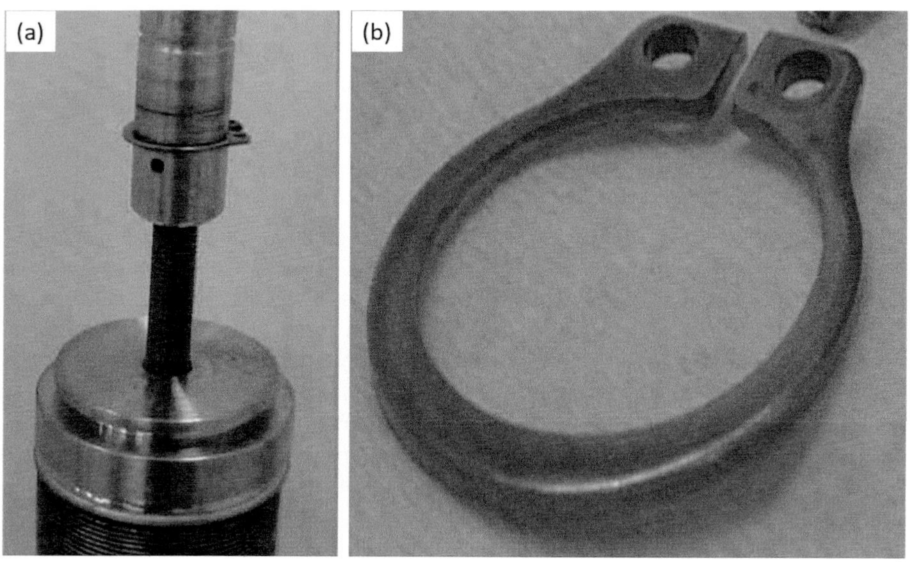

FIGURE 2.26 (a) Snap ring installed on the linear motion manipulator shaft. (b) Secure the snap ring to prevent it from opening. (Projects accomplished by authors at Janis Research Company. Image courtesy of FormFactor, Inc.)

3. Wheel diameter
   The wheel diameter is determined by the specific cryostat design. While a larger wheel diameter offers a higher angular resolution, it occupies more space within the system. Additionally, for the same rotation angle, a larger wheel requires greater spring deflection and a longer travel distance from the linear motion manipulator to achieve the desired rotation range.

2.5.3 Commercial Rotators

Several popular commercial rotators are currently available on the market. The nonmagnetic universal rotator from Oxford Instruments, also known as "Swedish rotator", is shown in Figure 2.27 (a)] [18]. This rotator, constructed from Torlon—a non-magnetic material—provides a 15 × 15 mm^2 sample space and allows up to 260° of polar rotation in high magnetic field environments. It is suitable for operation at millikelvin temperatures in high magnetic fields.

Other commonly used rotators include, but not limited to, the Dual Axis Rotator, the Compact Rotator with 360° Rotation Range from IceOxford [19], and the Attocube Compact Rotator with Horizontal Rotation Axis [20].

2.5.4 Customized Rotators by Research Groups

Numerous research groups have developed customized rotators tailored to their specific experimental requirements. Notable examples include a piezoelectric rotator designed to investigate quantum effects in semiconductor nanostructures under high magnetic fields and low temperatures [21]; a stable, five-axis cryogenic photoemission manipulator that functions without a differentially pumped rotary feedthrough [22]; and a rotating sample holder engineered without mechanical linkages [23], etc.

Two recent developments in rotating sample stages are introduced below.

2.5.4.1 Example #1: Rotating Sample Holder for Sub-K ARPES Cryostat

Different angles between the incoming beam line and the sample surface are key parametera for ARPES measurement.

Figure 2.28 shows a highly specialized 6-axis sample stage driven by piezo motors, designed for ARPES [24]. The sample on this stage can move along the X, Y, and Z directions. Simultaneously, it can rotate in the polar, tilt, and azimuthal angles, with individual range listed in Table 2.1.

FIGURE 2.27 Nonmagnetic universal rotator for high magnetic field and low temperatures (Reprint with permission from Oxford Instrument) [18].

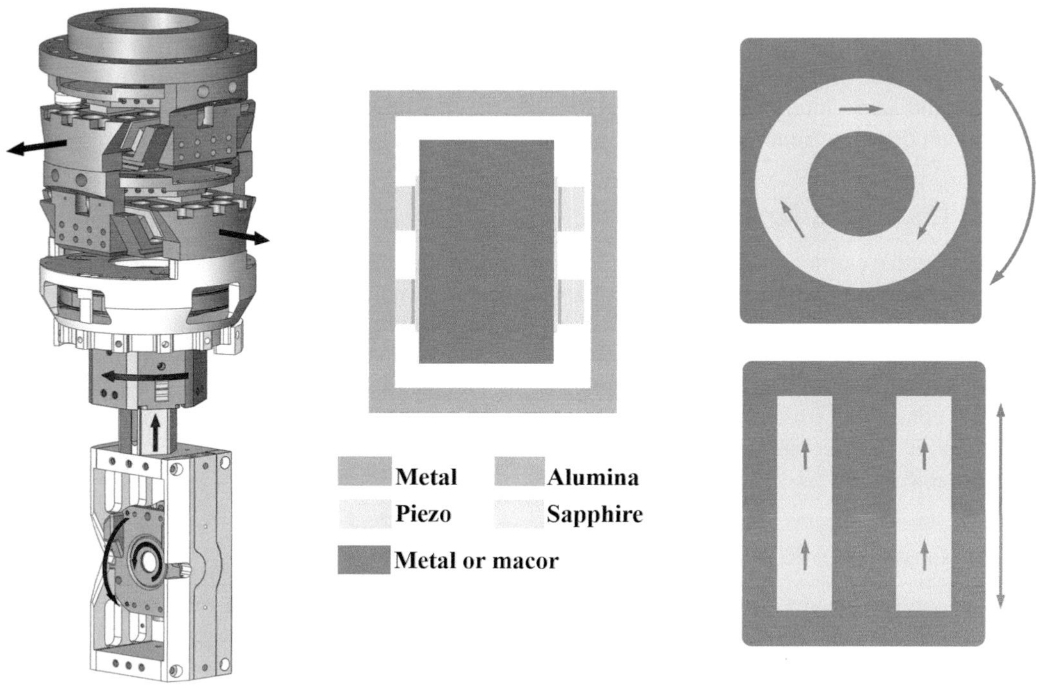

FIGURE 2.28 Six-axis sample stage driven by piezo motors, designed for ARPES [24].

TABLE 2.1
Moving/Rotation Range [24]

Axis	X	Y	Z	Polar	Tilt	Azimuthal
Motion range	±2.5 mm	±2.5 mm	±1.6 mm	165°	95°	65°

TABLE 2.2
Slot Locations and Dimensions [24]

Location	Width (mm)	Length (mm)
40K tail	3.8	22.2
3K tail	3.6	19.5

This sample stage is mounted at the bottom of a sub-Kelvin high-power He-3 system for ARPES measurement. This sub-K ARPES system has the slots as listed in Table 2.2. The analyzer is located right at the vacuum shroud at room temperature.

The sample temperature reaches 0.51 K with slots closed and 0.8K when the slots are open, respectively. The project originator [24] believes that there are rooms to improve, particularly to improve the thermal link.

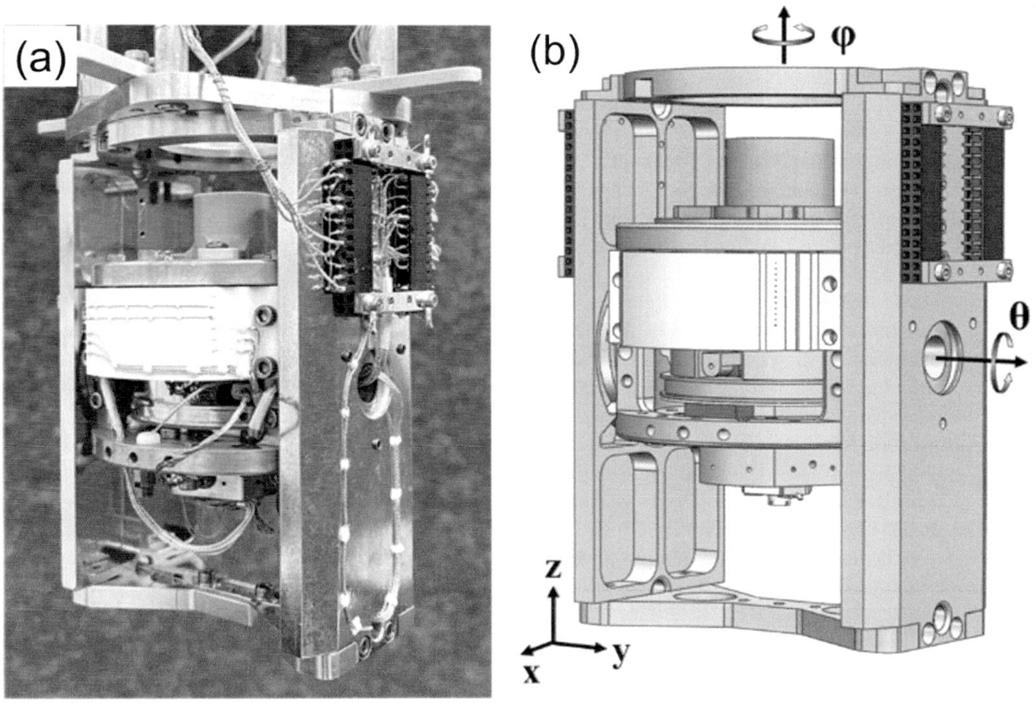

FIGURE 2.29 Two-axis piezoelectric rotating STM stage [24].

2.5.4.2 Example #2: Rotating STM Stage in a High-Field Solenoid

Measurements at varying angles between the sample and an applied high magnetic field have long been sought by STM users. The conventional solution involves the use of a vector magnet. However, most commercially available vector magnets are limited to a maximum rotating magnetic field of approximately 2 tesla.

To overcome this limitation, a two-axial piezoelectric STM rotating stage was developed [24], as illustrated in Figure 2.29, and installed inside an 18-tesla solenoid.

The key innovation is rotating the STM stage itself—rather than the magnetic field—within a fixed high-field solenoid. This novel approach enables STM measurements at arbitrary angles between the sample and the magnetic field, while preserving access to high magnetic field strengths.

The rotating STM stage, mounted on a wet DR, achieved a base temperature below 4 mK, enabling ultra-low-temperature experiments.

This advancement significantly expands the scope of STM-based research in high magnetic field environments.

REFERENCES

1. Chang-Ning Pan [a b], Xue-Kun Chen [a], Li-Ming Tang [a], and Ke-Qiu Chen [a], Orientation Dependent Thermal Conductivity in Graphyne Nanoribbons, *Physica E: Low-dimensional Systems and Nanostructures*, 64, 129–133, (November 2014).
2. Devin Todd[2,1], Maximilian Schwager[2,1], and Walter Mérida[3,1], Three-Dimensional Anisotropic Electrical Resistivity of PEM Fuel Cell Transport Layers as Functions of Compressive Strain, *Journal of The Electrochemical Society*, 162(3), F265–F272, (2015).
3. Jialu Wang, Haiyang Yang, Linchao Ding, Wei You, Chuanying Xi, Jie Cheng, Zhixiang Shi, Chao Cao, Yongkang Luo, Zengwei Zhu, Jianhui Du, Mingliang Tian, & Yuke Li, Angle-Dependent

Magnetoresistance and Its Implications for Lifshitz Transition in W_2As_3, *Quantum Materials*, 58, 1–7, (2019).
4. Paul Branch, Kozo Osamura, and Damian Hampshire, Weak Emergence in the Angular Dependence of the Critical Current Density of the High Temperature Superconductor Coated Conductor REBCO, *Superconductor Science and Technology*, 33(10), 104006, (3 September 2020).
5. Jeff Walter, Shameek Bose, Mariona Cabero, Maria Varela, and Chris Leighton, Giant Anisotropic Magnetoresistance in Oxygen-Vacancy-Ordered Epitaxial La0.5Sr0.5CoO3−δ Films, *Physical Review Materials*, 4, 091401(R) – Published, (17 September 2020).
6. E.C. Palms and T.P. Murphy, Very Low Friction Rotator for Use at Low Temperatures and High Magnetic Fields, *Review of Scientific Instruments*, 70(1), (January 1999).
7. A. Schenstrom, M.F. Xu, D. Bein, M. Levy, B.K. Sarma, S. Adenwalla, Z. Zhao, T. Tokoyasu, D. Hess, J.B. Ketterson, J.A. Sauls, and D.G. Hinks, Anisotropy of the Magnetic-Field-Induced Phase Transition in Superconducting UPt3, *Physical Review Letters*, 62, 332, (1989).
8. Xingjiang Zhou, *Lecture: Introduction to ARPES* as part of the China/US Joint Winter School on Novel Superconductors, Hongkong, (2013).
9. Bum Joon Kim, Hyeong-Do Kim, Deok-Yong Cho, Myonglin Kim, S.-J. Oh, and Changyoung Kim, Stable Five Axes Cryogenic Photoemission Manipulator without a Differentially Pumped Rotary Feedthrough, *Review of Scientific Instruments*, 76, 063910, (2005).
10. J. Fang, C. L. Bull, H. Hamidov, J. S. Loveday, M. J. Gutmann, R. J. Nelmes, and K. V. Kamenev, A rotator for single-crystal neutron diffraction at high pressure, *Review of Scientific Instruments*, 81, 113901, (2010).
11. B. Heacock, D. Sarenac, D.G. Cory, M.G. Huber, D.S. Hussey, C. Kapahi, H. Miao, H. Wen, and D.A. Pushin, Angular Alignment and Fidelity of Neutron Phase-Gratings for Improved Interferometer Fringe Visibility, *RESEARCH ARTICLE*, (AUGUST 19 2019).
12. Rob Delhez, E.J. Mittemeijer, Th H. de Keijser, and H.C.F. Rozendaal, Corrections for the Angle Dependence of Lorentz Polarization and Structure Factors in X-ray Diffraction Line Profiles, *Journal of Physics E Scientific Instruments*, 10(8), 784, (February 2001).
13. T.R.S. Prasanna, *Angle Dependence of the Integrated Intensity in Thin Film X-ray Diffraction*, Department of Metallurgical Engineering and Materials Science Indian Institute of Technology, Bombay Powai, Mumbai 400076, India. (2016). DOI:.48550/arXiv.1612.01805.
14. Zhaoyang Zhang, Yuan Feng, Feng Li, Sergei Koniakhin, Changbiao Li, Fu Liu, Yanpeng Zhang, Min Xiao, Guillaume Malpuech, and Dmitry Solnyshkov, Angular-Dependent Klein Tunneling in Photonic Graphene, *Physical Review Letters*, 129, 233901, (30 November 2022).
15. S. Adenwalla, S.W. Lin, Q.Z. Ran, Z. Zhao, J.B. Ketterson, J.A. Sauls, L. Taillefer, D.G. Hinks, M. Levy, and Bimal K. Sharma, Phase Diagram of Superconducting UPt3, *Physical Review Letters*. 65, 2298, (1990).
16. Swiss Jewel Company 555 E. City Avenue, Bala Cynwyd, PA, USA Email: sales@swissjewel.com Website: www.swissjewel.com
17. MDC Precision, LLC 111 Pinnacle Drive, Foothill Ranch, CA, USA Website: www.mdcprecision.com
18. Oxford Instruments plc Tubney Woods, Abingdon, Oxfordshire, UK Website: www.oxinst.com
19. Ice Oxford Eden House, Avenue Four, Witney, Oxford, UK Email: support@iceoxford.com; Website: www.iceoxford.com
20. Attocube Systems AG Eglfinger Weg 2, Haar, Germany Email: info@attocube.com Website: www.attocube.com
21. L.A. Yeoh, A. Srinivasan, T.P. Martin, O. Klochan, A.P. Micolich, and A.R. Hamilton, Piezoelectric Rotator for Studying Quantum Effects in Semiconductor Manostructures at High Magnetic Fields and Low Temperatures, *Review of Scientific Instruments*, 81, 113905, (2010).
22. Bum Joon Kim, Hyeong-Do Kim, Deok-Yong Cho, Myonglin Kim, S.-J. Oh, Changyoung Kim, Stable Five Axes Cryogenic Photoemission Manipulator without a Differentially Pumped Rotary Feedthrough, *Review of Scientific Instruments*, 76, 063910, (2005).
23. L.J. Azevedo, Rotating Sample Holder without Mechanical Linkages, *Review of Scientific*, 50, 231–232, (1979).
24. Private communication with Prof. Shuheng H. Pan, Institute of Physics, Chinese Academy of Sciences, Beijing, China. Email: span@iphy.ac.cn

3 Cryostat Design for High-Field Magnet with Room Temperature Bore

This chapter presents the design and construction of integrating a self-contained "wet" top-loading He-3 cryostat with sample-in-liquid into a high-field magnet with room temperature bore.

3.1 MULTIPLE TAILS FOR TIGHT SPACE FIT

3.1.1 Engineering Challenge

A high-field magnet, for example one with maximum field of 30 tesla, involves a complex design and large physical size, making it impractical to integrate directly with a cryostat inside a liquid helium dewar. Instead, the magnet and cryostat can be placed in two separated dewars and integrated together at room temperature. In this configuration, the self-contained cryostat, an insert coupled to its own liquid helium dewar, has a room-temperature vacuum shroud and tail, while the magnet installed in another dewar provides a room temperature bore (the bore of *the magnet dewar*). The cryostat tail is designed to fit into the bore of the magnet dewar, allowing the sample positioned inside the tail to remain at cryogenic temperatures while simultaneously being exposed to high magnetic fields. However, the room-temperature bore of such magnets dewar is typically very small, which requires a specialized cryostat tail design.

In this chapter, we use a top-loading He-3 cryostat with sample-in-liquid as the example to discuss how it can fit into a 30 tesla magnet dewar with a 32-mm diameter room temperature bore. The He-3 cryostat tail includes:

- Outer vacuum chamber (OVC) tail at room temperature
- Liquid nitrogen temperature tail (78K tail)
- Liquid helium reservoir tail (4K tail)
- Inner vacuum tail (IVC tail)
- He-3 sample tube (He-3 tail or He-3 pot)

None of the tails can touch the adjacent tails, except for the 4K tail and the IVC tail, which are allowed to be in direct contact because they are always maintained at the same temperatures. The diameter of OVC tail must be 30.7 mm or less to provide sufficient clearance within the 32 mm magnet dewar bore. The He-3 tail typically has an inner diameter larger than 17 mm, which leaves only a 6.5 mm radial clearance to accommodate the remaining fives tails with reasonable separations, as illustrated in Figure 3.1.

DOI: 10.1201/9781042000050-3

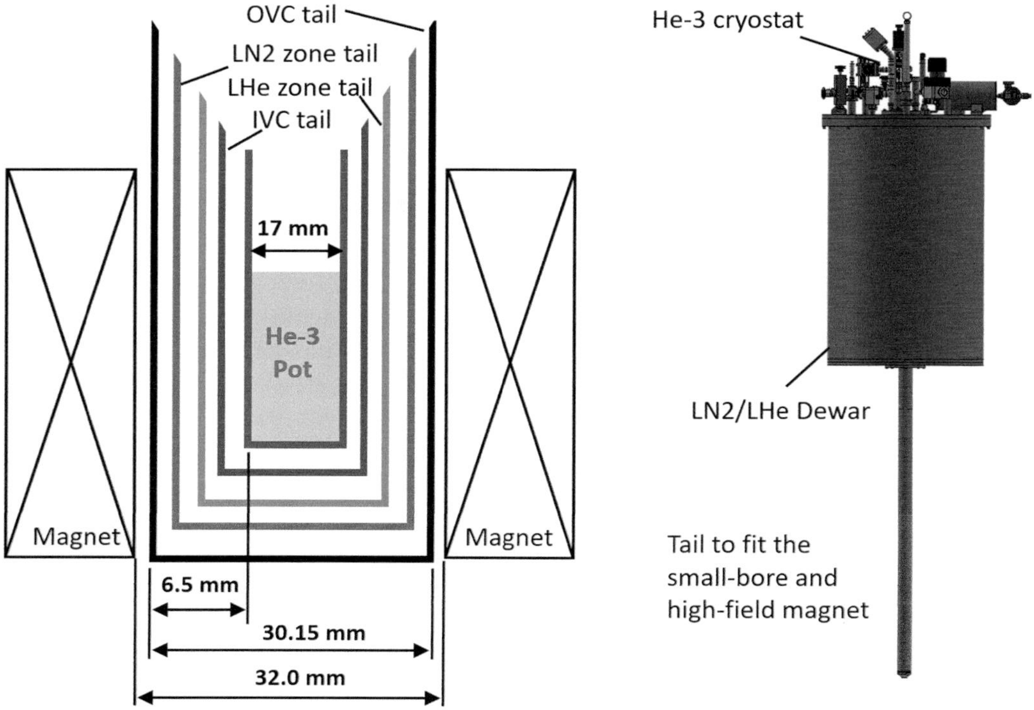

FIGURE 3.1 Typical top-loading He-3 cryostat integrated with a high filed magnet with a room temperature bore (dimensions in this sketch are disproportionate).

3.1.2 Tail Design Concept

The first step in this project is selecting the appropriate material and precise dimensions for each tail. Table 3.1 and Figure 3.2 provide an example of the tail design details, including materials, diameters, wall thicknesses, tail gaps, and overall dimensions. Since most of the tail tubes are not commercially available in the specified sizes, they must be custom machined from standard tubing or even solid rods.

Thermal shorts between different temperature zones must be prevented. The thermal isolation failure may severely compromise the performance of the cryostat and, in some cases, result in physical damage. The three most critical issues are summarized below.

- Risk of mechanical damage due to ice formation: Contact between the OVC tail and the 78K tail can lead to the formation of ice on the OVC tail (Figure 3.3 (a)). If the ice accumulates to a sufficient size, it may bond the tail to the magnet dewar, thereby posing a risk of mechanical damage to both components.
- High cryogen consumption: Direct contact between the 78K tail and the 4K tail introduces a substantial heat load to liquid helium reservoir, resulting in significant increase of the liquid helium boil-off rate.
- Degradation of He-3 performance: If the IVC tail contacts the He-3 sample tube, the heat load from liquid helium not only prevents the He-3 cooldown to the base temperature but also shortens the He-3 pot holding time.

TABLE 3.1
Set of Modified Tubes That Meet the Requirement

Item	Temperature	Material	O.D.	Wall Thickness	Gap	I.D.
OVC tail	300 K	Stainless steel	1.187"/30.15 mm Fixed OD	0.028"		1.131"
LN2 zone tail	78 K	Copper	1.095""	0.020"	0.0180" to OVC tail	1.055"
LHe zone tail	4.2 K	Copper	0.976"	0.020"	0.0395" to LN2 tail	0.936"
IVC tail	4.2 K	Stainless steel	0.845"	0.020"	0.0455" to LHe tail	0.805"
HE-3 Pot	0.3 K	Stainless steel	0.704"	0.012"	0.0505" to IVC tail	0.68"/17.27 mm Fixed ID

Source: Data from the authors' project.

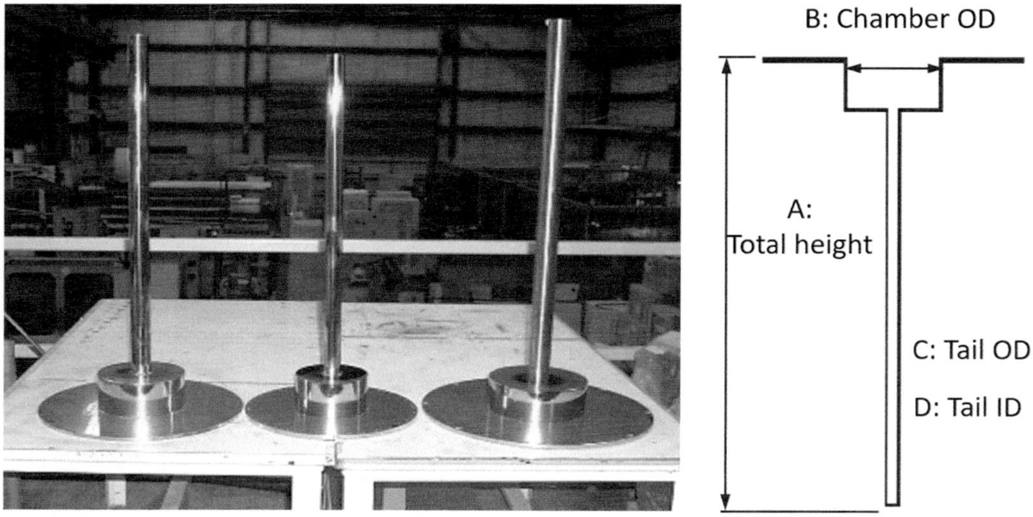

	RT Tail	LN2 Tail	LHe Tail	IVC Tail	He-3 Tail
A	45.76″	44.01″	42.26″	35.58″	34.27″
B	8.00″	7.00″	5.75″	NA	NA
C	1.187″	1.095″	0.976″	0.845″	0.704″
D	1.131″	1.055″	0.936″	0.805″	0.680″

FIGURE 3.2 Examples of the cryostat tails. (Projects accomplished by authors at Janis Research Company. Image courtesy of FormFactor, Inc.)

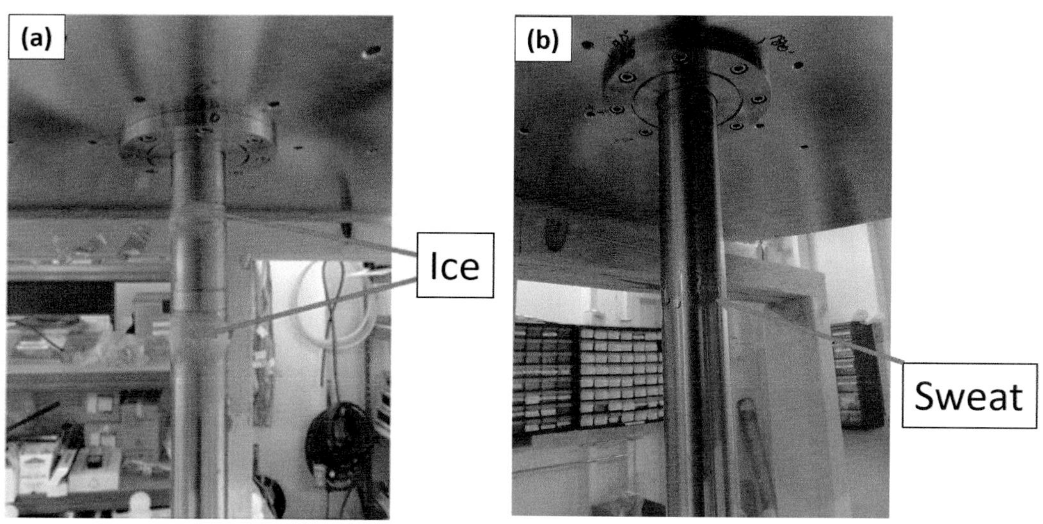

FIGURE 3.3 **(a)** Ice formed on OVC tail and **(b)** sweating on OVC tail. (Projects accomplished by authors at Janis Research Company. Image courtesy of FormFactor, Inc.)

3.1.3 Gaps between Different Temperature Zone Tails

The gaps between different temperature zones vary: they are smaller at higher temperatures (Table 3.1). Since the He-3 pot is protected by multiple radiation shields, the occasional contact between outer tails does not significantly impact cryostat performance, and the gap between the outer tails can therefore be relatively small. In contrast, a sufficiently large gap between the IVC tail and the helium reservoir (4K) tail is required to maintain adequate liquid helium around the IVC tail and to prevent the effect of helium magnetic levitation when the cryostat is operated in high magnetic fields. This phenomenon is discussed in more detail in the next section [1, 2].

The OVC tail and the 78K tail have the smallest gap but the largest temperature difference, making them highly susceptible to misalignment, particularly when the 78K tail reaches its operating temperature. The misalignment can lead to cold contact between the two tails, which in turn causes ice formation on the OVC tail. To avoid tail contact, all tails and flanges must be fabricated precisely and installed properly.

3.1.4 Tail Fabrication Tips

- All tubes must be straight or straightened on a lathe.
- The flanges on top of the tubes need to be faced after they are soldered onto the tube to ensure a 90-degree angle between the flange and the tube.
- The tube diameters should be highly uniform with a maximum tolerance of 0.002″.
- The typical lengths of tails range between 30″ and 40″, and the specified tolerance must be maintained over the entire length.
- The above design specification is applicable to both "wet" and "dry" systems.

3.1.5 Thermal Short Prevention

To maintain the tail concentricity and ensure adequate clearance, a special concentricity spacer, made of 0.011″ diameter monofilament nylon fishing thread [3], is wrapped around the tails to serve as a spacer for the small gaps (Figure 3.4 (a)). The thread is wrapped in a back-and-forth crossover pattern. When the thread spirals down and reaches the end of the tail, it spirals up in the opposite direction, as illustrated in Figure 3.4 (d). This configuration ensures that the effective spacing is at least twice the thread diameter. The diameter of the nylon threads may vary depending on the dimension of specific tail gaps. Additionally, a conventional triangle G-10 spacer is installed at the bottom of each tail (except for the OVC tail), as illustrated in Figure 3.4 (b).

The release of cryogenic gas from the reservoirs can change the temperature distribution on the dewar radiation shields and reduce the vent tube temperatures. Although these temperature

FIGURE 3.4 Concentricity spacers. (Projects accomplished by authors at Janis Research Company. Image courtesy of FormFactor, Inc.)

variations are subtle, the resulting thermal expansion or contraction of the metal components is sufficient to alter the gap between the tails. To mitigate this effect, multiple vent ports are typically incorporated into each reservoir. In case of contact, the user can adjust the cold-gas flow to relieve the contact. Due to the limited available clearance between the OVC tail and the 78K tail, condensed moisture rather than solid ice may remain after tuning.

3.2 HELIUM TAIL BUBBLIES AND MAGNETIC LEVITATION EFFECT

Small helium bubbles may form in the tail section as a result of the liquid helium boiling within the narrow space between the 4K and IVC tails. These bubbles remain uncondensed as long as boiling persists in the tail. When a magnetic field is applied, liquid helium experiences a force proportional to (B(dB/dx)) in a nonuniform magnetic field due to its diamagnetic susceptibility ($\chi = -1.8914\times10^{-6}$). Experiments revealed that once the magnetic forces exceed a threshold level of approximately 21 T^2/cm, liquid helium may be elevated, leading to trapping of helium bubbles near the magnetic field center within a narrow gap, as illustrated in Figure 3.5.

Due to the poor thermal conductivity of helium gas, the accumulation of trapped helium bobbles may prevent liquid helium from cooling the IVC tail, resulting in a local temperature increase by several kelvins in the bubble region [1]. Several approaches can be employed to mitigate this effect.

- Capillary extraction: Installing a capillary to actively pump the bubbles away.
- Isothermal copper tails: Fabricating the tail from high-resistive drawn copper to reduce the heating for bubble formation while minimizing eddy currents.
- Sub-lambda pumping: Lowering the liquid helium temperature below the lambda point to suppress bubble formation.
- Gap optimization: Increasing the spacing between the IVC and 4K tails to prevent bubble trapping.

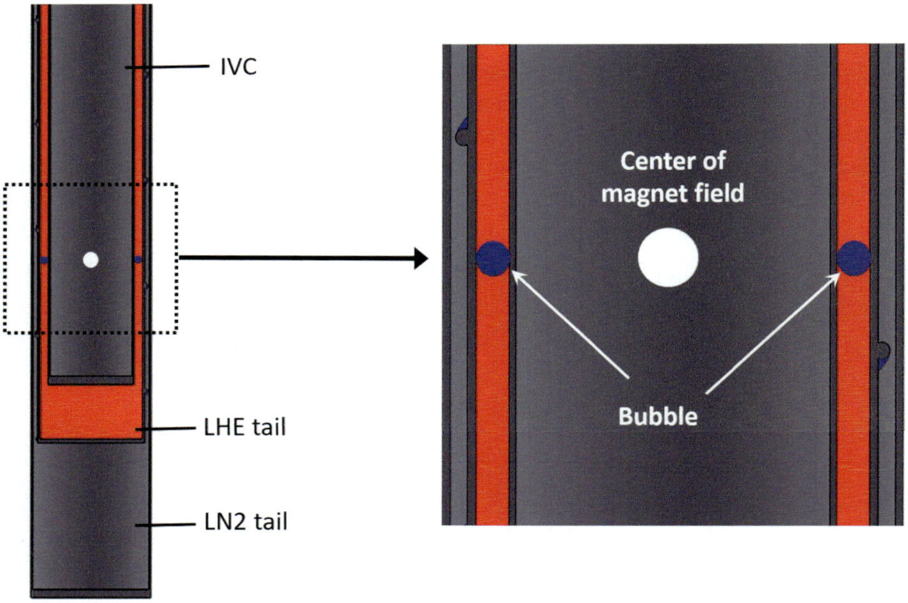

FIGURE 3.5 Magnetic levitation effect on helium.

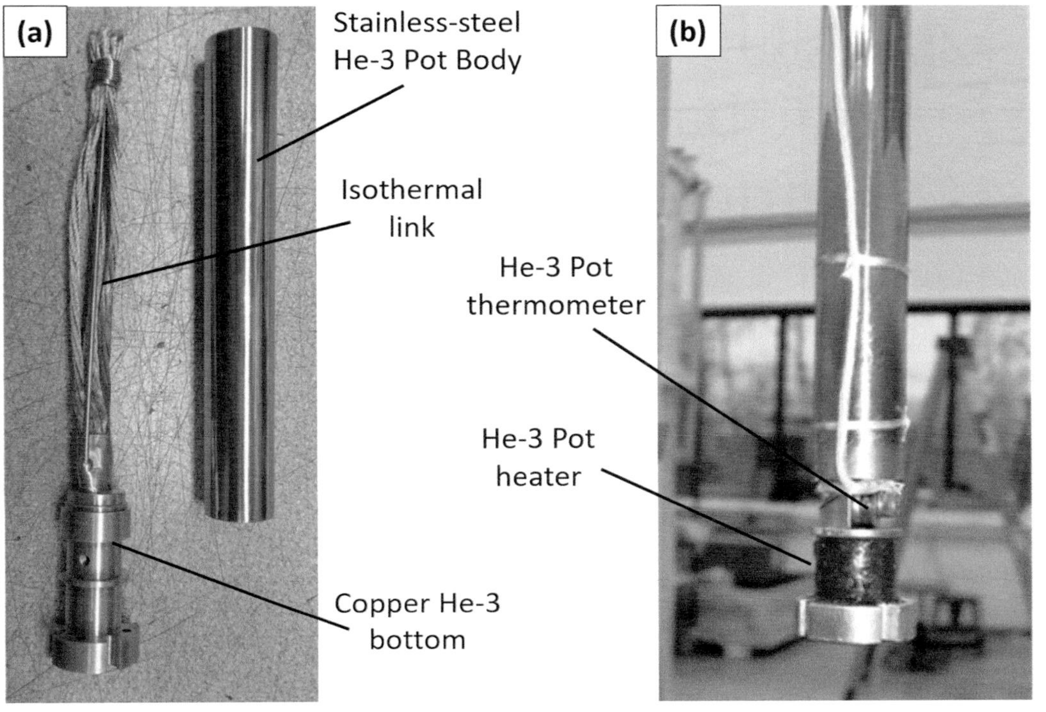

FIGURE 3.6 Isothermal zone made of copper wires. (Projects accomplished by authors at Janis Research Company. Image courtesy of FormFactor, Inc.)

3.3 ISOTHERMAL HE-3 POT IN SLIM SPACE

In certain applications, the He-3 pot must be fabricated from long stainless-steel cylindrical tube in order to fitting into the slim IVC tails. A temperature gradient may arise in the deep pot due to the hydrostatic pressure effect [4]. To mitigate this gradient, an isothermal zone along the He-3 pot can be generated by plating copper on its surface [5].

An alternative approach is to install copper wires inside the He-3 pot to enhance the conductive heat exchange within the liquid He-3, maintaining an effectively isothermal condition. Figure 3.6 shows a stainless steel He-3 pot made of a 5.0″ long cylindrical tube with the diameter of 0.75″, which consists of a copper bottom and copper thermal link. Due to the presence of copper wires in the He-3 pot, this design is suitable for the sample-in-vacuum applications only.

REFERENCES

1. E.J. McNiff, Jr., B.L. Brandt, S. Foner, L.G. Rubin, and R.J. Weggel, Temperature Anomalies Observed in Liquid 4He Columns in Magnetic Fields with Field-Field-Gradient Products >21 T^2/cm, *Review of Scientific Instruments*, 59, 2474–2476, (1988).
2. M.A. Weilert Dwight, L. Whitaker, H.J. Maris, and G.M. Seidel, *Magnetic Levitation and Noncoalescence of Liquid Helium*, *Physical Review Letters*, 77, 23, 4840–4843, (1996)
3. Berkley Trilene® XL® Monofilament Berkley (Pure Fishing, Inc.) 1900 18th Street, Spirit Lake, IA, USA Website: www.berkley-fishing.com
4. I. Takacs, J. Takacs, W.M. Reiff, and J.D. Ramsden, Hydrostatic Pressure Effect on the Minimum Temperature of a Helium-3 Cryostat, *Review of Scientific Instruments*, 57(10), 2605–2608, (1986).
5. Zuyu Zhao and Chao Wang (Eds.), *Cryogenic Engineering and Technologies: Principles and Applications of Cryogen-Free Systems*, CRC Press (Taylor & Francis Group), Boca Raton, FL, USA, (2020).

4 Salt Pill Growth for the Adiabatic Demagnetization Refrigerators

Single-shot Adiabatic Demagnetization Refrigerator (ADR) is a popular testbed for observational millimeter-wave, X-ray astronomy and dark matter, etc. Interested readers may find excellent descriptions of ADR systems and its construction in references [1–3], particularly for the commercial two-stage ADR cryostat employing Gadolinium Gallium Garnet (GGG) ($Gd_3Ga_5O_{12}{\cdot}24H_2O$) and Ferric Ammonium Alum (FAA) ($Fe_2(SO_4)_3{\cdot}(NH_4)_2SO_4{\cdot}24H_2O$) paramagnetic salt pills. The lowest operating temperature of ADR depends on the paramagnetic material. For example, a two-stage ADR cryostat with GGG and FAA can reach 50 mK or lower. In this chapter, we share our own experience of growing the FAA salt pill.

4.1 INTRODUCTION

4.1.1 Entropy Conservation Process

Entropy in a thermodynamic system is defined by the number of accessible microscopic states consistent with the macroscopic constraints. A reversible adiabatic process is therefore an isentropic process.

In ADR, we use a special paramagnetic material whose magnetic entropy depends mainly on the ratio of the applied magnetic field strength to the temperature (B/T) when the material follows Curie's law. During adiabatic demagnetization, this ratio tends to remain approximately constant, so the temperature decreases in proportion to the magnetic field. This approximation breaks down as the material gets colder and approaches its magnetic ordering temperature, where spin–spin interactions become significant, Curie's law fails, and the entropy is no longer determined solely by B/T.

4.1.2 The Lowest Temperature of ADR

Paramagnetic materials contain unpaired electrons, most commonly in the 3d orbitals of transition elements or the 4f orbitals of rare earth elements. These atoms or ions are often referred to as *magnetic ions*. In the absence of an applied magnetic field at room temperature, the magnetic moments of these ions are randomly oriented.

In an ADR system, the paramagnetic material (commonly called a *salt pill*) is first magnetized isothermally in an applied magnetic field [4]. During this process, the magnetic moments become partially aligned, reducing the spin entropy while the released heat is transferred to a thermal bath.

DOI: 10.1201/9781042000050-4

The system is then thermally isolated, and the process proceeds approximately adiabatically. Because essentially no heat is exchanged with the surroundings, the total entropy of the salt pill remains approximately constant. As the magnetic field decreases, the spin entropy increases, and the temperature correspondingly decreases in order to maintain this constant entropy condition.

At sufficiently low temperature, known as the magnetic ordering temperature, weak dipole–dipole or exchange interactions between the magnetic ions become significant. These interactions can drive long-range magnetic ordering, producing an effective internal magnetic field B_{int}. This internal field prevents the temperature from decreasing indefinitely and therefore sets a fundamental lower limit to the temperature achievable by adiabatic demagnetization, as described by:

$$T_f = T_i \frac{\sqrt{B_f^2 + B_{int}^2}}{B_i} \tag{4.1}$$

where

B_i: Initial applied magnetic field ($B_i >> B_{\text{int}}$)
B_f: Final applied magnetic field
T_i: Initial temperature
T_f: Final temperature

Therefore, salt pill temperature cannot reach absolute zero even when the final applied magnetic field becomes zero.

4.1.3 Salt Pill Candidates

Figure 4.1 shows the reduced entropy of several paramagnetic salt candidates as functions of temperatures. Materials such as CMN, which has a small intrinsic magnetic field, display lower magnetic ordering temperatures. This property makes them suitable for low-temperature thermometry (refer to Chapter 9). In contrast, salts such as GGG and FAA, which possess a high total spin angular momentum (J), exhibit significant magnetic entropy changes with temperature, making them promising candidates for demagnetization refrigeration applications. In addition, FAA has a reasonable low intrinsic field (Table 4.1) [2].

4.2 HOME-GROWN FAA PILL

While GGG can be purchased as a commercial product, FAA pill needs to be grown in-house with the following properties.

Chemical symbol:	$Fe(NH_4)_2(SO_4)_2.12H_2O$
Solubility at 20 °C:	1.24 g per c.c.
Solubility at 35 °C:	4.0 g per c.c.
Crystal density at 20 °C:	1.71 g per c.c.
Melting temperature:	39 ~ 41 °C:
Saturated solution density at 20 °C:	1.385 g per c.c.
Spin angular momentum:	J = 5/2
Internal field (B_{int})	~ 50 mT
Ordering temperature:	35 mK

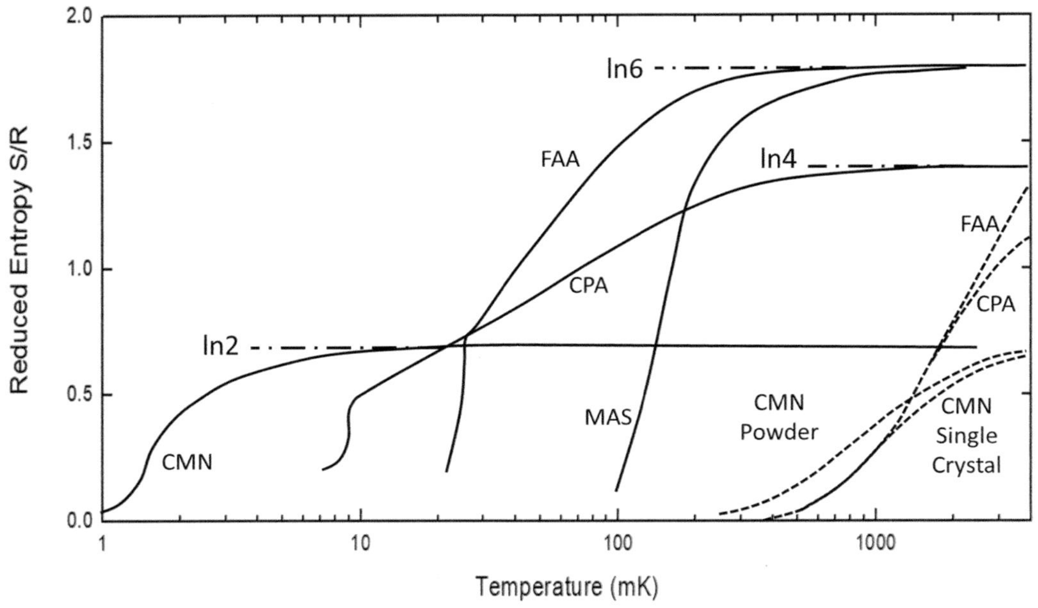

FIGURE 4.1 Reduced entropies (S/R) of four salt candidates for ADR as a function of temperature in zero magnetic field (full lines) and 2 tesla (dashed lines) (Information from Frank Pobell, Matter and Materials at Low Temperatures, Third, Revised and Expended Edition, Springer-Verlag Berlin Heidelberg 2007) [2].

TABLE 4.1
Behavioral Comparison of FAA and CMN at Low Temperatures [4]

Material	*m* (2*J*+1)	B_{int}	T_f	B_i	T_i
Ferric Ammonium Alum (FAA) $Fe(NH_4)_2(SO_4)_2 \cdot 12H_2O$	6	50 mT	30 mK	4.0 T	4.2 K
Cerium Magnesium Nitrate $Ce_2Mg_3(NO_3)_{12} \cdot 24H_2O$	2	4.2 mT	2.0 mK	N/A	1.0 K

Source: Data from Frank Pobell, *Matter and Methods at Low Temperatures*, Third, Revised and Expended Edition, Springer-Verlag Berlin Heidelberg 2007.

FAA undergoes a chemical transition above 40 °C, which alters its spin state. Home-grown FAA pills may be produced using the following procedures:

- Preparation: Gather the required materials, including polycrystal FAA and deionized water
- Equipment Setup: Assemble and calibrate the apparatus required for the growth process.
- FAA Solution Preparation: Mix the appropriate chemicals under controlled conditions to create an FAA solution.
- Crystallization: Allow the solution to crystallize under defined environmental parameters.
- Harvesting: Carefully isolate the resulting FAA crystals.
- Pill Formation: Compress the harvested crystals into pill for subsequent applications.
- Quality Control: Evaluate the final product for purity and uniformity to ensure reproducibility and reliability.

4.2.1 FAA Growth Setup

Because FAA and its solution have poor thermal conductivity and are highly corrosive, proper thermalization of component and careful selection of apparatus materials are required for the FAA growth. The suitable materials include gold, stainless steel, Stycast epoxy 1266, and acrylic.

The growth setup consists of two main parts:

Part I—FAA Pill Harvest Housing (Stainless-Steel Housing, SPH)

The SPH is constructed from stainless steel and designed as a cylindrical tube approximately 1.0 inch in diameter and 5.5 inches in length. An OFHC copper cold finger is silver-soldered to the bottom plate of the SPH. To enhance thermal conduction between the FAA pill and the copper cold finger, approximately 200 thin gold wires are silver-soldered to the top of the cold finger.

After soldering, the solder flux must be thoroughly cleaned with trichloroacetic acid (TCA) as shown in Figure 4.2(a). The solder joints must be coated with Stycast epoxy 1266 to protect the copper from corrosion by the FAA solution (see Figure 4.2(b)). The stainless-steel bottom plate should then be welded to the SPH, and the gold wires should be terminated approximately 0.25 inch below the top of the housing to allow installation of the SPH cap later.

Part II—FAA Solution Container

The FAA solution container (Figure 4.3) is constructed from two concentric acrylic tubes. The inner tube, approximately 2.0 inches in diameter, holds the oversaturated FAA solution. The SPH is hermetically bonded to the bottom of the tube to prevent leakage of the FAA solution. Copper wires are wound around this tube and form a homemade resistive heater. The outer tube is positioned concentrically around the inner tube and serve as a thermal buffer to reduce the impact of ambient temperature fluctuations on the FAA solution.

Additional thermal stabilization is achieved by surrounding the SPH with Styrofoam insulation. A glass thermometer is inserted into the inner acrylic tube to monitor the solution temperature.

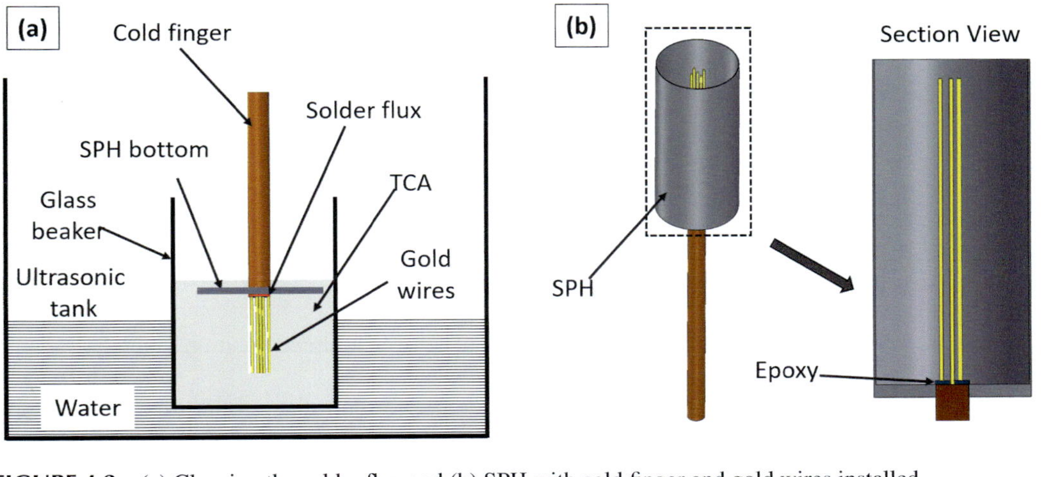

FIGURE 4.2 (a) Cleaning the solder flux and (b) SPH with cold finger and gold wires installed.

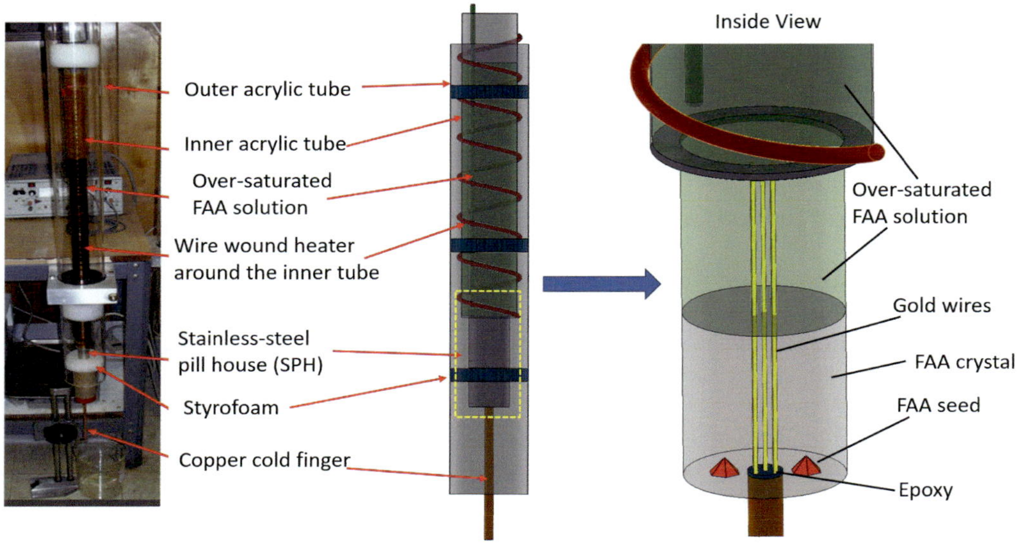

FIGURE 4.3 FAA growth setup. (Projects accomplished by authors at Janis Research Company. Image courtesy of FormFactor, Inc.)

Finally, a thin acrylic rod is prepared for gently poking the top of the solid FAA pill during the growth process.

4.2.2 Grow FAA Pill

1. Initial Measurement
   - Accurately weigh and record the total weight of the SPH, including both end caps, the cold finger, the gold wires, as well as the FAA seeds.
   - Accurately measure the capacity of the SPH.
2. Pre-warming and Stabilization
   - Pre-warm the inner acrylic tube and stabilize its temperature at approximately 38 °C.
3. Seed Crystal Placement
   - Place two small FAA seed crystals (few millimeters or smaller in diameter) at the bottom of the SPH. These will serve as nucleation seeds for crystal growth.
4. Solution Preparation
   - Measure 125 ml of deionized water and pour it into a clean beaker, and then place the beaker on a hot plate.
   - Set and maintain the water temperature at ~37 °C, which is approximately 1 °C lower than the temperature of the inner acrylic tube. Monitor continuously with a glass thermometer.
   - Add 500 mg of commercial FAA polycrystal to the warmed water to create an oversaturated FAA solution.
   - Maintain the solution at a steady temperature between 35 °C and 37 °C while all solids dissolve.
   - Repeat the steps above to prepare 500–550 ml of oversaturated solution in total.
5. Solution Transfer
   - Carefully pour the oversaturated FAA solution into the inner acrylic tube using a glass funnel.
   - Prevent any undissolved FAA polycrystals from entering the inner tube during the pour.

FIGURE 4.4 Home-grown FAA pill. (Projects accomplished by authors at Janis Research Company. Image courtesy of FormFactor, Inc.)

6. Temperature Regulation
   - Measure the temperature of the solution inside the inner acrylic tube using a glass thermometer.
   - Regulate the temperature to stay between 35.5 °C and 36.0 °C using an appropriate temperature controller.
   - Inspect the seal between the SPH and the inner acrylic tube to ensure there is no solution leakage.
7. Crystal Growth
   - Once the solution is stabilized, FAA crystals will begin to grow from the seed crystals at the bottom of the SPH.
   - Gently inspect the growing crystal using a thin acrylic stick every 12 hours, depending on the growth rate of the FAA pill.
   - Monitor the crystal more frequently as it is close to the top of the SPH.
8. Termination of Growth
   - Stop the crystal growth by turning off the heater power supply when the FAA pill reaches approximately 0.05–0.10 inches below the top edge of the SPH.
9. Crystal Harvesting
   - Carefully decant the remaining FAA solution into a clean beaker.
   - Allow the entire assembly to cool gradually to room temperature.
   - Detach the SPH from the inner acrylic tube carefully to avoid damaging the crystal.
   - A home-grown FAA pill, similar to that shown in Figure 4.4, should now be obtained.

4.2.3 End Cap Installation

Sealing the SPH with an end cap is critical for preserving the FAA pill's crystallization water, especially because the pill will remain in ADR vacuum for most of its operational time. If the installation of the end cap is delayed, the FAA pill must be stored in a high-humidity environment. A practical method is to place the SPH in a sealed plastic bag together with an open container of water (see Figure 4.5). This setup helps maintain sufficient moisture and prevents the FAA crystal from drying out.

FIGURE 4.5 Keep the FAA pill in high humidity. (Projects accomplished by authors at Janis Research Company. Image courtesy of FormFactor, Inc.)

The end cap can be installed with the following procedure:

1. Trimming the Overgrown Crystal
   - Prepare a cup of lukewarm water maintained at approximately 36 °C.
   - Moisten a sponge with the water and gently rub and rotate it against the overgrown FAA crystal near the top of the SPH.
   - Continue until the crystal top surface is leveled with the ends of the gold wires, ensuring that the stainless-steel end cap can be fitted securely.
2. End Cap Installation
   - Weld the stainless-steel end cap in place as soon as possible to prevent the FAA pill from drying out.
   - During welding, wrap a wet towel around the SPH to prevent the FAA crystal from overheating.
   - Stycast 1266 epoxy can be used to seal the joint if the welding leaves a small gap.

4.2.4 Estimate the FAA Filling Factor

The filling factor represents the fraction of the internal volume of the SPH that is occupied by solid FAA, $f = \frac{V_{solid}}{V_{total}}$. The remaining space may either be filled by FAA solution or simply remains as a void with air. This parameter is of critical importance, as it directly affects the performance of the adiabatic demagnetization cryostat.

The filling factor can be estimated by measuring the weight increase of the complete salt pill after crystal growth. Using the known internal volume of the SPH along with the densities of FAA crystal (1.71 g/cm³, NIST) and FAA saturated solution (1.385 g/cm³, NIST), the filling factor can be calculated based on the mass of FAA that solidified, the remaining solution volume, and the total capacity of the SPH. For reliable operation in an ADR cryostat, the filling factor should typically exceed the empirically established threshold of 97%.

REFERENCES

1. Zuyu Zhao and Chao Wang (Eds.), *Cryogenic Engineering and Technologies: Principles and Applications of Cryogen-Free Systems*, CRC Press (Taylor & Francis Group) Boca Raton, FL, USA, (2020).
2. Frank Pobell, *Matter and Materials at Low Temperatures*, Third, Revised and Expended Edition, Springer-Verlag, Berlin Heidelberg, (2007).
3. Grant W. Wilson and Peter T. Timbie, Construction Techniques for Adiabatic Demagnetization Refrigerators using Ferric Ammonium Alum, *Cryogenics*, 39(4), 319–322, (April 1999).
4. O.V. Lounasmaa, *Experimental Principles and Methods Below 1 K*, Academic Press Inc., (London) LTD, (1974).

5 Joule–Thomson Condensing Assembly Unit

5.1 INTRODUCTION

The Joule–Thomson (JT) effect is one of the fundamental phenomena in physics of real fluids. A JT expansion is an adiabatic and isenthalpic process in which a nonideal gas may either cool or warm, depending on its thermal dynamic state when the expansion occurs. The details of the JT expansion were discussed in reference [1] and can be summarized as follows:

- JT coefficient (μ_{JT}) is a key parameter to describe the change in temperature of a fluid with respect to pressure during an isenthalpic process, defined as

$$\mu_{JT} = (\partial T/\partial P)_H \tag{5.1}$$

  A positive value of the coefficient indicates that the gas cools upon expansion, while a negative value means that the gas experiences a temperature rise as it expands. The sign and magnitude of the JT coefficient depend on both the temperature and pressure of the gas.

- JT inversion curve is the envelop of the turning points ($\mu_{JT}=0$) of the JT coefficients (Figure 5.1). Along this curve, a fluid experiences no temperature change during isenthalpic expansion. It separates the regions of cooling ($\mu_{JT} >0$) and heating ($\mu_{JT} <0$). The maximum temperature along the curve is called the maximum inversion temperature, above which a gas will always warm upon expansion, regardless of pressure.
- The JT effect applies only to nonideal gas. For an ideal gas, where enthalpy is exclusively a function of temperature, isenthalpic expansion does not alter the temperature.

5.2 JOULE–THOMSON INVERSION CURVE OF HE-3 AND HE-4

The JT expansion has been widely employed in the continuous-flow helium cryostat to achieve lower temperatures. The essential requirement is that the expansion occurs under conditions where the JT coefficient is positive to ensure that the gas cools with expansion. Figure 5.2 (a) shows the inversion curves of He-4, and the corresponding numerical data are listed in Table 5.1.

A full-range JT inversion curve for He-3 is constructed using a virial-type equation of state for this quantum fluid [2]. Based on data and analysis reported in Refs. [2] and [3], a representative JT inversion curve for He-3 is illustrated in Figure 5.2 (b), and some critical temperatures and pressures are summarized in Table 5.2.

The maximum inversion temperature of He-3 is approximately 34 K, indicating that no cooling is produced through JT expansion above this temperature. In the low-pressure limit,

DOI: 10.1201/9781042000050-5

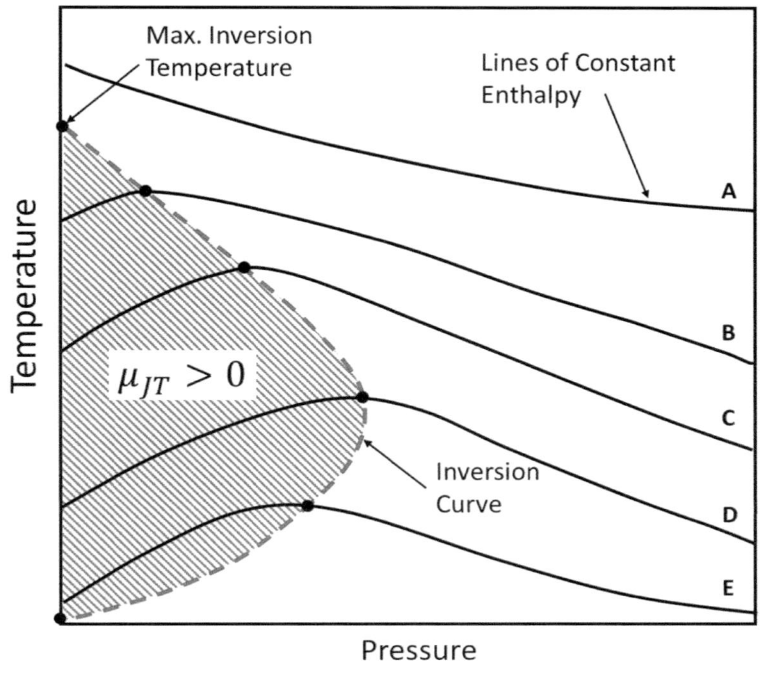

FIGURE 5.1 Joule–Thompson (JT) inversion curve and JT coefficient. (Reprint with permission of Taylor & Francis Group, *Cryogenic Engineering and Technologies: Principles and Applications of Cryogen-Free Systems*, Taylor & Francis Group, LLC, (2020), edited by Zuyu Zhao and Chao Wang.)

the minimum inversion temperature approaches approximately 2.8 K. The maximum-pressure point on the inversion curve, commonly referred to as the "peak," occurs at approximately 22–25.9 bar and 19.6–22 K for He-3. In our practical designs, the precooling stages in cryogen-free He-3 systems and dilution refrigerators employing commercial two-stage cryocoolers typically operate in the temperature range of 2.8–3.15 K at pressures between 0.5 and 1 bar prior to JT expansion. In contrast, substantially higher inlet pressures (2–4 bar) are required when the precooling stage operates near 4.2 K in a wet refrigerator configuration before JT expansion (see Section 5.3.1). Based on the information listed in Table 1.1, some He-3 in a cryogen-free He-3 cryostat or DR could be partially liquefied before entering the JT. The He-3 inversion curve shown in Figure 5.2(b), however, represents the behavior of gaseous He-3 only. Therefore, Figure 5.2(b) and Table 5.2 should be considered representative references rather than exact operating conditions.

In cryostats utilizing JT expansion with helium isotopes or mixtures as refrigerants, achieving condensation requires maintaining temperature and pressure within specific ranges characteristic of each isotope. Operating outside these ranges prevents effective condensation and reduces cooling efficiency. Therefore, precise control of thermodynamic conditions relative to the inversion curve is essential for optimal performance of helium-based JT cryostats.

5.3 FABRICATION

A typical JT expansion assembly used for gas condensation and liquid cooling in the continuous-flow cryostats is illustrated in Figure 5.3, which is referred to as the Joule–Thomson condensing unit (JTCU). The following discussion focuses its applications on the continuous-flow He-3 cryostat.

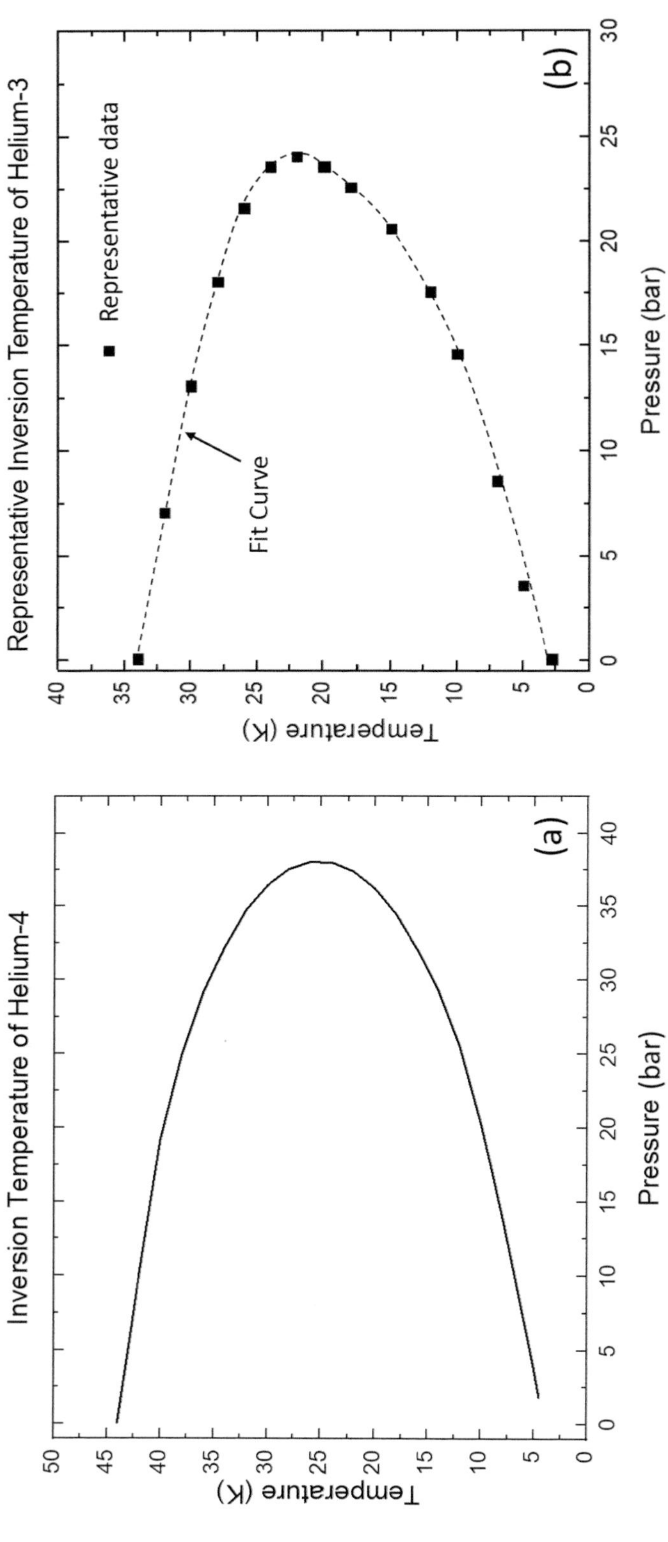

FIGURE 5.2 (a) Joule–Thomson inversion curve of He-4. (Reprint with permission of Taylor & Francis Group, *Cryogenic Engineering and Technologies: Principles and Applications of Cryogen-Free Systems*, Taylor & Francis Group, LLC, (2020), edited by Zuyu Zhao and Chao Wang.) (b) Representative Joule–Thompson inversion curve of He-3 interpreted from [2–5].

TABLE 5.1
Inversion Temperature of He-4 [1]

He-4 Inversion Temperature (K)	Pressure (atm)	He-4 Inversion Temperature (K)	Pressure (atm)
4.5	1.821	22	37.33
5	3.768	24	37.93
6	7.266	26	37.98
7	10.74	28	37.48
8	14.10	30	36.40
9	17.31	32	34.71
10	20.36	34	32.22
12	25.57	36	29.13
14	29.29	38	24.89
16	32.07	40	19.11
18	34.44	42	9.8
20	36.18	44	0.03

TABLE 5.2
Representative Data of Inversion Temperature and Pressure of He-3 [2, 3]

He-3 Inversion Temperature (K)	Pressure (atm)	He-3 Inversion Temperature (K)	Pressure (atm)
2.8	0	19.6–22.0	22.0–25.9
34.0	0		

Note: According to information listed in Table 1.1, some He-3 should in our cryogen free He-3 cryostat or DR should have been liquefied before JTCU, while the He-3 inversion curve as illustrated in Figure 5.2(b) applies to gaseous He-3 only. In addition, both Figure 5.2(b) and Table 5.2 provided representative information for reference only.

A complete JTCU typically consists of a compressor, a recuperative heat exchanger, and a JT valve that may be either a fixed impedance or a throttling valve. Since the unit is often used for cryogen condensation, a receiver (such as a He-3 pot or a mixing chamber) is commonly installed below the JT valve to collect the condensed liquid. JTCUs are now widely employed in continuous-flow He-3 cryostats and dilution refrigerators, in both wet and cryogen-free" systems.

5.3.1 Compressor

When a JT expansion occurs within the inversion curve, a larger pressure drop generally leads to more effective cooling. Particularly, in wet systems, He-3 gas is typically pre-cooled to 4.2 K by liquid helium before entering the counter-flow heat exchanger in gaseous state. A compressor is usually incorporated into our wet DR systems to increase the JT inlet pressure to approximately 4 bar for the initial condensation of the He-3/He-4 mixture. During normal operation, the inlet pressure can be reduced. The compressor should meet the following minimum criteria for the optimal performance:

i. Hermetic seal with a leak rate in the lower range of 10^{-8} mbar·L/s. or less
ii. Pressure generation of up to four atmospheric pressures.
iii. Gas throughput of up to 2 millimole per second.

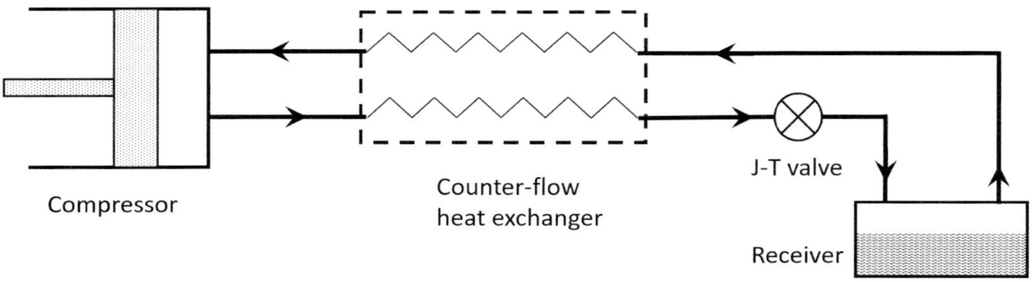

FIGURE 5.3 Joule–Thomson condensing unit.

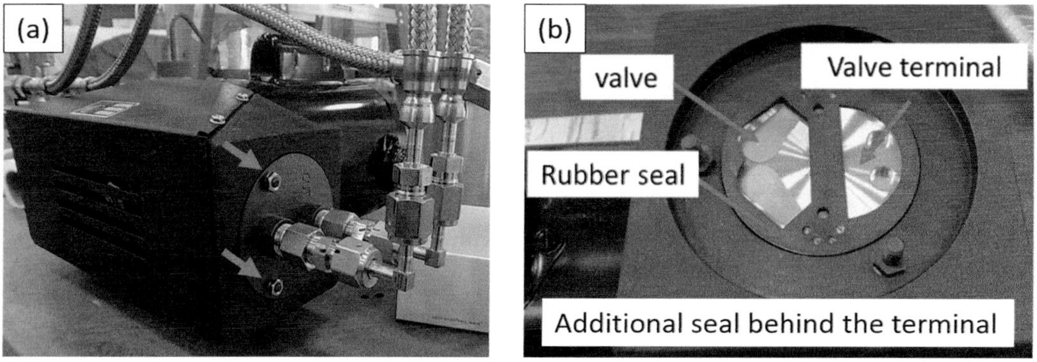

FIGURE 5.4 Model MB-602 compressor. (Projects accomplished by authors at Janis Research Company. Image courtesy of FormFactor, Inc.)

One of the compressor candidates is the Aerospace model MB-602 [6] (Figure 5.4 (a)), a metal bellows compressor. Similar to other commercial compressors, this model may occasionally experience small helium leaks. The following approaches may be taken to improve the sealing:

1. Remove the screws and take off the compressor cover.
2. Carefully remove the valve terminal, which contains rubber seals. *Note:* The valves are the thin metal films that cover the holes on the terminal.
3. Remove the two rubber seals—one from each side of the valve terminal.
4. Apply a thin layer of vacuum-compatible grease to both seals. Reinstall the seals into the valve terminal, then mount the terminal back onto the compressor body.
5. Reattach the cover and perform a helium leak check around the rubber seals to ensure proper sealing before operating the compressor.

In a dry He-3 cryostat, the pulse tube refrigerator (PTR) can cool He-3 gas to approximately 2.5 K or lower. At these temperatures, He-3 condenses into liquid under a pressure of about 330 mmHg (0.44 atm), as shown in Table 1.1, which is lower than the typical return pressure in a dry He-3 cryostat operating without a compressor. In most cryostats, the returning He-3 gas has already condensed into liquid before reaching the JTCU. After the JT expansion, the liquid He-3 is further cooled and finally collected in the He-3 pot. Therefore, no compressor is required in dry He-3 systems utilizing a PTR.

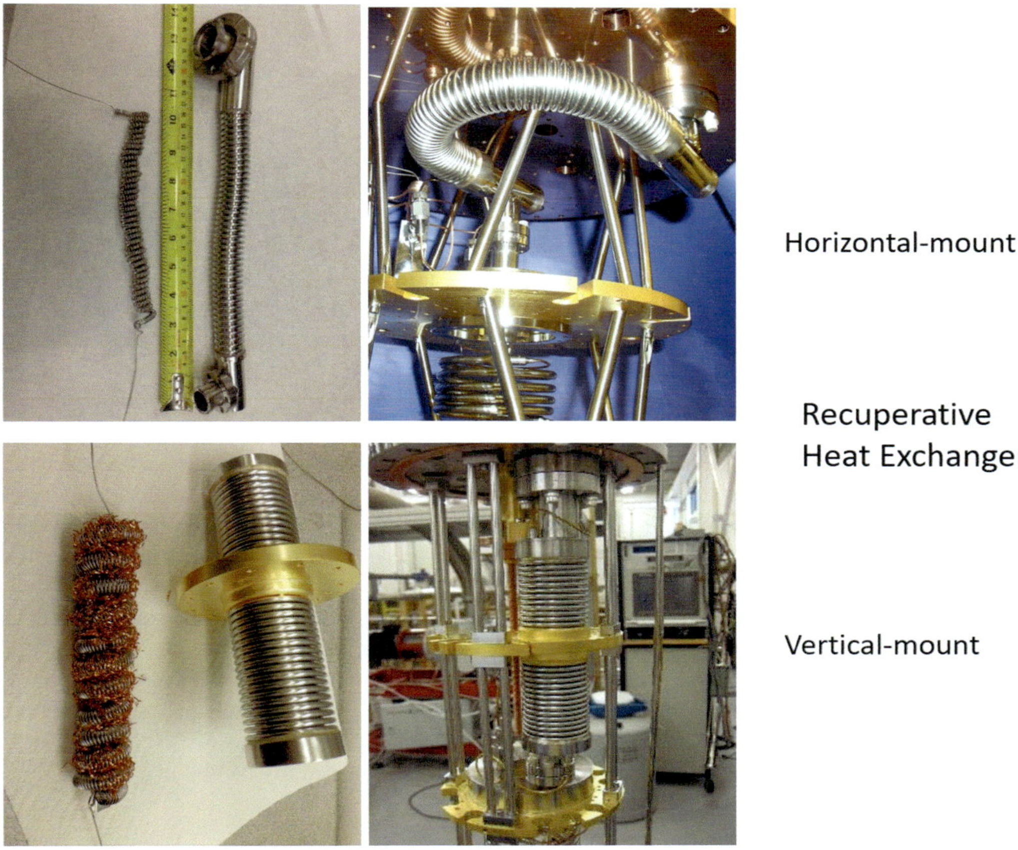

FIGURE 5.5 (a) Recuperative heat exchanger installed horizontally and (b) Recuperative heat exchanger installed vertically. (Projects accomplished by authors at Janis Research Company. Image courtesy of FormFactor, Inc.)

A similar situation applies to dry dilution refrigerators, where the He-3/He-4 mixture can also be condensed by a PTR without the need for a compressor.

5.3.2 Recuperative Heat Exchanger

A counterflow heat-exchanger consists of an outer stainless-steel flexible hose and an internal stainless-steel capillary. For example, the outer hose has an inner diameter of approximately 2 cm and a length of approximately 20 cm, while the internal capillary is roughly 10 m long (Figure 5.5). The outgoing gas from the He-3 pot flows through the outer flexible hose and precools the incoming gas inside the inner capillary. During normal operation, the typical temperature of the outer flexible hose in a dry-cryostat is around 2.5 K or higher.

It is generally desired to install the recuperative heat exchanger horizontally since this configuration minimizes the IR radiation entering JTCU directly. However, due to the limited space in certain cryostats, the JTCU is often installed vertically, as illustrated in Figure 5.5. In this vertical arrangement, IR radiation may pass the recuperative heat exchanger and introduce unwanted heat load to the cryogenic components below. To mitigate this effect, copper wool or similar baffles can be inserted inside the recuperative heat exchanger to reduce IR transmission.

5.3.3 JT Valve—A Fixed (Condensing) Impedance

The JT valve has two primary functions:

i. To establish the proper head pressure of the incoming gas inside the recuperative heat exchanger
ii. To restrict the flow rate in order to optimize the system performance.

A lower impedance corresponds low flow restriction on the incoming fluid, which results in lower head pressure and higher circulation rates. This can enhance the cooling power as more gas circulates through the system. However, the high flow of incoming warm fluid may also lead to elevated base temperature. In contrast, a higher impedance imposes high restriction on fluid flow, generating higher head pressure but reducing circulation rates, which limits the cooling power. Therefore, an optimal impedance requires balancing these competing factors to optimize the performance for each specific system.

In most cryogenic systems, the JT valve has a fixed condensing impedance, as illustrated in Figure 5.6, and is typically made of small CuNi capillaries. The impedance can be further increased by installing a fine wire into the capillaries, which can also reduce the required overall length of the impedance assembly. For the continuous-flow cryostats, the typical fixed impedance ranges from 10^9 (1/cm^3) to 10^{10} (1/cm^3), depending on the desired flow rates.

Due to the extremely small diameter of the inner capillary, impurities such as residual air, nitrogen, or hydrogen tend to solidify at low temperatures and block the impedance. In this case, the user typically needs to stop the circulation and warm up at least part of the system to clear the blockage. This process is time-consuming and labor-intensive, and, for wet systems, incurs additional cost of cryogen. A practical solution is to apply localized heating directly to the clogged impedance, rather than warming the entire system. The heating assembly can be installed using the following procedure:

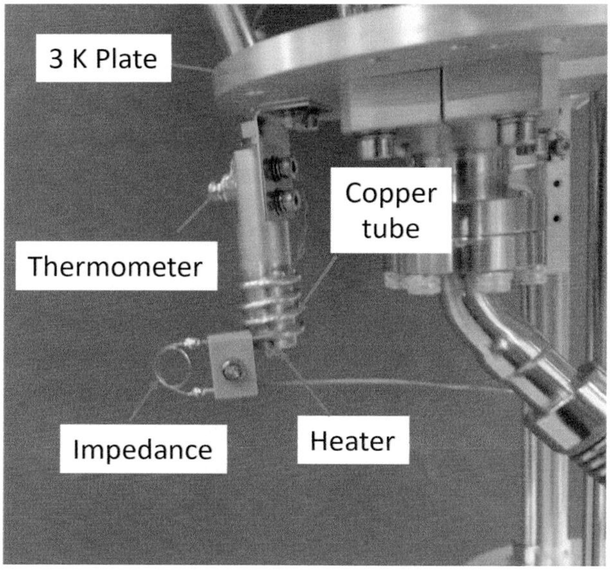

FIGURE 5.6 Condensing impedance. (Projects accomplished by authors at Janis Research Company. Image courtesy of FormFactor, Inc.)

- Solder a small copper capillary onto the inlet end of the impedance.
- Coil the copper capillary on a copper cylinder, as illustrated in Figure 5.6.
- Install a heater and thermometer on the copper cylinder.

By heating the copper cylinder to a preset temperature, the blockage can be cleared efficiently. This approach not only simplifies the cryostat recovery but also helps diagnose the blockage root cause.

- If the blockage is removed at ~20 K, hydrogen is the likely cause.
- If it unblocks at ~80 K, nitrogen is the probable contamination source.
- If water vapor is the cause, the impedance must be heated to room temperature for full recovery.

For effective and rapid heating, the copper block should be thermally isolated from the rest of the system. The impedance assembly is terminated at both ends with VCR fittings, facilitating easy replacement or substitution with components of different impedance values.

Interested readers may refer to the Appendix for more details on impedance construction, measurement techniques, and specific examples.

5.3.4 Receiver

The receiver, such as the He-3 pot or the mixing chamber, is located at the bottom of the JTCU to collect the condensation.

5.3.5 Complete Setup of Joule–Thompson Mixture Condenser in Cryogen-Free Dilution Refrigerator

The JT condenser assembly for the Cryogen-Free Dilution Refrigerator is basically identical to that of cryogen-free He-3 cryostat as illustrated in Chapter 1 (Figure 1.10).

REFERENCES

1. Zuyu Zhao and Chao Wang (Eds.), *Cryogenic Engineering and Technologies: Principles and Applications of Cryogen-Free Systems*, CRC Press (Taylor & Francis Group Boca Raton, FL, USA, (2020).
2. B. Z. Maytal, Joule–Thomson inversion curves of helium isotopes, *Cryogenics*, 36(11), 843–847 (1996).
3. Xi Lin, *Introduction of Low Temperature Experiments*, Peking University Press, Beijing, China, (2025).
4. J.G. Daunt, Preliminary Thermodynamic Data for the Inversion Curve of He^3, *Cryogenics*, 10, 473–475 (1970).
5. J. Kraus, et al., Enthalpy–pressure (H–P) Diagram of He^3 in the Range $1.0\ K \leq T \leq 4.17\ K$ and $0 \leq P \leq 6.5$ atm and Inversion Curve for $T \leq 4.17$ K, *Cryogenics*, 14 (1), 29–35, (1974).
6. Senior Aerospace Metal Bellows 1075 Providence Highway, Sharon, MA, USA Website: www.metalbellows.com/aerospace

6 Excessive Cooling Power from Pulse Tube Refrigerators

6.1 PULSE TUBE REFRIGERATORS (PTRS)

The PTR can be implemented in three primary configurations:

- **U-tube configuration**: Consists of two parallel tubes—the pulse tube and the regenerator tube—connected by heat exchanger to form a U-shape configuration. In this configuration, the user has direct access to the regenerator tube.
- **Coaxial configuration**: Uses a coaxial design, where one tube is positioned concentrically inside another one.
- **In-line configuration**: Aligns the pulse tube and regenerator tube along a single axis, forming a straight, continuous structure.

6.2 TYPICAL COMMERCIAL PTR

Commercial PTRs typically employ the U-tube configuration, which provides user access to the regenerator tubes and enables the utilization of their excess cooling capacity. In this chapter, we focus on the second-stage regenerator, discussing the extraction and applications of excess cooling power.

Two types of excess cooling power can be extracted from the regenerator: distributed cooling power and intermediate cooling power.

6.2.1 Distributed Cooling Power

The distributed cooling power refers to the cooling available along the entire length of the regenerator tube. To extract the distributed cooling power, a small metal tube—typically made of copper, brass, or stainless steel with an outer diameter of 0.094–0.125"—can be coiled and soldered onto the second-stage regenerator of the PTR, as illustrated in Figure 6.1. In our discussion, this small tube is referred to as the "Special Heat Exchanger".

This setup is commonly used to cool helium (either He-4 or He-3) flowing through the Special Heat Exchanger, with the exhaust helium typically reaching around 4 K [2]. If the incoming helium gas is fully thermalized by the first stage at a temperature below 40 K, this result implies that nearly 90% of the helium enthalpy is removed by the Special Heat Exchanger.

DOI: 10.1201/9781042000050-6

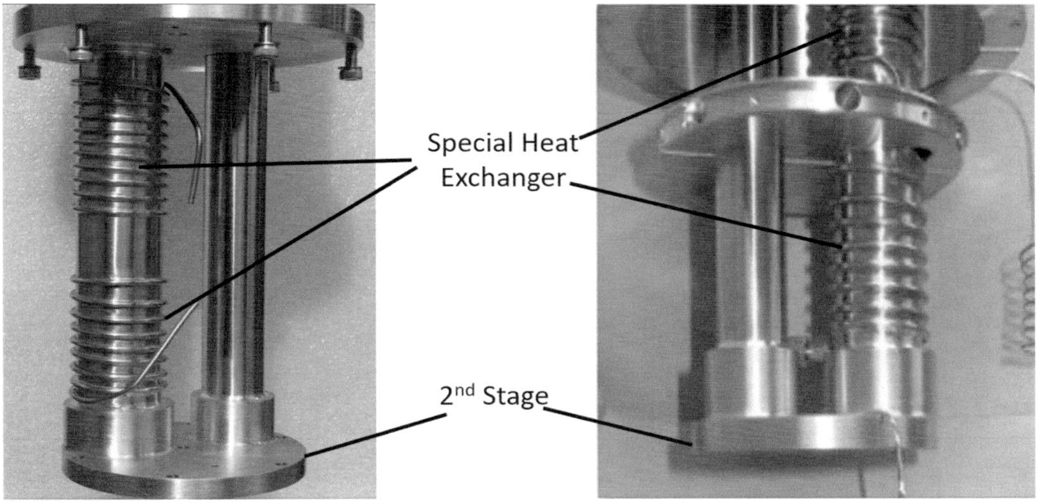

FIGURE 6.1 Extract the distributed cooling power. (Courtesy of Bluefors Cryocooler, [1].)

6.2.2 Intermediate Cooling Power

The intermediate cooling power refers to the cooling capacity available at a specific location along the regenerator. To extract intermediate cooling power, an OFHC cold plate is usually clamped or soldered onto the regenerator at a desired position. This cold plate is frequently used as a thermal anchor stage, providing stable cooling at intermediate temperatures.

A systematic study performed on a Bluefors Cryocooler Technologies Model PT410 cryocooler revealed the following results [2]:

- The temperature distribution along the second stage regenerator is non-linear, spanning between the first- and the second-stage temperatures.
- Several watts of distributed cooling power and a few watts of intermediate cooling power can be extracted, respectively, without significantly impacting the performance of the second stage.
- Slightly higher cooling powers can also be extracted from the pulse tubes than that from the regenerator tubes using the similar method.

6.3 APPLICATIONS

6.3.1 "10K" Cold Plate—Utilization of Intermediate Cooling Power

An OFHC copper plate can be installed near the midpoint of the second-stage regenerator (Figure 6.2). Depending on the PTR model and the exact placement of the plate, it typically reaches a temperature range of 5–10 K. This plate is commonly referred to as "10K plate".

The 10K plate serves as a general-purpose thermal anchoring stage, primarily used to intercept heat from wires, RF cables, and other components thermally connected to higher-temperature stages. By utilizing intermediate cooling power, the plate can effectively reduce the heat load on the colder stage and improve overall system performance. An example of the installation of multiple high-Tc current leads using the 10K plate is presented in Section 6.4.3.

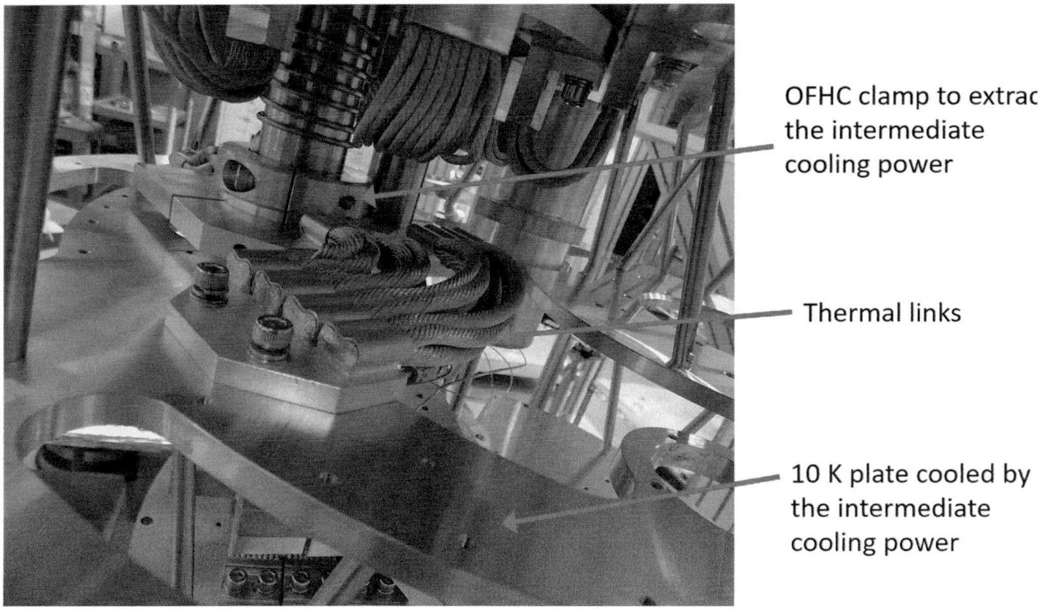

FIGURE 6.2 Example of specific application of the intermediate cooling power. (Projects accomplished by authors at Janis Research Company. Image courtesy of FormFactor, Inc.)

6.3.2 Efficient Refrigerant Condensation—Utilization of Distributed Cooling Power

As discussed above, by utilizing distributed cooling power, this Special Heat Exchanger can remove nearly 90% of enthalpy of the refrigerant after its initial cooling at the PTR first stage. For cryogen-free cryostats, this significantly reduces the thermal load on the PTR second stage, which typically operates around 2.8 K, thereby facilitating the refrigerant condensation process.

This approach is especially beneficial in cryogen-free DRs and continuous-flow He-3 cryostats equipped with JT condensers. When sufficiently cooled around 2.8 K at the PTR second stage, both He-3 and He-4 exhibit positive JT coefficients ($\mu > 0$) in JT assemblies, even when the head pressure is below 1 bar. The capability to operate at low head pressure enables the cryostat to function without an external compressor, which effectively reduces costs, vibration, and noise, and improves overall user-friendliness.

6.4 QUANTUM CRYOGENICS

Qubit coherence is the critical factor for the functioning of a quantum computer [3]. Even minute thermal fluctuation, EM disturbance, or mechanical vibrations at cryogenic temperatures may disrupt qubit coherence and lead to decoherence, whereby fragile quantum superposition states lose their phase relationships and evolve into classic-like statistical mixtures.

6.4.1 Challenges on Heat Management

In order to protect the fragile quantum states from thermal noise and ensure sufficiently long coherence times for qubit manipulation, devices such as Traveling Wave Parametric Amplifiers (TWPA), Josephson Parametric Amplifiers (JPA), and High Electron Mobility Transistors (HEMT) need to be operated at cryogenic temperatures with minimal thermal noise.

For example, HEMT is an important device in quantum computers, often used as the first stage of amplification in the readout chain, providing low-noise amplification of weak microwave readout signals before transmission to room-temperature electronics. HEMTs are often installed on the second stage of the PTR and maintained at ~3 K or lower during normal high-frequency operation. Their heat dissipation during normal operation is in the order of a watt, which requires the PTR second stage to deliver at least 1 W of usable cooling power [4]. After combined with additional contributions from wiring, RF cables, and other components, the total heat load from an operating quantum computer presents a significant challenge to the cryostat, including the PTR cryocooler. A detailed analysis of heat management strategies in quantum computing systems can be found in Reference [5].

6.4.2 Next-Generation PTR

To meet the growing thermal demands, advanced PTR plays a critical role in effective heat management for quantum computers [1, 6]. It must provide sufficient cooling power to ensure all devices operate at their required cryogenic temperatures. For example, it is desirable to operate some devices such as HEMT on the second stage at 3 K or lower temperatures.

The recently developed next-generation PTR, model Q3G (3 K) and Q4G (4 K) (Figure 6.3) from Boston Cryogenics LLC [6] directly addressed this need. The Q4G, optimized for 4 K operation, delivers 2.4 W at 4.2 K and more than 0.9 W at 3 K on the second stage, along with 70 W at 50 K on the first stage. The Q3G provides even better performance below 3.8 K, supplying at least 1 W of usable cooling power at 3 K on the second stage (Figure 6.3). The overall load map of the model Q3G PTR is shown in Figure 6.4.

Higher circulation rates are essential for achieving larger mixing chamber cooling power in a dilution refrigerator. In a cryogen-free DR, to reduce heat load to low temperature zones, it is critical to maximum the precooling of the return mixture on the PTR. To address this requirement, Boston Cryogenics developed novel heat exchangers to enhance the PTR precool efficiency (Figure 6.5). For example, an improved design implemented on a Q4G cold head provides progressive cooling when circulating refrigerant through the heat exchange assembly, including first-stage heat exchanger, intermediate precooling heat exchangers, and the second-stage heat exchanger. The performance measurements were conducted to characterize the Helium-4 gas temperature after the second-stage heat exchanger as a function of circulation rate while maintaining the first-stage temperature at 50 K. The corresponding data are listed in Table 6.1. As an example, with a high circulation rate of 3.1 SLM (2,148 μmol/s), the second-stage temperature is still able to stabilize at 2.4 K, potentially providing sufficient mixture precooling for high cooling power dilution refrigerators.

In addition, the Q4S and Q3S PTR models offer further operational advantages, such as reduced vibration and lower electromagnetic noise, making them particularly suitable for applications requiring exceptional mechanical and electrical stability.

6.4.3 Proper Thermal Anchor

DC wiring, RF cables, attenuators, amplifiers, and other microwave components must be installed across both stages of the PTR to ensure proper thermal anchoring. Thermal leakage from wiring and RF cables has been discussed in detail in Reference [7]. Figure 6.6 illustrates a typical assembly of RF cables and auxiliary devices, together with their thermal anchoring scheme.

Although hundreds or even thousands of qubits can now be fabricated on a single quantum processor and cooled to millikelvin temperatures [8], scaling up the corresponding control and readout electronics remains a significant challenge. This challenge arises primarily from the limited cooling power available in cryogenic systems, which constrains not only the number of physical connections but also the ability to dissipate the heat loads associated with dense cabling.

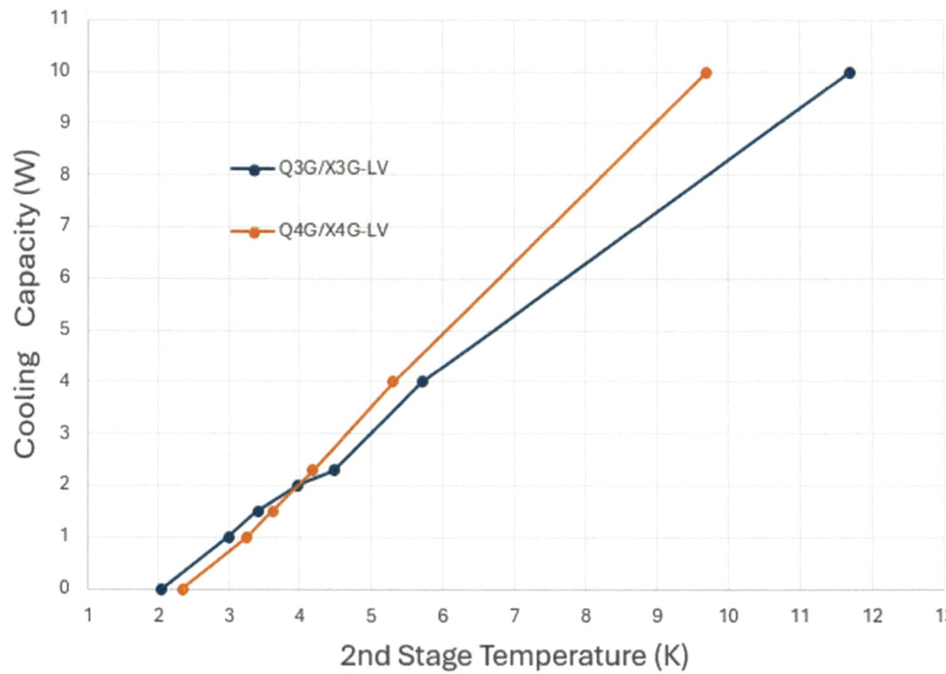

FIGURE 6.3 (a) Novel PTR from Boston Cryogenics and (b) the second stage temperature versus cooling power of PTR provided by Boston Cryogenics. (Reprint with permission from communication with Wang, C. at Boston Cryogenics LLC, 12 Dunham Rd, Unit 1, Billerica, MA 01821 info@boscryo.com] [6].)

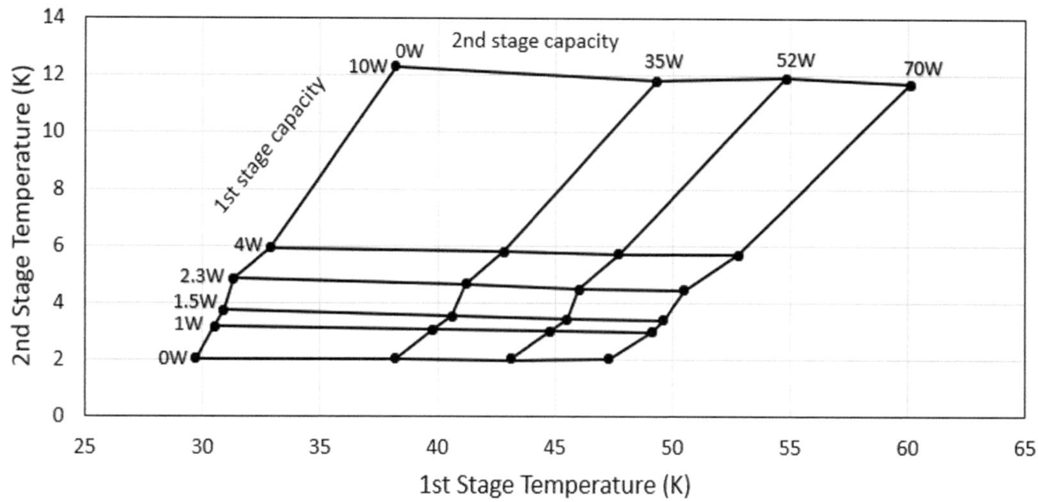

FIGURE 6.4 Overall load map of the model Q3G PTR. (Reprint with permission from communication with Wang, C. at Boston Cryogenics LLC, 12 Dunham Rd, Unit 1, Billerica, MA 01821 info@boscryo.com.)

TABLE 6.1
Temperature on the Second Stage of Model Q4G PTR as a Function of Helium-4 Circulation Rate

He-4 Gas Flow Rate in SLM	He-4 Gas Flow Rate in μmole/s	First Stage Temperature (K)	Second Stage Temperature (K)
1.5	1,039.5	50	~ 2.25
3.1	2,148.3	50	2.4
7.0	4,851.0	50	< 3.1

Source: Wang, C. at Boston Cryogenics LLC, 12 Dunham Rd, Unit 1, Billerica, MA 01821, info@boscryo.com

One promising approach to overcome these challenges involves integrating the system with multiple high-power PTRs. In an optimized configuration, one PTR can be dedicated to maintain the DR operation, while additional PTRs are assigned to handle the thermal load from microwave components and cryogenic wiring. This architecture distributes the cooling capacities more effectively and enables the scaling of cryogenic environments to support large-scale quantum processors.

6.4.4 Minimizing the Photon Perturbation

Photons can interact with qubits and potentially disrupt their coherence. To mitigate this effect, several approaches are employed, including designing a light-tight cryostat and blackening the inner surface of the mixing chamber radiation shield [9], thereby reducing photon-induced decoherence.

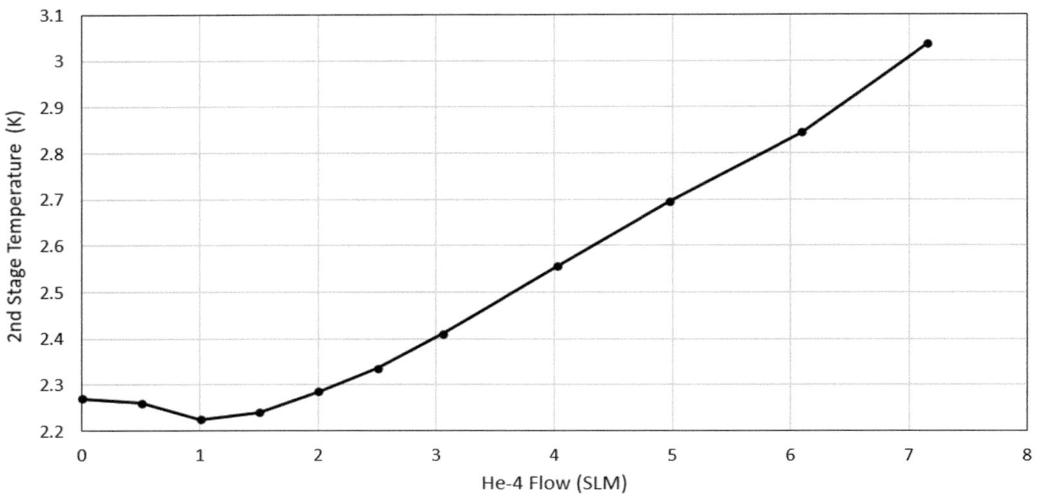

FIGURE 6.5 Helium-4 gas temperature after the second-stage Heat Exchanger vs helium-4 flow rate. (Reprint with permission from communication with Wang, C. at Boston Cryogenics LLC, 12 Dunham Rd, Unit 1, Billerica, MA 01821 info@boscryo.com.)

6.4.5 UHV-Compatible Cryostat

Unwanted impurities around qubits present another source of decoherence and performance degradation. For this reason, quantum DR users require the system to be UHV-compatible. This ensures a clean environment around the quantum devices and minimizes impurity-induced noise. Readers may refer to Chapter 12 for a detailed discussion of UHV technology and its implementation in cryogenic systems.

6.4.6 Example of Hybrid Quantum System Setup

Figure 6.7 illustrates a UHV-compatible platform for the hybridization of atomic and superconducting quantum systems [10]. In this setup, ultracold atoms can be transferred and trapped in a millikelvin cryogenic environment, where "*interactions between atomic and superconducting qubits may be established, paving the way for hybrid quantum systems*" [10]. Interested readers are referred to Reference [10] for further details.

From a cryogenic engineering perspective, this highly customized millikelvin cryostat differs from commercial dilution refrigerators in two key aspects:

1. **Integration of superconducting coils**
   Superconducting coils were installed at the bottom plate of second stage (3K stage) and still shields (at 0.8 K) to collimate cold atoms after injection from the bottom of the cryostat. ***During factory testing***, eight high-Tc current leads were initially installed between the first-stage (50K stage) and the second stage cold plate to power the coils. However, their combined conductive heat load raised the second-stage temperature above 4.0 K, which in turn increased the still temperature and reduced the mixing chamber cooling power.

   To mitigate this issue, an intermediate cold plate was added to the second-stage regenerator of the PTR above the 3K plate. This plate operates at approximately 8.2 K, referred to as the 8K plate, and serves as the thermal termination point for the lower ends of the high-Tc leads, as shown in Figure 6.8. From this point downward, low-Tc superconducting wires were used

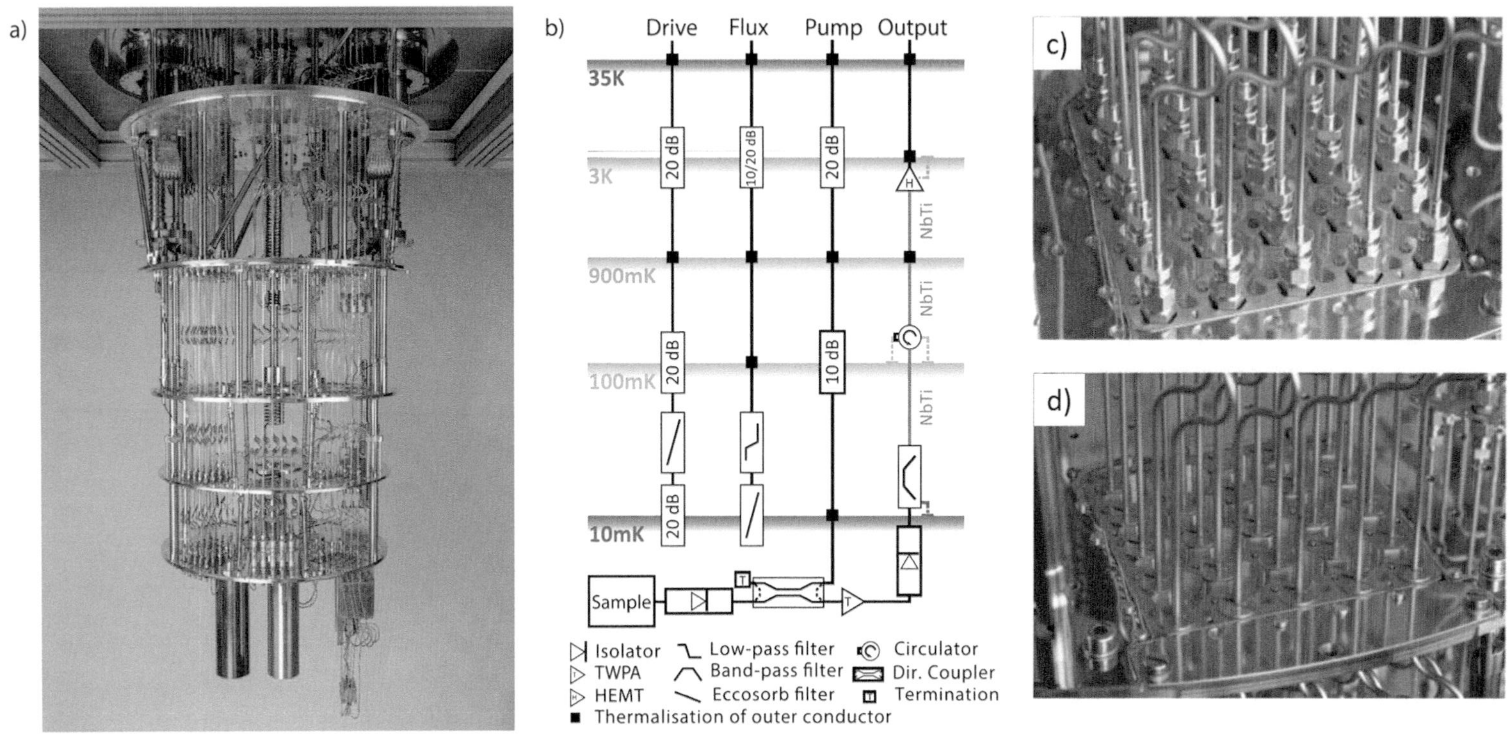

FIGURE 6.6 Cabled dilution refrigerator (DR) for quantum computer research. (Reprint with permission from S. Krinner, S. Storz, P. Kurpiers, P. Magnard, J. Heinsoo, R. Keller, J. Lütolf, C. Eichler and A. Wallraff, Engineering cryogenic setups for 100-qubit scale superconducting circuit systems, *EPJ Quantum Technology* (2019, [7].)

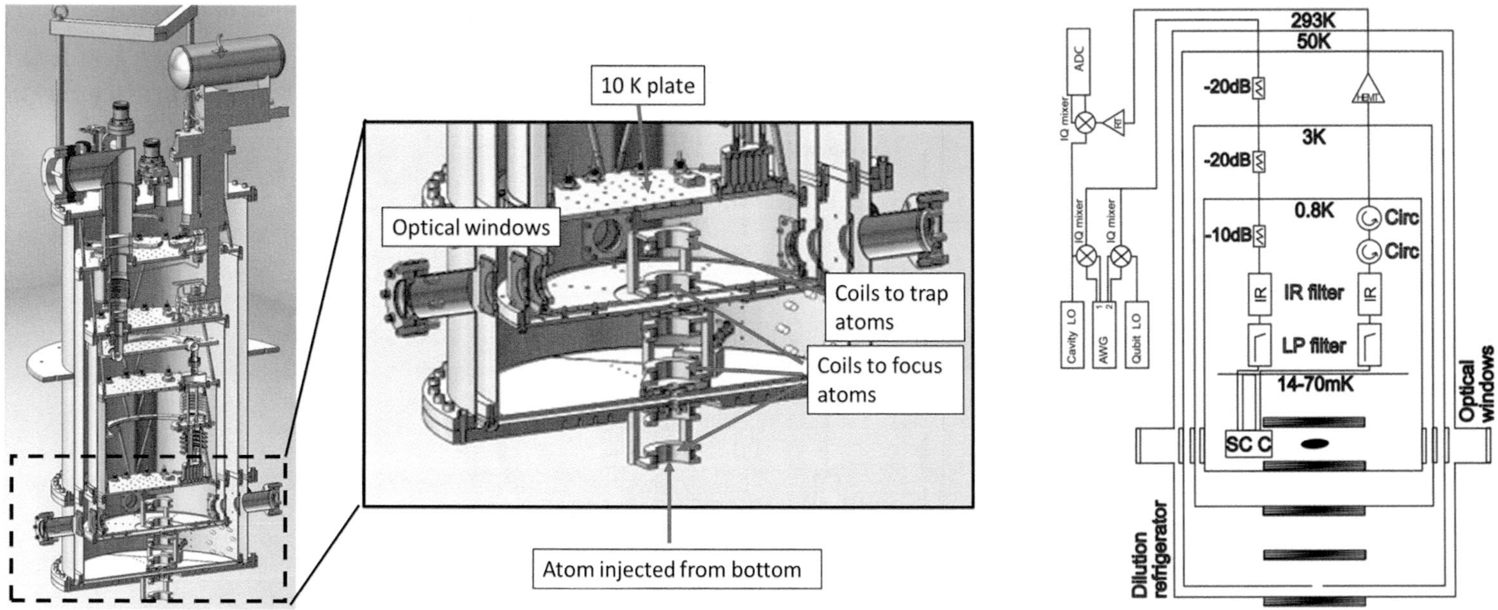

(a) (b)

FIGURE 6.7 (a) Experimental platform for hybridization of atomic and superconducting quantum systems. (Projects accomplished by authors at Janis Research Company. Model courtesy of FormFactor, Inc.) (b) Schematic of the room temperature and cold microwave electronics for probing superconducting circuits. Attenuators, low pass filters (LP filters), custom-made infrared filters (IR filters), circulators (Circ), HEMT, and low noise room temperature (RT) amplifiers have been installed. Superconducting quantum circuits (SCC) can be probed and measured with arbitrary waveform generator (AWG), local oscillators (LO), and analog to digital converter (ADC). (Reprint with permission from Alessandro Landra, Christoph Hufnagel, Lim Chin Chean, Thomas Weigner, Yung Szen Yap, Long Hoang Nguyen, and Rainer Dumke, The design of an experimental platform for hybridization of atomic and superconducting quantum systems, *Physical Review A*, Volume 99, Article 053421, on May, 2019, [10].)

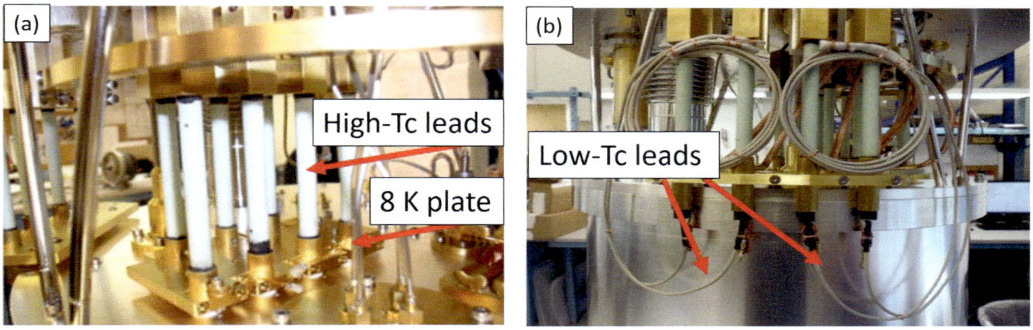

FIGURE 6.8 (a) Installation of high-Tc current leads between the "50K" and "8K" cold plates and (b) Low Tc leads installed from the bottom of 8K plate. (Projects accomplished by authors at Janis Research Company. Image courtesy of FormFactor, Inc.)

to connect the high-Tc leads to the superconducting coils. In this setup, the excess cooling power of the PTR second-stage regenerator is utilized to absorb the conductive heat load from multiple high-Tc leads, thereby restoring the normal DR operation.

2. **Optical cryostat with window ports**
   This system is also configured as an optical cryostat equipped with four CF40 window ports. The optical measurements were performed under several different configurations.
   - Configuration #1: Factory Test
     All window ports—including those on the OVC, the 50K radiation shield, the 3K radiation shield, and the still radiation shield—were sealed with blank flanges. The cryostat reached the flowing results:
     - The 8K cold plate stabilized at temperature of 7 K.
     - The DR system reached a base temperature of 14 mK.
     - The mixing chamber achieved a cooling power of 400 μW at 100 mK.

     The similar cryostat performance was replicated during the on-site test.
   - Configuration #2: User Measurements [11]
     Two sets of quartz windows with CF40 window retainer were installed along one viewing axis (i.e., two windows on each OVC, 50K, and 3K shield), while ZnSe windows were installed on the still radiation shield. The window ports along the other viewing axis were sealed with blank flanges. In this configuration, the DR system reached a base temperature of 80 mK.
   - Configuration #3: Heat Absorbing Windows [11]
     This configuration is identical to Configuration #2, except the ZnSe windows on the Still radiation shield that were replaced with Model FSQ-KG2 "heat absorbing" windows from Newport [12]. In this setup, the DR system achieved a base temperature of approximately 65 mK.
   - Configuration #4: Dual Viewing Axes [11]
     The window materials are same as Configuration #3, except that the windows were installed along both viewing axes—resulting in four windows on each radiation shield. During this test, the DR reached a base temperature of approximately 100 mK.

These test results are summarized in Table 6.2, providing reference data on the performance of the optical dilution refrigerator system in the millikelvin temperature range. A small opening is

TABLE 6.2
Performance of Optical DR

Test	Window Setup	Windows on OVC, 50K, and 3K Shields	Still Shield Window	Viewing Axis with Windows	MC Temperature	Equivalent Cooling Power without Windows (Circulation Rate: 800 μmole/s)
1	All ports blocked	None	None	None	14 mK	0 μW
2	2 sets	Quartz	ZnSe	One	80 mK	279 μW
3	2 sets	Quartz	FSQ-KG2	One	65 mK	182 μW
4	4 sets	Quartz	FSQ-KG2	Two	100 mK	400 μW

Source: Data from communication with Christoph Hufnagel and Rainer Dumke, Centre for Quantum Technologies (CQT), National University of Singapore, Singapore 117543 [10].

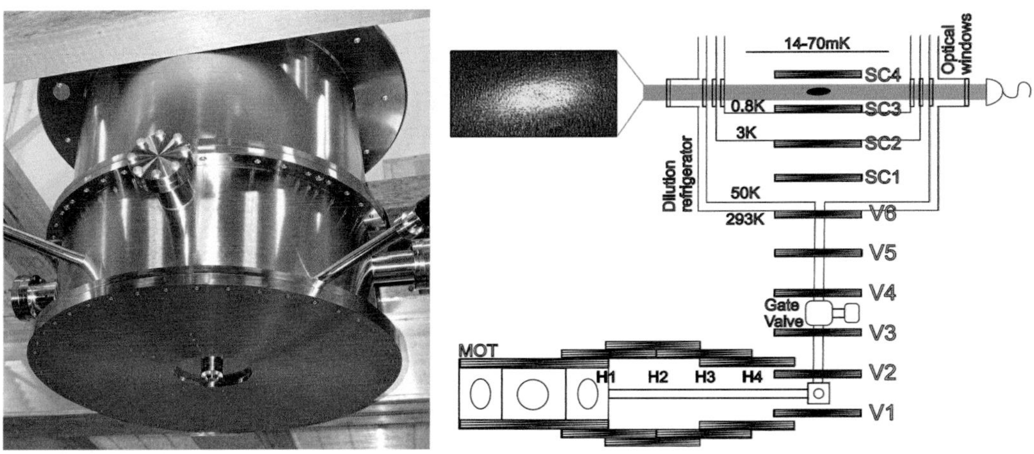

FIGURE 6.9 (a) Bottom view of the cryostat. (Projects accomplished by authors at Janis Research Company. Image courtesy of FormFactor, Inc.) (b) Loading sample (atoms) from bottoms. (Reprint with permission from Alessandro Landra, Christoph Hufnagel, Lim Chin Chean, Thomas Weigner, Yung Szen Yap, Long Hoang Nguyen, and Rainer Dumke, The design of an experimental platform for hybridization of atomic and superconducting quantum systems, *Physical Review A*, Volume 99, Article 053421, on May 2019, [10].)

installed at the bottom of the cryostat for sample (atom) injection from a separate vacuum chamber as illustrated in Figure 6.9. A window will be installed on the opening when the vacuum chamber for the cold atoms is not connected to the cryostat.

REFERENCES

1. Bluefors Cryocooler Technologies Inc. 6682 Moore Rd, Syracuse, NY, USA Website: www.bluefors.com
2. C. Wang, Extracting Cooling From Pulse Tube and Regenerator in a 4K Pulse Tube Cryocooler, Proceedings of the 15th International Cryocooler Conference, 177–184, (2009).

3. Orchestrating a Quantum Leap using Cold Atoms, *Nature Research Custom Media* www.nature.com/articles/d42473-023-00439-4
4. Chu-Dan Qiu, Yuan-De Jin, Jun-Xiang Zhang, Gang-Qin Liu, and Wen-Long Ma, How Coherence Measurements of a Qubit Steer Its Quantum Environment, *Physical Review B*, 110, 024311, (2024).
5. Muhaimin Haziq, Shaili Falina, Asrulnizam Abd Manaf, Hiroshi Kawarada, and Mohd Syamsul, Challenges and Opportunities for High-Power and High-Frequency AlGaN/GaN High-Electron-Mobility Transistor (HEMT), *Applications: A Review Micromachines*, 13(12), 2133, (2022).
6. Boston Cryogenics LLC Dunham Rd, 1, Billerica, MA, USA Email: info@boscryo.com Website: www.boscryo.com
7. S. Krinner, S. Storz, P. Kurpiers, P. Magnard, J. Heinsoo, R. Keller, J. Lütolf, C. Eichler, and A. Wallraff, Engineering Cryogenic Setups for 100-qubit Scale Superconducting Circuit Systems, *EPJ Quantum Technology*, 6, 1–29, (2019).
8. K. Padavic-Callaghan IBM unveils world's largest quantum computer at 433 qubits, New Scientist, (November 9, 2022) (www.newscientist.com).
9. Socomore Group ZI du Prat, Avenue Paul Duplaix, 56037 Vannes Cedex, France Website: www.socomore.com/en/brand/aeroglaze
10. Alessandro Landra, Christoph Hufnagel, Lim Chin Chean, Thomas Weigner, Yung Szen Yap, Long Hoang Nguyen, and Rainer Dumke, The Design of an Experimental Platform for Hybridization of Atomic and Superconducting Quantum Systems, *Physical Review A*, 99, Article 053421, (on May, 2019).
11. Private communication with Dr. Christoph Hufnagel, Centre for Quantum Technologies, Singapore.
12. FSQ-KG2-C is a KG2 glass IR cut-off filter with a protective coating for humidity resistance, used to block infrared and reduce heat in optical systems. Website: www.newport.com/p/FSQ-KG2-C

7 Special Cryostats

7.1 MICROWAVE SPECTROSCOPY CRYOSTAT AT MILLIKELVIN

Many important microwave experiments are conducted below 1 K, including electron paramagnetic resonance (EPR) on magnetically ordered systems, microwave surface impedance measurements of metals, studies of conventional and unconventional superconductors focusing on collective modes, and investigations of magnetic superconductors [1]. Microwave spectroscopy at millikelvin temperatures has also become a crucial tool in quantum computing and fundamental quantum physics, enabling qubit characterization in superconducting processors, spectroscopy of hybrid quantum systems, and exploration of quantum materials. In this section, we introduce a highly customized microwave spectroscopy cryostat operating below 100 mK.

In this system, an Oxford model TLE 400 top-loading dilution refrigerator has been adapted to build an X-band microwave spectrometer operating at millikelvin temperatures. The dilution refrigerator provides temperatures between 0.1 and 1 K, corresponding roughly to photon energies from 2.3 to 23 GHz, within the microwave regime.

Operating below 1 K suppresses thermal noise, producing a "quiet" environment where the system approaches its quantum ground state—particularly below 100 mK. At these temperatures, observed spectral features reflect intrinsic quantum responses rather than thermal excitations. This suppression of thermal decoherence enables high-resolution spectroscopy and long-lived quantum states.

7.1.1 Design and Construction

The millikelvin microwave spectrometer (mK-MWS) consists of six major components (Figure 7.1):

- Dilution refrigerator (DR): provides the millikelvin-temperature environment.
- Microwave generator (electronics): located outside the DR at room temperature and generates the microwave signals.
- Cylindrical waveguide: transfers microwaves from the microwave generator to the sample region.
- Microwave resonator (cavity): a resonant chamber that stores and manipulates energy at specific frequencies.
- Gordon coupler: regulates the coupling of energy between the waveguide and the resonator to control the power flow into and out of the resonator.
- Sample top-loader: allows samples to be loaded from ambient conditions into the millikelvin environment.

Since the DR and microwave generator are standard commercial products, our discussion will focus on the remaining four components.

DOI: 10.1201/9781042000050-7

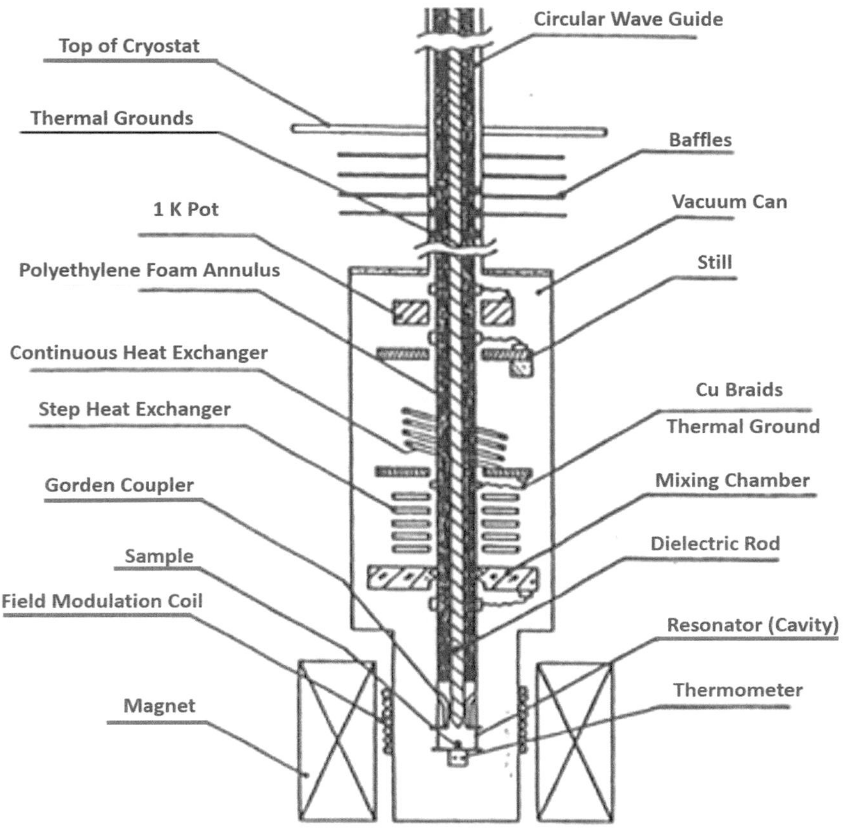

FIGURE 7.1 Schematic, overall view of the elements of microwave spectroscopy cryostat at mK. (Reprint from permission of Zuyu Zhao, *A top loading, very low-temperature, X-band, microwave spectrometer*, M. Z. Lin, Z. Zhao, Y. H. Shen, and J. B. Ketterson, Y. J. Qian and Bimal K. Sarma, Rev. Sci. lustrum. 61 (5), May 1990.)

7.1.1.1 Cylindrical Waveguide

A waveguide is a metallic transmission line used at microwave frequencies to connect the microwave generator (e.g. klystron) to the receiver (e.g. resonator). In this setup, instead of conventional coaxial cables, the waveguide is a stainless steel circular tube adapted from the central sample-loading access of the dilution refrigerator. This design offers several significant advantages.

- The circular wave guide tube offers MWS top-loading capabilities in the cryostat. The sample can be directly transferred from a room temperature zone to the cold microwave resonator (≤ 4 K) through the waveguide without warming up the low-temperature region.
- The Gordon coupler and microwave resonator are installed at the bottom of guide tube without the requirement of additional mounting structure.
- The circular waveguide has small energy loss and wide bandwidth. To further increase reflection and reduce energy loss, the tube inner surface can be coated with conductive materials, such as gold or superconductive metals.
- The simple circular guide tube is easy to be fabricated.

One of the most important properties of the waveguide is its cutoff frequency, the minimum frequency at which a particular propagation mode can travel through the waveguide without significant attenuation. For a circular wave guide, the cutoff frequency can be expressed as the following equations [2]:

For the TM_{mn} mode:

$$f_{mn} = \frac{c}{2\pi\sqrt{\mu\varepsilon}} \cdot \frac{x'_{mn}}{R} \tag{7.1}$$

and, for the TM_{mn} mode

$$f_{mn} = \frac{c}{2\pi\sqrt{\mu\varepsilon}} \cdot \frac{x_{mn}}{R} \tag{7.2}$$

where:

- m and n are the mode indices representing the function number and the root number, respectively.
- x_{mn} and x'_{mn} are the n-th roots of the cylindrical Bessel function of the first kind of order m, $J_m(x)$, and its derivative, $J'_m(x)$, respectively.
- ε is the relative permittivity (also referred to as the dielectric constant; $\varepsilon = 1$ in vacuum).
- μ is the relative permeability ($\mu = 1$ in vacuum)
- c is the speed of light in air or vacuum
- R is the tube inner radius.

To achieve a wider microwave bandwidth, a waveguide with a larger inner diameter is desired. However, space constraints in the DR limit the maximum allowable inner diameter to 0.667 in (Figure 7.2). For a vacuum-filled circular waveguide of this size, using the values of x_{01} and x'_{11} listed in Table 7.1, the cutoff frequencies are is 10.15 GHz for the TE_{11} (transverse electric) mode and 13.6 GHz for the TM_{01} (transverse magnetic) mode [2]. These values are close to the upper limit of the IEEE-defined X-band (8–12 GHz).

Because the waveguide provides a direct path from room temperature to the millikelvin stage, thermal radiation must be effectively blocked. In this design, the tube is filled with a stack of annular porous polyethylene foam rings with an inner diameter of 0.25 in. The foam blocks the majority of thermal radiation while maintaining a central opening for Gordon coupler tuning and sample loading, as discussed later. Although the effective dielectric constant of the porous polyethylene foam is not precisely known, it exhibits high transparency to microwave signals. During the experiments, no significant microwave attenuation or measurable shift in cutoff frequencies was observed. In order

TABLE 7.1
Roots of Cylindrical Bessel Functions

Function Number m	Root Number n	Roots of the Bessel Function J_m $x_{m,n}$	Roots of the Bessel Function Derivatives J'_m $x'_{m,n}$
0	1	2.405	
1	1		1.841

Source: Data from reference [2].

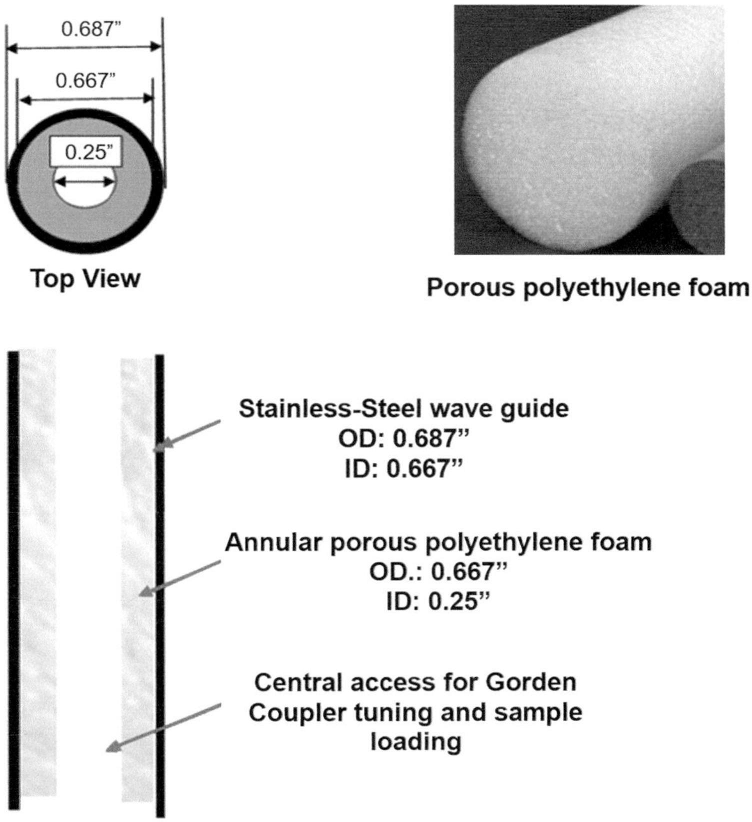

FIGURE 7.2 Stainless steel circular waveguide. (The sketch is disproportionate.)

to block thermal radiation, the guide tube is thermally anchored to the DR at three temperature stages—150 K, 78 K, and 4.2 K—using copper braids.

7.1.1.2 Microwave Resonator (also referred to as "Microwave cavity")

A microwave resonator is a chamber that supports standing waves at specific resonant frequencies (Figure 7.3). In this setup, a cylindrical resonator made of OFHC copper is installed at the bottom of the waveguide and thermally anchored to the DR mixing chamber using flexible copper braids. The resonator temperature is monitored by thermometers mounted at the base.

The resonant frequency depends on several factors, including the resonator dimensions, operating mode, temperature, and the dielectric constant of the material inside the resonator. For the initial design, the TE_{111} and TM_{010} modes were selected. These two modes are close in frequency but sufficiently separated to avoid ambiguity during measurement. The resonant frequencies for a cylindrical cavity resonator can be calculated using the following formulae [3].

For the TE_{mnp} mode:

$$f_{mnp} = \frac{c}{2\pi\sqrt{\mu\varepsilon}}\sqrt{\left(\frac{x'_{mn}}{R}\right)^2 + \left(\frac{p\pi}{h}\right)^2} \tag{7.3}$$

and, for TM_{mnp} mode

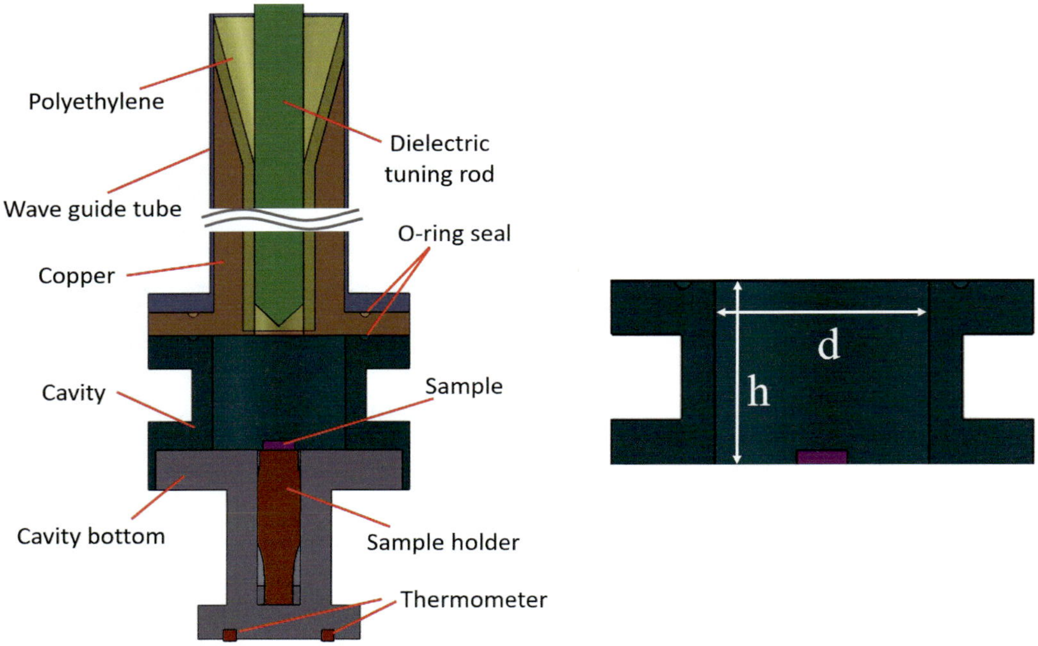

FIGURE 7.3 Schematic drawing of the microwave resonator.

$$f_{mnp} = \frac{c}{2\pi\sqrt{\mu\varepsilon}}\sqrt{\left(\frac{x_{mn}}{R}\right)^2 + \left(\frac{p\pi}{h}\right)^2} \tag{7.4}$$

where:

- p is the number of half-wave variations along the cavity axis (p=0, 1, 2…).
- h is the length of cylindrical cavity.

For the TE_{111} mode, $m = 1$, $n = 1$, and $p = 1$, whereas for TM_{010} mode, $m = 0$, $n = 1$, and $p = 0$ (See Table 7.1 for the values of the roots of the cylindrical Bessel functions and their derivatives).

The optimal resonator dimensions are chosen to place the resonant frequency slightly above the cutoff frequency of the circular waveguide. This ensures efficient propagation of microwave signals into the resonator while filtering out higher-frequency noise. It also prevents impedance mismatch and sensitivity issues that may occur if the cutoff and resonant frequencies are too close.

In this set-up, the resonant frequency of the resonator was measured to be 11.420 GHz when the Gordon coupler was tuned to its critical position (see Section 7.1.1.3, Gordon Coupler). Several points are noteworthy:

- The resonator interior is initially under vacuum ($\varepsilon = 1.0$).
- The TE_{111} mode exhibits its strongest electric field at the center of the resonator bottom, making it ideal for electron spin resonance (ESR) studies.
- In a cylindrical resonator, the TE_{111} mode actually consists of two degenerate modes corresponding to two possible orthogonal electric field polarizations. This degeneracy can be lifted by introducing a perturbation, such as a radial protrusion on the resonator bottom, allowing polarization control by rotating the resonator about its axis.

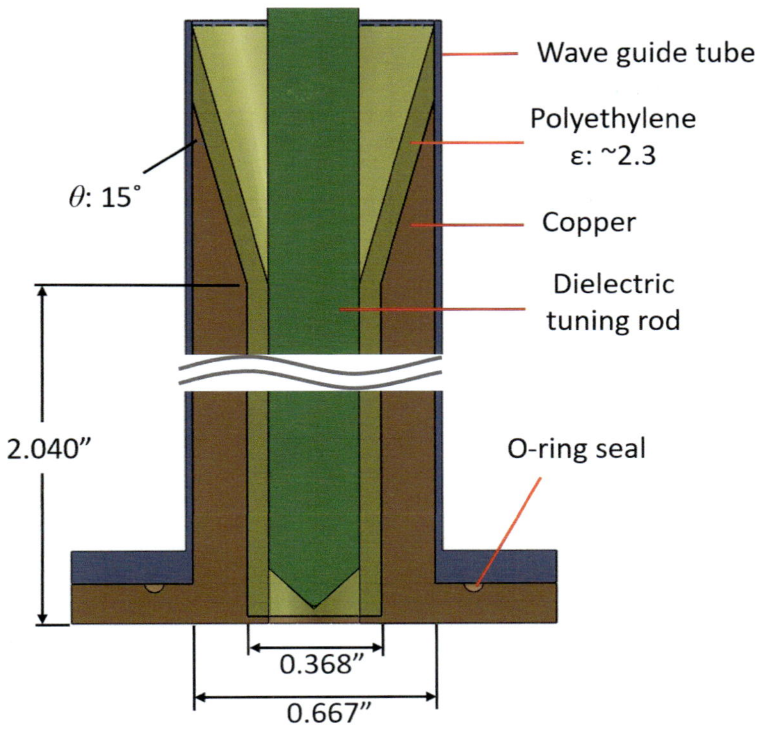

FIGURE 7.4 Gorden coupler.

7.1.1.3 Gordon Coupler

The Gordon coupler is installed between the bottom of the waveguide and the top of the resonator (see Figure 7.4) to optimize signal coupling between the two components [4]. It consists of a cylindrical copper insert that fits precisely within the waveguide tube (Figure 7.4). The inner diameter of this copper section is chosen so that its cutoff frequency lies above the resonant frequencies of the resonator. The interior of the copper insert is filled with a dense polyethylene dielectric rod ($\varepsilon = 2.3$) featuring a central access hole of 0.25 in. diameter.

By inserting a dielectric rod into this narrow copper insert, the cutoff frequency of the section is lowered below the operating frequency, enabling controlled coupling to the resonator. Fine adjustment of the coupling strength can be achieved by changing the insertion depth of the rod until the optimal (critical) performance is reached (Figure 7.5). A detailed characterization of the Gordon coupler can be found in Reference [1].

When the dielectric rod is fully withdrawn, the access hole becomes a top loading channel, enabling sample loading or removal without interrupting the cryogenic environment (see Section 7.1.1.4.b). The tapered shape at the top of the Gordon coupler serves as a guide for the sample loader, ensuring precise insertion into the resonator.

The resonator performance is commonly evaluated using DPPH (2,2-diphenyl-1-picrylhydrazyl), an organic compound with a strong ESR signal arising from unpaired electrons with a g-value of 2.0036. The DPPH adsorption peak is reliably detected unless the excitation frequency deviates significantly from the cavity resonance. Figure 7.6 illustrates an ESR measurement of a Cu(tatbp)I powder sample with a small amount of DPPH placed on top as a reference. In the spectrum, peak "A" represents the sharp and strong absorption peak of DPPH, while peaks "B" and "C" correspond to the $g_\perp$ and $g_\parallel$ resonances, respectively, of the Cu(tatbp)I sample [5].

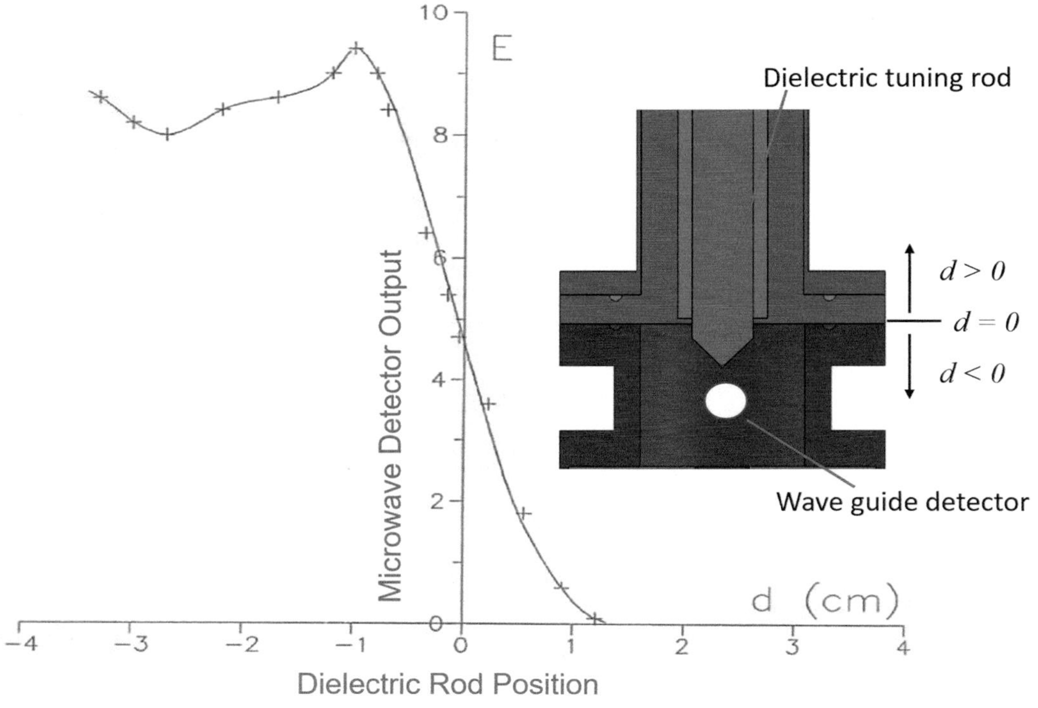

FIGURE 7.5 Detector output (optimized at each point) as a function of the position of the dielectric rod. X-axis: Position of the dielectric rod (Coupler tuner); Y-axis: Microwave detector output (arbitrary unit).

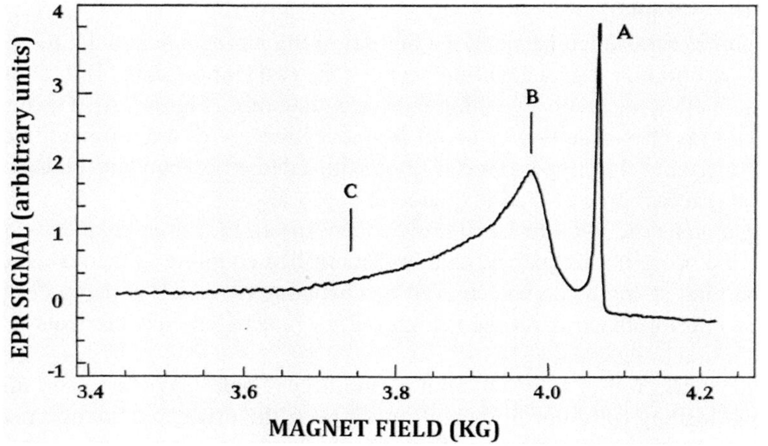

FIGURE 7.6 EPR signal from DPPH. *Source:* Information from Zuyu Zhao, Ph.D. Thesis, 1990, Northwestern University, Evanston, IL.

7.1.1.4 MWS Top Structure

Two critical structures are integrated with the DR at the top of the millikelvin microwave spectrometer [1, 4]:

a) **Connection between the Microwave Generator and the MWS**
   The room-temperature microwave generator (a Klystron) is connected to the MWS via a rectangular waveguide (Figure 7.7). A clear Mylar (polyester) sheet is clamped between the O-ring–sealed flanges. The Mylar sheet not only provides a hermetic seal between the waveguide and the Klystron but also allows the microwave signal to pass through with negligible loss.
b) **Top-Loading Feature**
   This MWS top-loading features allow samples to be directly loaded into the microwave resonator from room temperature without warming the cryostat above approximately 4 K [1, 5]. Figure 7.8 illustrates the structure of the top-loading assembly. A ball valve and a pop-off valve are installed on the waveguide adapter, with the in-line ball valve opened during sample exchange.

The sample is attached to a stainless steel sample holder using silver epoxy or GE 7031 varnish and transferred by a stainless steel loading rod. The rod features a left-hand thread at its bottom end, which mates with the matching thread on the sample holder. The lower end of the sample holder contains a tapered section followed by a right-hand thread (Figure 7.9), designed to match the threaded resonator bottom. The tapper improves the thermal contact between the sample holder and the resonator by increasing the contact surface area and the normal force at the interface.

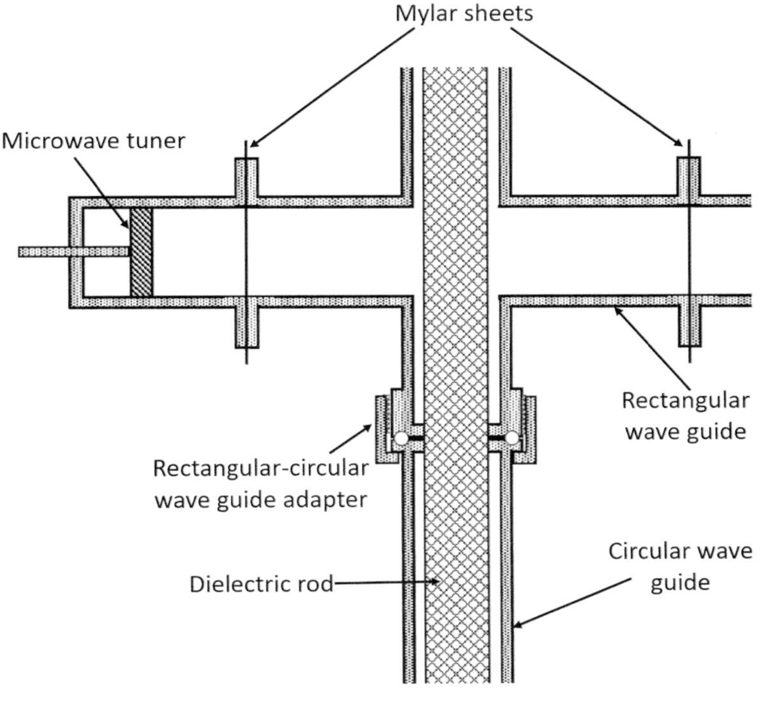

FIGURE 7.7 The rectangular to circular waveguide adapter. *Source:* Information from Zuyu Zhao, Ph.D. Thesis, 1990, Northwestern University, Evanston, IL.

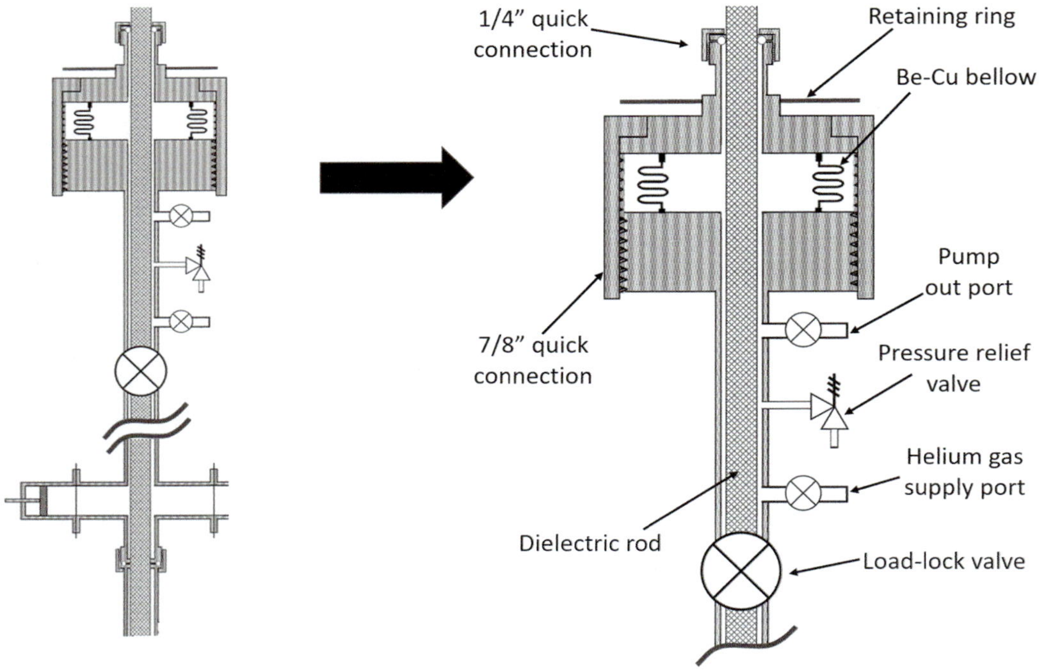

FIGURE 7.8 Top-loading sample set-up. *Source:* Information from Zuyu Zhao, Ph.D. Thesis, 1990, Northwestern University, Evanston, IL.

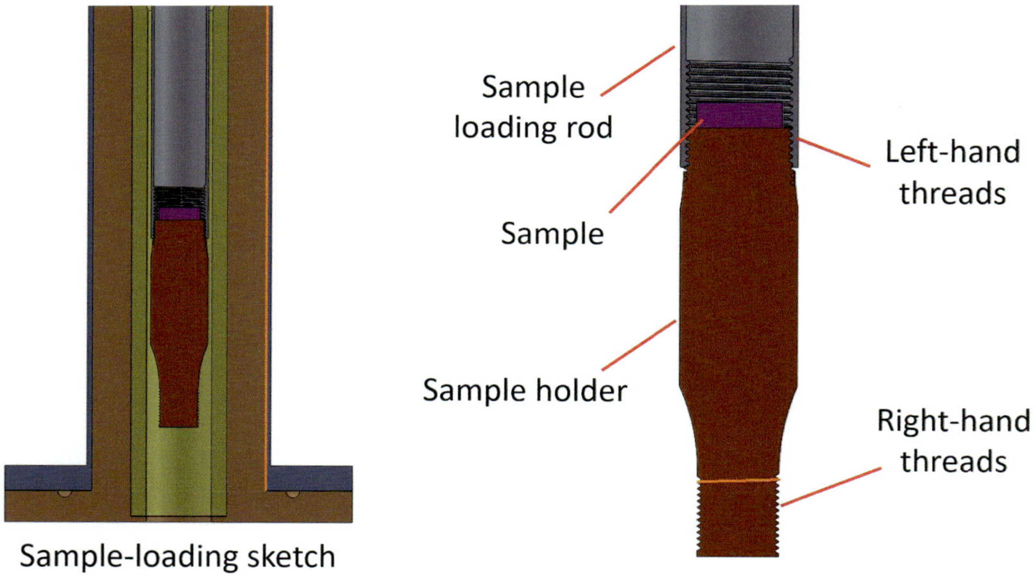

FIGURE 7.9 Sample loading rod with sample holder.

During loading, the rod carrying the preloaded sample holder is inserted into the cryostat and threads the sample holder into the resonator. Once sufficient torque is reached, the holder automatically detaches from the rod. This reverse process is used for the sample removal. The limitation of this design is the relatively moderate contact force between the resonator and the sample, which may result in non-negligible thermal boundary resistance at the contact interface. Consequently, a temperature gradient between the sample holder and the resonator is expected. Mitigation strategies for this issue are discussed in the following section.

After the sample is secured, the sample-loading rod is withdrawn and replaced with the dielectric rod for the Gordon coupler tuning. Fine and continuous adjustment of the dielectric rod position can be achieved via the bellows assembly at the top of the cryostat, which incorporates a threaded linear motion.

7.1.1.5 Cryogenics

To ensure that the sample can reach millikelvin, the heat load to the microwave resonator must be minimized. This requirement is addressed with three main approaches [1, 5].

a) **Intercepting the Conductive Heat Load:**
The low-temperature section (below 4 K) of the stainless steel waveguide tube is thermally anchored to the dilution refrigerator at multiple locations (Figure 7.1). Copper rings are clamped around the tube, and silver-soldered copper braids connect these rings to small copper blocks at various thermal stages on the DR, such as the 1K pot, Still, and intermediate cold plate (ICP). The lowest section of the waveguide, including the copper section of the Gordon coupler, is thermally anchored to the mixing chamber using the same method. The flexible copper braids offer limited rotation of the microwave tube along its central axis.

b) **Blocking the IR Radiation:**
As discussed above, the interior of the circular waveguide tube is filled with a stack of polyethylene foam annuli, extending from the point just below the circular-to-rectangular waveguide adapter down to the top of Gordon coupler. This foam stack attenuates thermal radiation while preserving a central access channel.

c) **Filling the Resonator with Liquid Helium**
When the Klystron was turned off, the resonator and sample cooled below 50 mK. Activating the microwave generator introduced additional heating, raising the sample temperature. To mitigate this effect, liquid helium was directly condensed into the resonator, thereby ensuring thermal equilibration between the sample and the resonator. Initial experiments with liquid He-4 resulted in a base temperature of 210 mK, as superfluid He-4 creeping along the waveguide created a thermal short to higher-temperature stages. Substituting He-4 with He-3 eliminated this problem, enabling both the resonator and the sample to reach base temperatures below 50 mK (Klystron off) or approximately 70 mK (Klystron on).
The dielectric constant of liquid He-3 at sub-Kelvin temperatures is slightly higher than that in vacuum, but the difference produces no measurable shift in the resonator's resonant frequencies. However, because the surface vibrations of liquid can destabilize the resonance frequency, a partially filled resonator should be avoided. A compact gas-handling system (Figure 7.10) is employed, with a dipstick serving as a pump to supply or extract He-3.

d) **Additional Consideration**
Indium O-rings were initially considered for the hermetic seals between the waveguide tube and the resonator to provide reliable electrical contact and preventing microwave leakage. However, indium becomes superconducting below 3.4 K in zero magnetic field with a critical field of 293 Gauss [6]. This superconducting transition can cause a dramatic change in the resonator Q-factor (Figure 7.11), which could be misinterpreted as sample signals. To avoid this artifact the indium seals were replaced with custom-made Teflon O-rings. In addition, residual

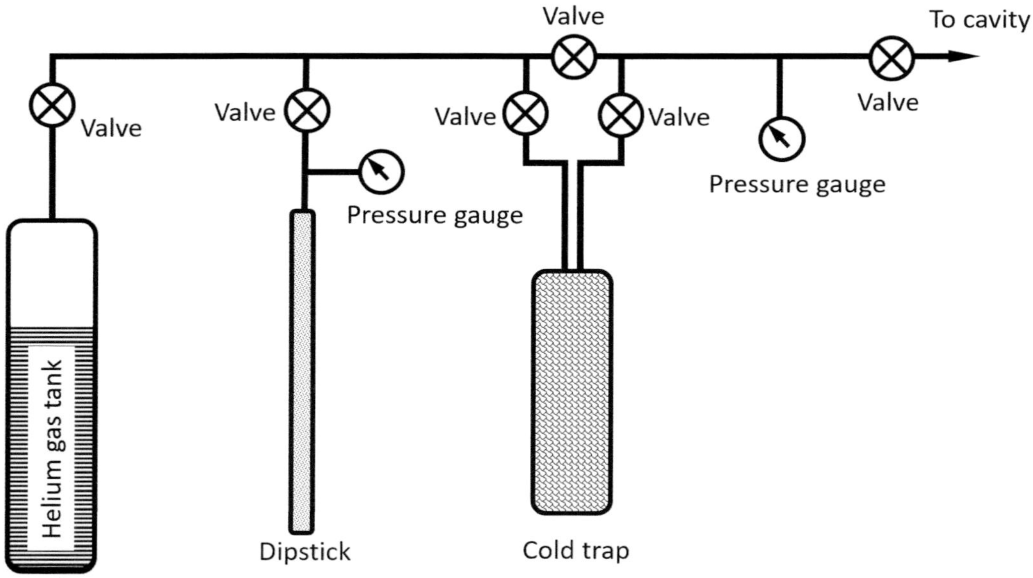

FIGURE 7.10 Schematic diagram of helium-3 gas handling system.

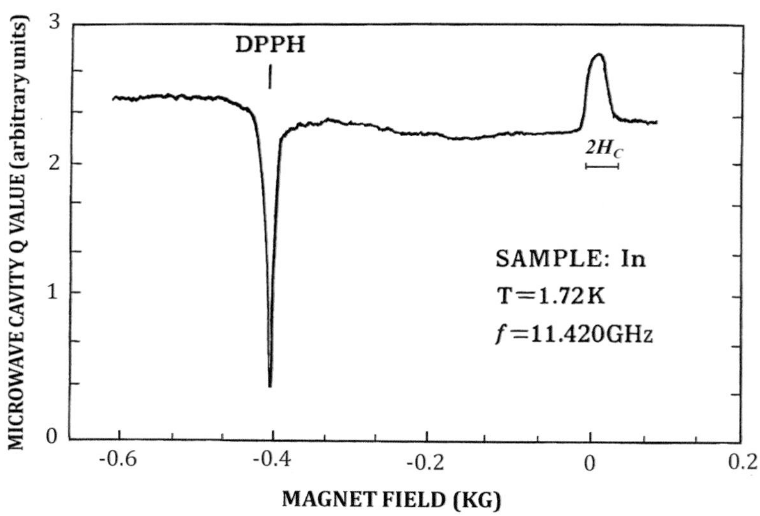

FIGURE 7.11 Microwave absorption arising from the Indium O-ring seal during a field sweep at 1.72K. A sharp increase of Q factor of the resonator is observed below the Indium critical field. *Source:* Information from Zuyu Zhao, Ph.D. Thesis, 1990, Northwestern University, Evanston, IL

magnetic flux can remain trapped in the superconducting magnet even when the current is set to zero. A bipolar power supply was therefore used to determine the precise location of zero magnetic field.

7.1.2 MWS Performance

The MWS performance was characterized through two types of tests: surface impedance measurements and ESR measurements. This section focuses on the discussion of surface impedance measurements.

Microwave surface impedance measurements on metals probe the interaction between microwaves and metallic surfaces to provide insights into material properties. Preliminary studies were carried out on conventional s-wave superconductors (Sn, Zn, and In) and the heavy fermion superconductor UPt_3 at 70 mK [1]. We use zinc (Zn) as an example to discuss the MWS performance. Zn exhibits a superconducting transition temperature of 0.875 K in zero magnetic field and a critical field of 53 Gauss at zero temperature [6]. The experimental sample was a spherical piece of 59-grade Zn with a diameter of approximately 4.1 mm.

Figure 7.12 illustrates a sharp increase of the resonator's Q-factor as the sample undergoes its superconducting transition during a temperature sweep in zero magnetic field. Figure 7.13(a) illustrates a sharp increase of the resonator Q-factor when the sample enters the superconducting state during a field sweep at 0.53 K. The critical field of Zn was measured as the function of sample temperatures, as shown in Figure 7.13(b).

For all measurements, a small speck of pure DPPH was attached to the Zn sample using GE 7031 varnish, serving as a calibrated reference. DPPH ensures that the resonator operates at its resonant frequency and provides a well-known g-factor for calibrating the magnetic field of the superconducting solenoid. Detailed procedures for these two measurements are given in References [1] and [5].

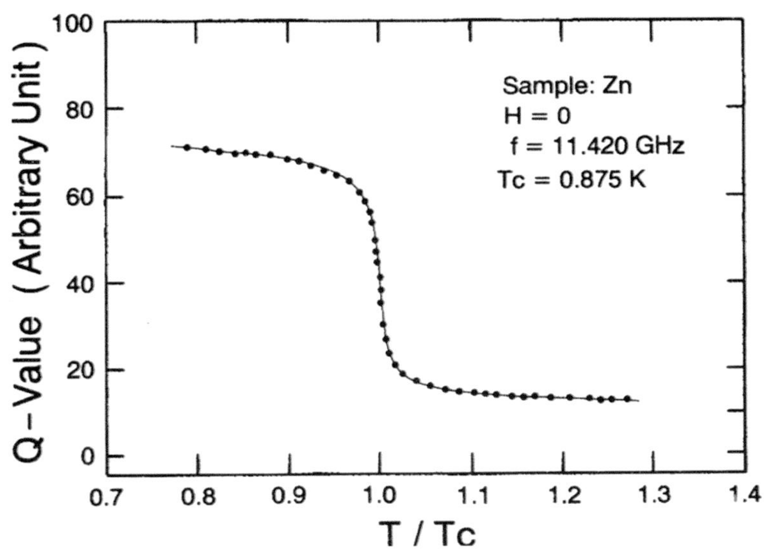

FIGURE 7.12 Q-factor of resonator with Zn vs. temperature sweep in zero magnetic field. *Source:* Information from Zuyu Zhao, Ph.D. Thesis, 1990, Northwestern University, Evanston, IL

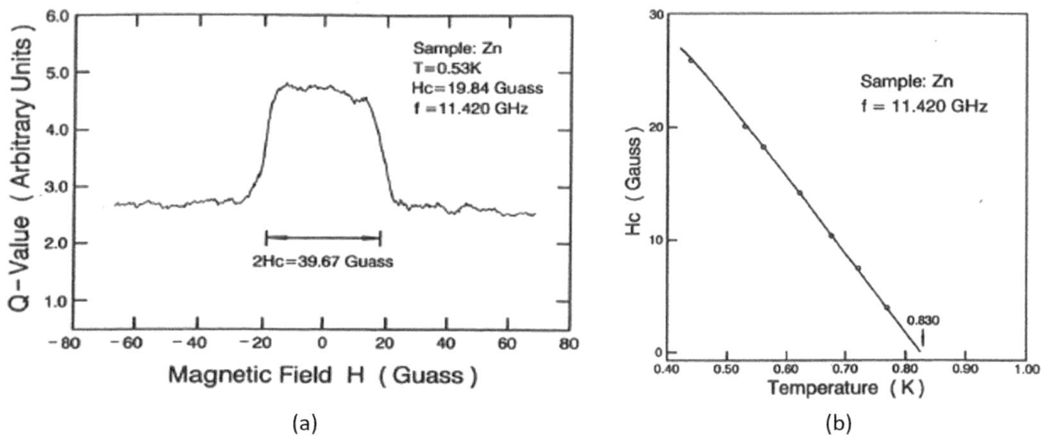

FIGURE 7.13 (a) Q-factor of resonator with Zn vs. magnetic field sweep at 0.53K. X-axis: Magnetic fields; Y-axis: Resonator Q-factor in arbitrary unit. (b) Critical field of Zn vs temperature. X-axis: Temperature; Y-axis: Critical field (Hc) of Zn. *Source:* Information from Zuyu Zhao, Ph.D. Thesis, 1990, Northwestern University, Evanston, IL.

7.1.3 Discussion

The Q-values of both the waveguide and the resonator can be further enhanced by optimizing the following components:

- Resonator surface: Gold plating or coating the inner surface with a superconducting material can reduce surface losses and increase the Q-value.
- Waveguide tube material: Using materials with higher electrical conductivity, such as copper, while incorporating short stainless steel sections to block conductive heat loads, can minimize signal loss.
- Alternative microwave transmission: High-frequency coaxial cables may be considered as an alternative to waveguides.

7.2 STRUCTURAL DESIGN OF LIGHT WEIGHT LONG DURATION FLIGHT CRYOSTAT

Cryostats designed for airborne or space-based missions must operate reliably at liquid helium temperatures and simultaneously meet strict requirements on mass, mechanical integrity, and thermal performance in space. These systems face unique challenges such as high launch and landing loads, limited cryogen capacity, liquid slosh during flight, and demanding hermetic sealing. This section presents general design strategies used to address these challenges.

7.2.1 Minimizing Joint Stress during Landing

A primary structural concern in flight cryostats is the mechanical load experienced during launch and landing. Rocket payloads typically experience sustained accelerations of 3–6 g for several minutes during launch and intense shocks of up to 9 g for ~0.1 s during landing. The cryostat joint structures require specialized design to withstand these loads, particularly the large shock forces generated during landing.

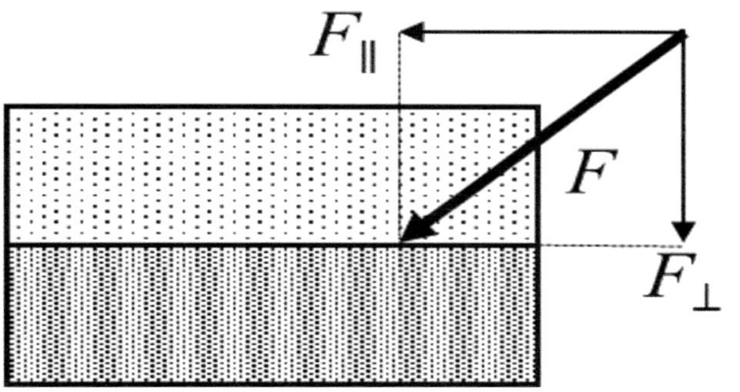

FIGURE 7.14 Decompose the applied force, (F_{SH}), into $F_{\perp}$ and $F_{\parallel}$.

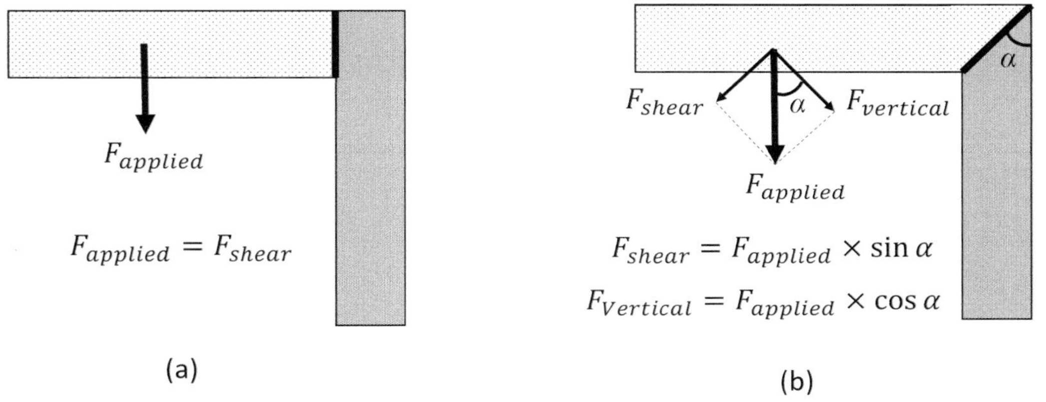

FIGURE 7.15 Shock force at the joints.

When two objects form a face-to-face joint (interface), any applied force (F) can be decomposed into two components: one perpendicular to the interface (**F⊥,** normal force) and one parallel to the interface (**F∥**, shear force), as illustrated in Figure 7.14. The shear force is the primary cause of the joint structure damaging. To mitigate shear stress, miter joints are often employed to redirect impact forces along multiple orientations. These joints are created by cutting each connecting part at an angle, allowing them to meet neatly at a corner (Figure 7.15).

A miter joint distributes mechanical loads across the joint more evenly than a simple butt joint, which reduces stress concentrations that could cause cracks or failures under the high g-force shocks. The miter design also provides a larger contact surface, providing better mechanical interlocking and enhancing resistance to shear and bending under impact. Importantly, the load distribution advantages apply regardless of the impact direction. For right-angle corners, a 45° miter joint is commonly considered an optimal design (Figure 7.16).

7.2.2 Maximize Usable Cryogen Reservoir Capacity

Because flight cryostats are volume and mass-limited, it is essential to maximize the effective cryogen reservoir capacity and ensure the precise measurement of the cryogen amount. A practical design approach involves opening both the fill and vent ports at the highest possible positions within the reservoir and tilting the helium level probe at an angle inside reservoir. Figure 7.17 illustrates a

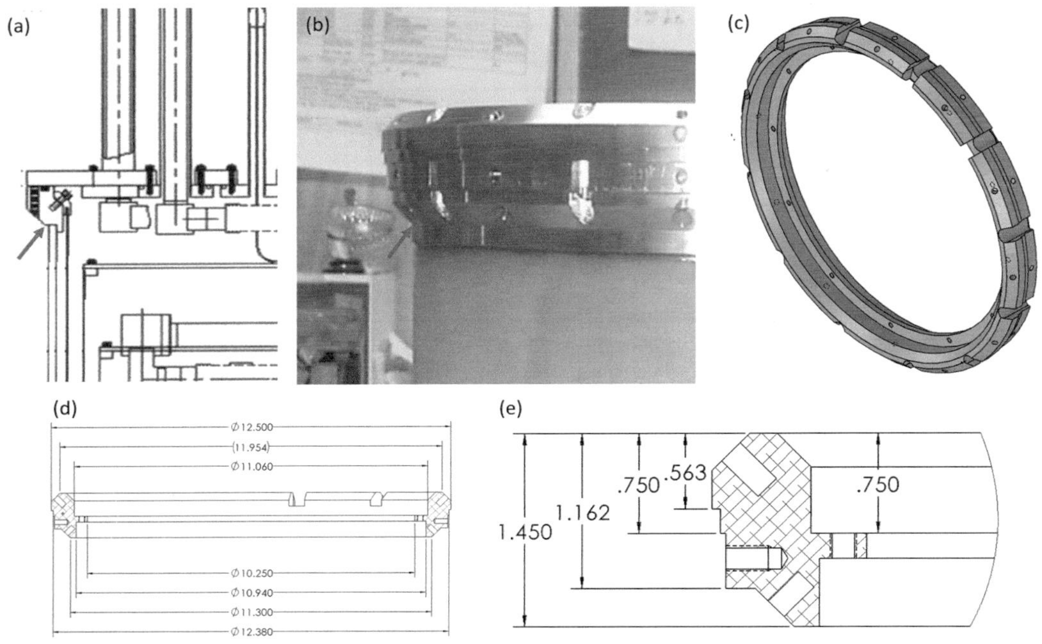

FIGURE 7.16 Design of 45 degree supporting structure.

custom configuration of the liquid helium fill and vent ports, the helium transfer line, and the helium level sensor on a horizontal long-duration airborne helium cryostat.

With this design, the usable liquid helium capacity exceeds 90% of the total reservoir volume.

7.2.3 Minimize Liquid Slosh during Flight

Liquid cryogen can move or slosh inside the cryostat during flight, generating unevenly distributed forces that destabilize the cryostat and raise safety concerns. The following approaches can be implemented to minimize sloshing:

- Incorporate internal baffles or dividers to reduce the free liquid surface. These structures can damp sloshing motion and reduce the oscillation amplitude (Figure 7.18).
- Install mechanical or passive slosh suppressors, such as foam inserts and flexible membranes, to absorb kinetic energy and decrease sloshing.
- Implement optimized geometries, such as cylindrical or spherical reservoirs rather than irregular shapes, to minimize free-surface area for the cryogen.

7.2.4 Reliable Vacuum Seal and Hermetic Joints

Compared to cryostats operated on the ground, flight cryostats have much higher requirements for vacuum integrity, especially in the cryogen reservoir. Even a tiny cryogen leak could cause a catastrophe. Since the entire cryostat is exposed to cryogenic conditions during flight in space, all hermetic seals must withstand thermal cycles between room temperature and cryogenic temperatures, and all sealing joints must accommodate thermal contraction and expansion.

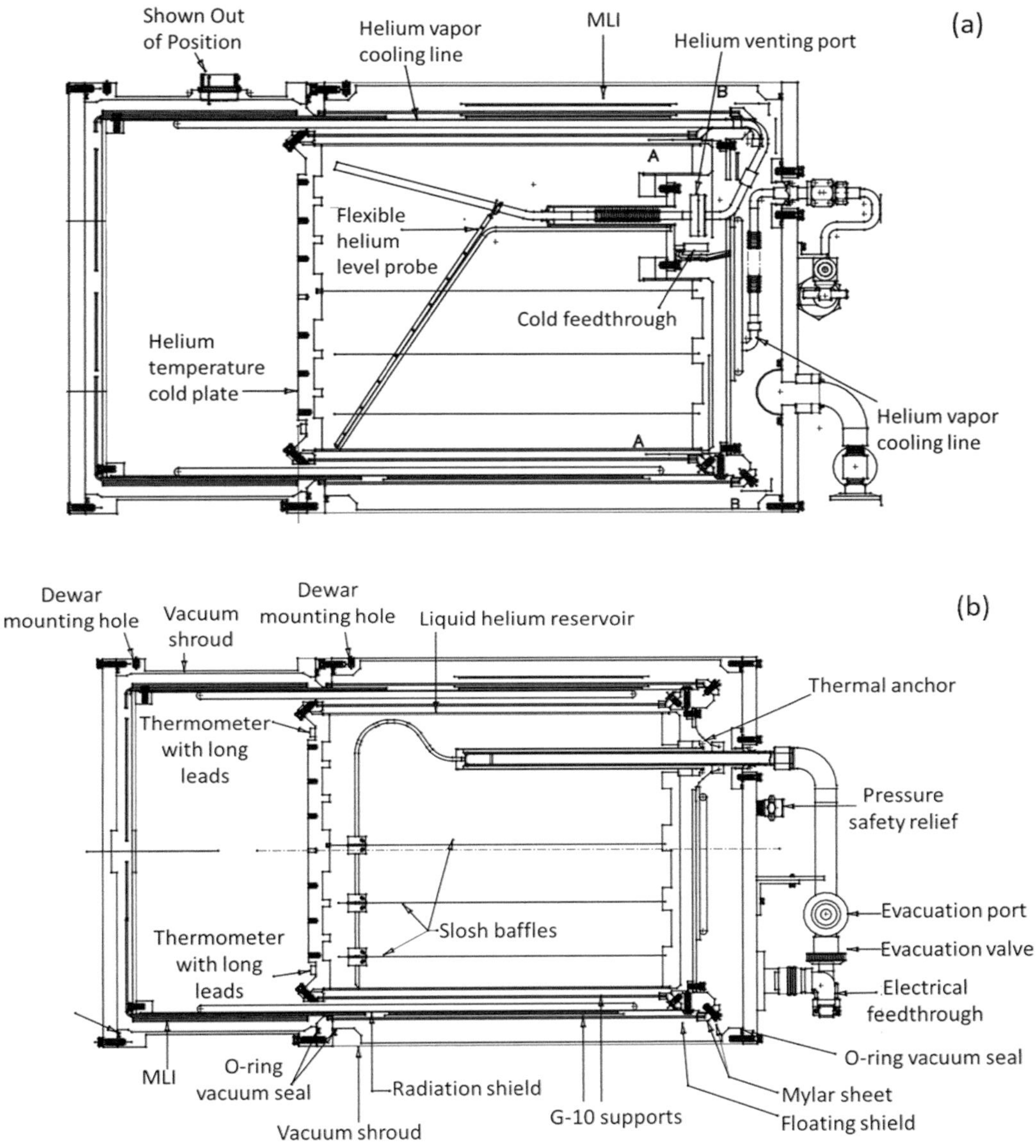

FIGURE 7.17 Maximize the effective helium reservoir capacity.

7.2.4.1 O-Ring Seals

Outer vacuum cans (OVCs) are often O-ring sealed. However, conventional sealing materials are typically made from polymeric and elastomeric materials, which may undergo a glass transition and become brittle at cryogenic temperatures, leading to cracking the rings. Additionally, since vacuum grease solidifies at cryogenic temperatures, O-ring materials in flight cryostats should be self-lubricating and avoid stick-slip behavior. To address these issues, fluorinated ethylene propylene (FEP)–coated O-rings with a fluoroelastomer (FKM/Viton) core are recommended [7–9]. This combination offers good rubber-elastic properties and a low compression set. These seals exhibit:

FIGURE 7.18 Baffle assembly to minimize the liquid slosh during the flight. (Projects accomplished by authors at Janis Research Company. Photos courtesy of FormFactor, Inc.)

- A wide operating temperature range (~22–473 K)
- Low compression set, allowing quick recovery from continuous strain and pressure cyclings
- High corrosion resistance against a wide range of chemicals and nonflammable materials
- Self-lubricating properties with extremely low friction and no stick-slip behavior

7.2.4.2 Reliable Aluminum Welding

Aluminum is an ideal material for flight cryostats construction due to its low cost and low density although aluminum welding presents unique challenges [10–12]. Tungsten inert gas (TIG) welding is most commonly used. To achieve reliable joints, several factors need to be considered for aluminum welding:

- **Right Joint Design**
  Typical aluminum welding joints are illustrated in Figure 7.19 for the reference.
- **Preparation and Cleanliness**
  Proper surface preparation and thorough cleaning are essential to ensure strong adhesion at welding joints. Aluminum exposed to ambient air rapidly forms thin oxide layers that hinder proper bonding and must be removed as thoroughly as possible. Stainless steel bristle brushes can be used to mechanically scrub and remove aluminum oxides from the workpiece surface. When feasible, acid solutions are recommended for more effective oxide removal.
  After oxide removal, the workpiece should be cleaned with solvents such as butanol, acetone, or paint thinner to eliminate dirt, grease, and other contaminants. All residual impurities must be removed during preparation, as they can burn into the weld pool and degrade the integrity of the final weld.
- **Temperature and Thermal Anchor**
  a) Melting and Preheating: Pure aluminum melts at approximately 1,200°F (650°C). Maintaining the proper temperature is crucial for achieving high-quality aluminum welds. Preheating the workpiece to around 275°F (135°C) helps achieve a uniform temperature distribution within the weld zone and can improve overall weld quality. However, excessive preheat should be avoided because it can degrade aluminum's mechanical properties and lead to distortion or loss of strength.

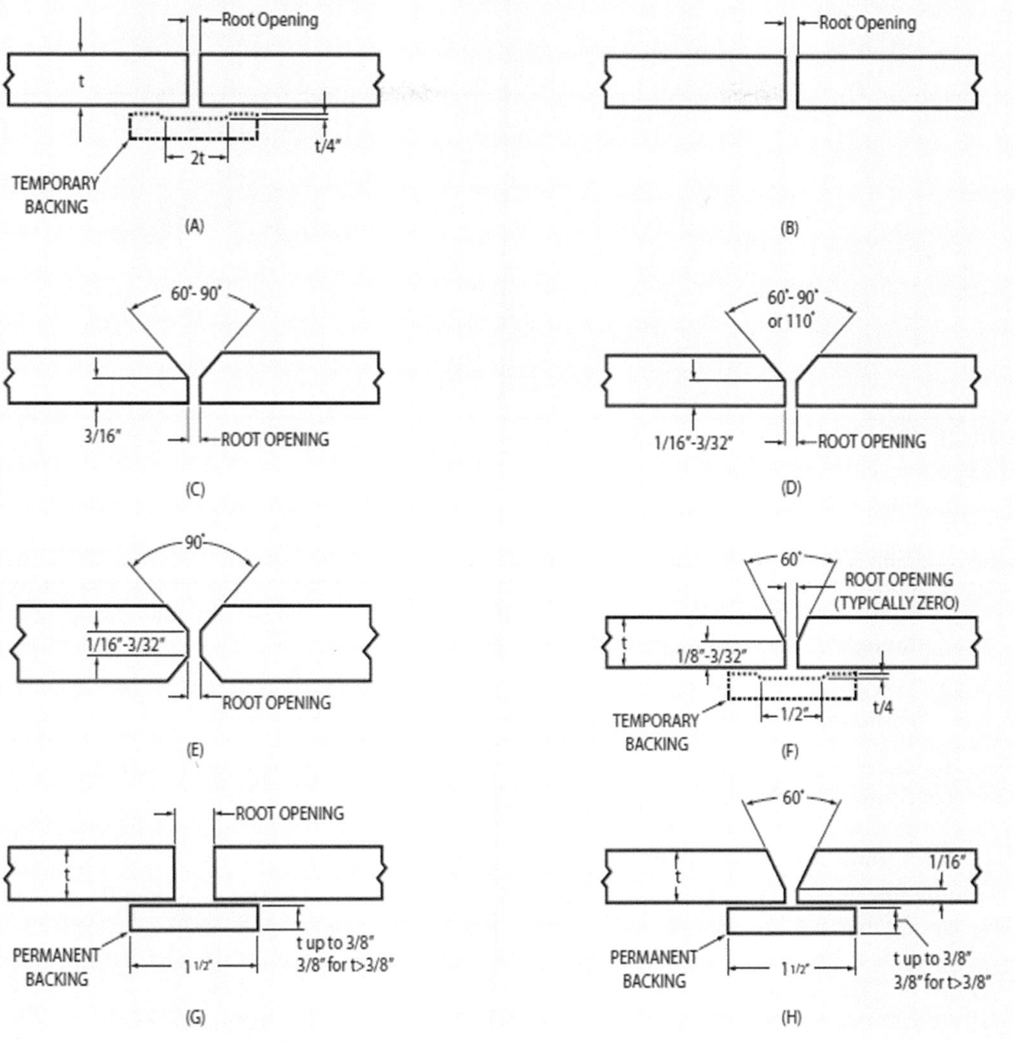

FIGURE 7.19 Examples of aluminum welding joint design. (Information from Hobart Brothers: *Guide for Aluminum Welding*, www.hobartbrothers.com; public domain.)

b) Oxide Layer Considerations: Completely removing the oxide layer during surface preparation is challenging. Since aluminum oxide melts at 3,700°F (2,040°C)—significantly higher than the base metal—the actual welding temperature must be sufficient to penetrate this layer. The optimal welding temperature may also vary depending on the filler rod composition and alloying elements.
c) Heat Management and Thermal Anchors: Proper heat management is essential when welding aluminum. Rapid temperature changes should be avoided to prevent warping, cracking, or residual stress. Heat sinks or thermal anchors made from highly thermally conductive materials, such as copper, are often attached to the workpiece—especially for parts with low thermal mass. These heat sinks help drain excess heat from the workpiece during welding and maintain stable thermal conditions.

FIGURE 7.20 A good aluminum welding. (Projects accomplished by authors at Janis Research Company. Model courtesy of FormFactor, Inc.)

High-quality aluminum welds typically exhibit a uniform "stack of dimes" pattern, also known as the *fish scale pattern*, shown in Figure 7.20. This characteristic pattern results from consistent torch movement, proper heat control, and steady filler metal addition throughout the welding process.

7.2.5 Minimizing Heat Load to Cryogens

Since flight cryostats have limited cryogen reserves, minimizing parasitic heat load is essential to extend operational duration.

7.2.5.1 Conductive Heat Load Reduction

Due to its low thermal conductivity and high mechanical strength, G-10 is a commonly used structural support material in cryostats, particularly in long-duration flight applications. To further reduce conductive heat load, a multipass (or *zig-zag*) support design is often employed, as illustrated in Figure 7.21. This design effectively increases the thermal path length, making the total length of the G-10 support nearly three times the direct distance between the two temperature zones. As a result, the conductive heat load to the low-temperature zone is reduced by approximately a factor of three compared with a straight support of equal span.

7.2.5.2 Radiation Heat Load Reduction

To reduce the radiative heat load, meticulous application of multilayer superinsulation (MLI) is essential for achieving the required thermal performance. The design, installation, and performance considerations of MLI are discussed in detail in Chapter 15.

7.2.6 Lightweight Design

Finally, minimizing mass is a key requirement. In addition to the extensive use of lightweight materials such as aluminum, unnecessary material is removed to further reduce the overall mass

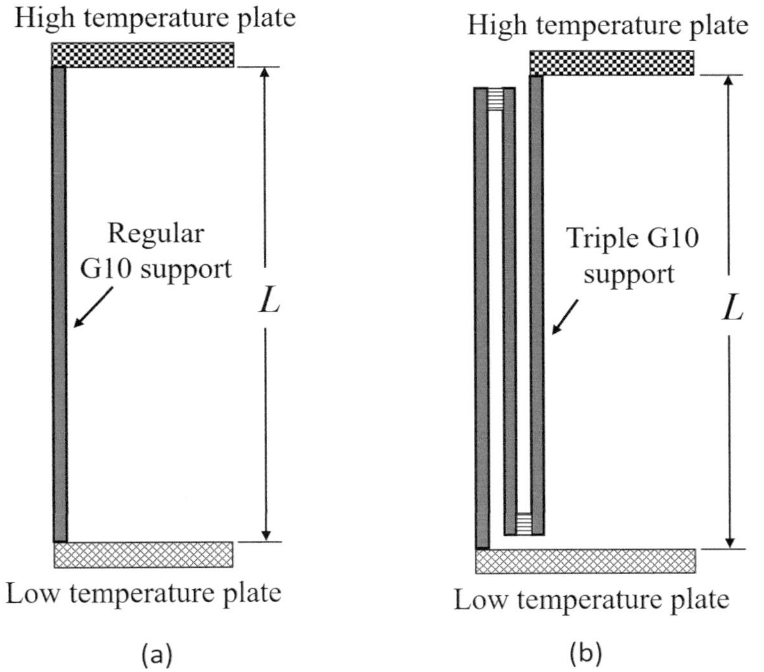

FIGURE 7.21 Folded support.

7.3 EXAMPLE #1: LIGHTWEIGHT HELIUM-ONLY AIRBORNE CRYOSTAT

This type of cryostat is used to cool infrared (IR) detectors to liquid helium temperatures (see Figure 7.17).

7.3.1 Requirement

- Maximum total weight: ≤ 75 lb
- Cold plate temperature for IR detector: ≤ 5.0 K
- Duration at target temperature: ≥ 24 h

Due to constraints on overall weight and dimensions, a conventional liquid-nitrogen shield cannot be used in this design. Instead, cold helium vapor is employed to cool the radiation shield, including the sidewalls, top, and bottom plates. The helium vapor can be treated as an ideal gas for convenience (within a few percent error). The molar enthalpy change with temperature is given by Equation 7.5:

$$\Delta h = \frac{5R}{2}\Delta T \tag{7.5}$$

where R is the molar gas constant, 8.31 J·K^{-1}·mol^{-1}, and ΔT is the temperature change. Thus, each mole of helium vapor can absorb approximately 20.78 J of heat for one Kelvin increase in temperature.

7.3.2 Cryostat Structure

OVC Shroud

Material: Aluminum
Diameter: 14.0 in
Height: 22.0 in

Liquid Helium Reservoir

Material: Aluminum
Temperature: 4.2 K
Diameter: 10.0 in
Height: 12.77 in
Effective helium reservoir capacity: 12.5 ~ 13.0 L.

Radiation Shields

Material: OFHC copper
Diameter: 11.62 in
Height: 15.58 in

40-layer MLI is installed on the radiation shield. Copper tubes (0.25" in diameter) are soldered with 60/40 solder onto the inner wall, top and bottom plates of the radiation shields (Figure 7.22).

7.3.3 Performance Estimate (Finite Element Analysis, FEA)

The vapor cooling of the radiation shield is a self-consistent process that continues until the floating radiation shield reaches an equilibrium temperature, where the parasitic heat load from the OVC is balanced by the cooling power of the helium vapor.

FEA simulated this process through iterative calculations and yielded the following results:

- Radiation shield temperature: 86 K
- Cold plate temperature: 4.2 K
- Heat load on helium reservoir: 0.1625 W
- Calculated liquid helium boil-off rate: 230.4 cc/h (≈ 0.002 mol/s)
- Predicted holding time (with 12.5 L effective liquid helium): 54 h

where

- Liquid helium density at 4.2 K: 0.031 mol/cc
- Liquid helium latent heat near 4 K: 81.6 J/mol

7.3.4 Actual Performance

- Total cryostat weight: ~71 lb
- Cold plate temperature for IR detector: 4.2–4.6 K
- Duration at required cold plate temperature: ~ 45 h
- Measured liquid helium loss rate: 272–288 cc/h
- Copper radiation shield temperature: ~ 86 K

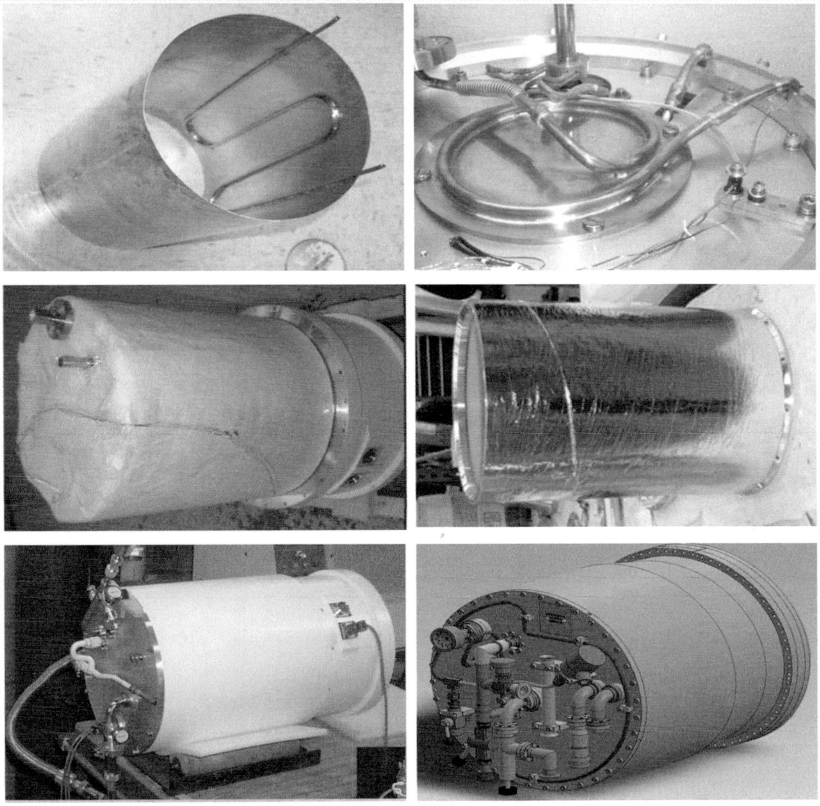

(a)
Vapor-cooled radiation shield (side and top)

(b)
MLI wrapping

(c)
Horizontal light-weight LHE-only airborne cryostat

FIGURE 7.22 Horizontal light weight LHE-only airborne cryostat. (Projects accomplished by authors at Janis Research Company. Photos courtesy of FormFactor, Inc.)

7.3.5 Discussions

7.3.5.1 Efficiency of Convection Cooling in the Vapor-Cooled Radiation Shields

Assuming the following conditions:

- The helium vapor temperature increases from 4.2 K to 86 K
- The helium vapor is treated as an ideal gas
- The measured liquid helium loss rate is 288 cc/h (≈0.0025 mol/s), implying that 0.2 watt heat load from the radiation shield is not intercepted
- According to FEA the expected heat load from the radiation shield to liquid helium without vapor cooling (Q_R):$Q_R = 3.337\ W$. Note: The conductive heat load is sufficiently small to be neglected.
- Total heat load intercepted by the vapor cooling is

$$Q_i = 3.337 - 0.2 = 3.137\ \text{W} \tag{7.6}$$

- From Eq. 7.5, the total enthalpy changes of the helium vapor when the helium valor warms from 4.2 K to 86 K should be:

$$Q_c = \dot{n}\frac{5R}{2}\Delta T = 4.248\ W \tag{7.7}$$

where $\dot{n}$ is the helium loss rate expressed in the mole per hour.

The convection cooling efficiency of the vapor-cooled radiation shield is:

$$\eta = \frac{Q_i}{Q_c} = 73.8\% \tag{7.8}$$

This result indicates that approximately 73.8% of the available cooling power from the helium vapor is effectively utilized by the radiation shield. The efficiency can be further improved by increasing the total length of the copper cooling tubes soldered onto the radiation shield, which enhances heat exchange between the vapor and the shield surface.

7.3.5.2 Alternative Configurations for Vapor Cooling Tube Installation

Two practical approaches for integrating vapor cooling tubes are illustrated in Figure 7.23:

- Vapor cooling tube epoxied to the aluminum radiation shield in areas where soldering is not feasible (Figure 7.23(a))
- Stainless steel cooling tubes soldered directly onto the stainless steel chamber wall (Figure 7.23(b))

(a) (b)

FIGURE 7.23 (a) Clamp, then epoxy the copper tube with Stycast 2850 onto the radiation shields; (b) Solder the colling tube onto the UHV sample tube. (Projects accomplished by authors at Janis Research Company. Photos courtesy of FormFactor, Inc.)

7.4 EXAMPLE #2: FAST ALTERNATIVE CRYOGENIC EXPERIMENTAL TESTBED

The Fast Alternative Cryogenic Experimental Testbed (FACET, Figure 7.24) was developed as a proof-of-concept payload to demonstrate cryogenic support for low-temperature scientific investigations within the constraints of the Hitchhiker siderail (HH-S) carrier on the Space Shuttle [14]. It consists of a hybrid solid neon–superfluid helium cryostat, a multi-channel thermometer/heater control system, and a multi-stage thermally isolated "probe" for precise thermal control of helium samples.

7.4.1 Required Performance

- **Base Temperature of the Cold Plate:** ≤ 2.17 K.
  In the vacuum of space, the external environment serves as a natural pump, continuously pumping on the liquid reservoir and converting the liquid helium into superfluid.
- **Operating Time:** ≥ 65 h of unattended operation at superfluid helium temperatures in orbit.
  This requirement implies that the total liquid helium holding time must be 160 hours or longer, providing at least 96 additional hours of helium storage prior to the Space Shuttle launch.
- **Space and Weight Constraints**
  The cryostat must fit within a compact envelope of 0.64 m × 0.64 m × 0.99 m, with a total weight not exceeding 216 pounds, which limits the liquid helium volume.
- **Launch Acceleration:** Typically between 3 g and 9 g
- **Operational Requirement**
  The flight cryostat must maintain reliable performance in any spatial orientation during all mission phases.

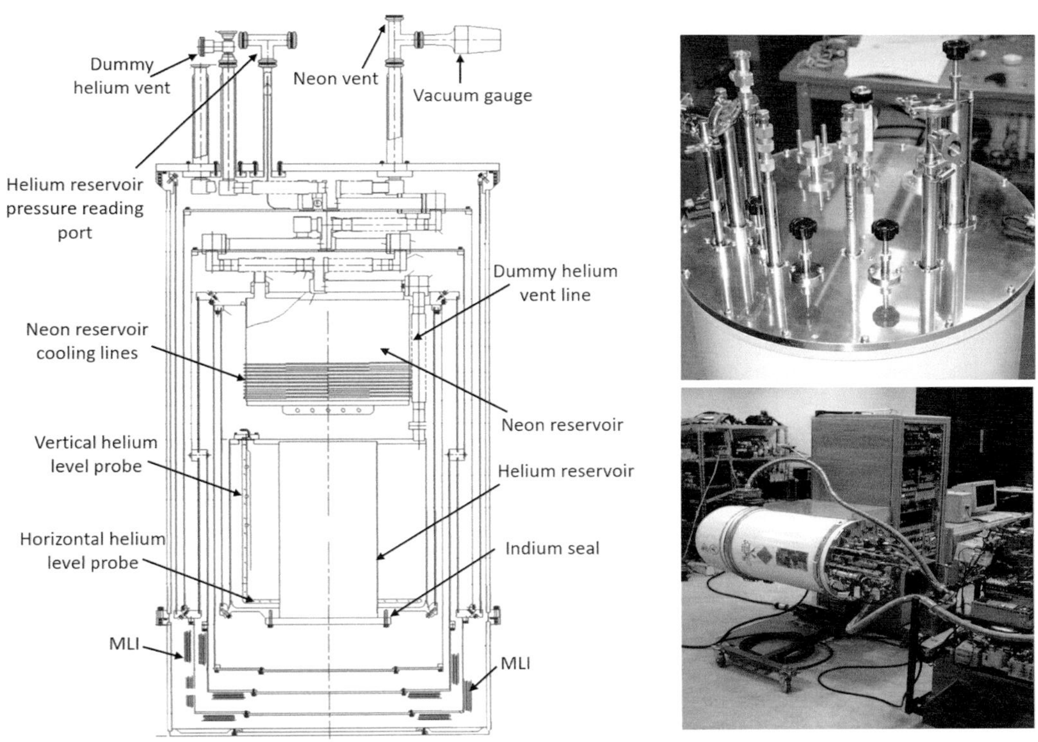

FIGURE 7.24 Fast Alternative Cryogenic Experimental Testbed (FACET). (Projects accomplished by authors at Janis Research Company. Model courtesy of FormFactor, Inc.) [13, 14].

7.4.2 Key Design Concepts

7.4.2.1 Neon-Temperature Radiation Shield

After evaluating all candidate materials, the traditional liquid nitrogen radiation shield was found to be less optimal. In contrast, a radiation shield using solid neon demonstrated perfect thermal performance.

As shown in Table 7.2, the characteristic temperature of neon is approximately one-third that of nitrogen in both its liquid and solid phases. The radiation heat load from the neon reservoir to the superfluid helium reservoir is expected to be near two orders of magnitude lower than that from a liquid nitrogen reservoir. Additionally, solid neon can be further cooled to even lower temperatures using liquid helium, which can further reduce the heat load to superfluid helium and increase the total holding time of the cryostat.

As illustrated in Figure 7.25, the neon reservoir and radiation shield are integrated into a single assembly. An aluminum tube is welded onto the neon reservoir as a cooling loop, through which liquid helium flowed to cool the reservoir. Aluminum foam is installed inside the neon reservoir and attached to the inner surface of the bottom plate to enhance heat exchange between the liquid neon and the reservoir. During the operation, liquid neon is first solidified at its triple point (~24 K). The solid neon is further cooled by liquid helium to approximately ~20 K to maximize the holding time.

7.4.2.2 Double Vapor-Cooled Radiation Shields between OVC and Neon Reservoir

To further reduce the heat load on the neon reservoir, two layers of vapor-cooled radiation shields were installed between the OVC and the neon reservoir. This multilayer configuration follows the design principles used in the long-duration lightweight cryostat described in Section 7.2. Each shield intercepts partial incoming thermal radiation, substantially lowering the net radiative heat load to the neon reservoir.

7.4.2.3 Holding Time of the Neon Reservoir

The neon undergoes the following thermal process before depletion:

- Solid neon warming: absorbs heat while increasing from the initial temperature to the melting point.
- Melting process: absorbs heat of fusion until the solid is fully melted.

TABLE 7.2
Parameters of Nitrogen and Neon

	Base Temperature (K)	Specific Heat of Solid (J/g.K)	Melting Point (K)	Heat of Fusion (J/g)	Specific Heat of Liquid (J/g.K)	Boiling Point (K)	Heat of Evaporation (J/g)
Neon	20*	1.03~1.20	24.56	16.6	1.03	27.1	85.8
Nitrogen	58	1.1	63.2	25.7	2.04	77.3	198.3
	*Approximate obtained temperature	Under normal pressures	Under normal melting temperatures	Under normal melting temperatures	Under normal pressures	Under 1.0 bar	Under normal boiling temperatures

Source: Data from Flynn, T., *Cryogenic Engineering*, Second Edition, (2005); R. D. McCarty, *Interactive FORTRAN IV Computer Programs for the Thermodynamic and Transport Properties of Selected Cryogens (Fluids Pack)*, NBS Technical Note 1025, National Bureau of Standards, Boulder, CO, 1980.

Note: Values are temperature sensitive.

FIGURE 7.25 Neon reservoir. (Projects accomplished by authors at Janis Research Company. Photos courtesy of FormFactor, Inc.)

- Liquid neon warming: absorbs heat while increasing from the melting point to the boiling point.
- Boiling and evaporation: absorbs heat of evaporation until all liquid neon depleted.
- Sublimation: negligible under normal operating conditions.

In this design, the reservoir contains over 2 kg of solid neon, which can absorb approximately 228,400 joules of heat before depletion. Given the measured heat load below 0.2 W, the expected holding time of the neon reservoir exceeds 317 hours.

7.4.2.4 Brief Summary of Special Features

This flight cryostat is a hybrid solid neon–liquid helium system. Special design features, such as the "folded G-10 tube" support, lightweight machining, and 45-degree joints, are incorporated into this cryostat. A porous plug is installed at the helium vent port to allow evaporated helium gas to escape from the dewar while retaining the superfluid liquid helium inside. Additionally, aluminum foam installed inside the helium reservoir helps thermalize the cold plate to the superfluid helium temperature. These design elements collectively improve thermal efficiency, reduce mass, and enhance operational reliability in varying orientations during flight.

7.5 GROUND TEST AND CONCLUSION

The following cool-down procedure was implemented during the ground performance test:

- Fill the neon reservoir with liquid neon.
- Fill the helium reservoir with liquid helium.

- Flow liquid helium through the cooling loop surrounding the neon reservoir to solidify the neon inside.
- Continue flowing liquid helium to further cool the solid neon to the lowest achievable temperature.

According to the report [13–15] "...... all requirements (performance, hitchhiker payload (7) and development constraints) were met or exceeded. Test data from the prototype indicated that flight cryostat that could provide superfluid helium cooling to an instrument for the full duration of even the longer shuttle missions (16 days). ..."

"The analysis and tests of these subsystems have proven the feasibility of a cryogenic HH-S carrier payload."

Interested reader can refer to the [13] to [15] for more detailed information.

REFERENCES

1. M.Z. Lin, Z. Zhao, Y.H. Shen, and J.B. Ketterson, Y.J. Qian and Bimal K. Sarma, A Top Loading, Very Low-Temperature, X-band, Microwave Spectrometer, *Review of Scientific Instruments*, 61(5), 1436–1441, (May 1990).
2. Neo Huang, *Circular Waveguide Cutoff Frequency Calculator (TE11 Mode)* www.calculatorultra.com/en/tool/circular-waveguide-cutoff-frequency-calculator-te11-mode.html
3. LearnEMC, LLC (Public Domain) 208 Industrial Drive, Stoughton, WI, USA Website: www.learnemc.com
4. J.P. Gordon, Variable Coupling Reflection Cavity for Microwave Spectroscopy, *Review of Scientific Instruments*, 32, 658–661, (1961).
5. Zuyu Zhao, Studies of Non-Conventional Superfluids: Ultrasound Propagation in 3he-B and the Microwave Surface Impedance of the Heavy-Fermion Superconductor upt3, Ph.D. Thesis, Northwestern University, Evanston, IL, (1990).
6. Charles Kittel, *Introduction to Solid State Physics*, 8th edition, John Wiley & Sons, Inc., New York, (2005).
7. Jens Schuster, Johannes Lutz, Yousuf Pasha Shaik, and Venkat Reddy Yadavalli, Recycling of Fluoro-Carbon-Elastomers – A Review, *Advanced Industrial and Engineering Polymer Research. Recycling of Rubbers*, 5(4), 248–254, (2022).
8. Research Advisory Board for Mechanical Engineering Frankfurt am Main, Germany Website: www.fkm-net.de (public domain)
9. FKM/FPM are standards for a broad family of chemically different fluoroelastomer formulations, while Viton® is a proprietary subset with its own graded product lines and performance differences. Website: www.fkm-rubber.com/uk/viton/competence/fluoroelastomer/properties/difference-viton-fkm-fpm.html (Public Domain).
10. Hobart Brothers LLC. Guide for Aluminum Welding. Website: www.hobartbrothers.com/wp-content/uploads/2020/09/Aluminum_Welding_Guide.pdf
11. Jogn D. Buckley and Paul G. Sandefur, Jr., *Low Temperature Solder for Joining Large Cryogenic Structures*, Scientific and Technical Information Branch, National Aeronautics and Space Administration (NASA), Washington, D.C. USA, (1980).
12. John H. Moore, Christopher C. Davis, Michael A. Coplan, with a chapter by Sandra C. Greer, *Building Scientific Apparatus*, Cambridge University Press, 10th printing, Cambridge, England, UK, (2019).
13. A. Nash, P. Shields, R. Abbott, J. Craig, and W. Holmes, The Fast Alternative Cryogenic Experiment Testbed, *Cryogenics*, 39(12), 1029–1035, (1999).
14. Alfred Nash, Peter Shields, Munir Jirmanus, and Zhao Zuyu, Development of the FACET cryostat, *The 1999 Cryogenic Engineering Conference. Part A.*
15. A Space Shuttle "small payload" flight program created to provide low-cost, modular, and rapidly integrated experiments. Website: nasa.fandom.com/wiki/Hitchhiker_Program (Public Domain).

8 Thermal Boundary Conductance
Modelling and Examples

8.1 THERMAL BOUNDARY CONDUCTANCE AND THERMAL BOUNDARY RESISTANCE

When heat flows across the interface between two materials, a thermal boundary resistance—also known as Kapitza resistance, thermal contact resistance, or interfacial thermal resistance—causes a temperature discontinuity (ΔT) at the interface (see Figure 8.1) [1]. Macroscopically, the thermal boundary resistance (R_B) is defined the ratio of the temperature change to the heat flow across the interface:

$$R_B = \frac{\Delta T}{Q}$$

where Q is the total heat flow (W). The reciprocal, thermal boundary conductance (κ_B), represents how well heat is transferred across the interface and is expressed as

$$\kappa_B = \frac{Q}{\Delta T} = \frac{1}{R_B}$$

R_B is measured in the unit of K/W, and κ_B is expressed in W/K.

The thermal boundary resistance depends on various physical parameters, with the interface condition playing a particularly important role. In practical applications, the actual contact area between two solid surfaces is typically only about 8% or less of the nominally "flat" interface, and in many cases, it is as low as 1–3% (Figure 8.2). Because heat must flow through these limited contact regions, the local heat flux—that is, the heat flow per unit area (W/m²)—can become very high. According to Fourier's law:

$$q = -\kappa \Delta T \tag{8.1}$$

where q is the heat flux, and κ is thermal conductivity in the unit of W/(m·K).
The high heat flux leads to a significant temperature difference (ΔT) across the contact interface in a cryogenic system.

DOI: 10.1201/9781042000050-8

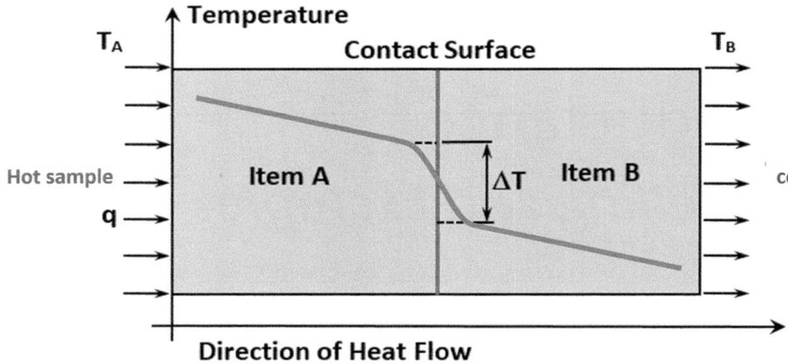

FIGURE 8.1 Temperature "discontinuity", ΔT, at the interface on a macroscopic scale caused by thermal boundary resistance at the interface.

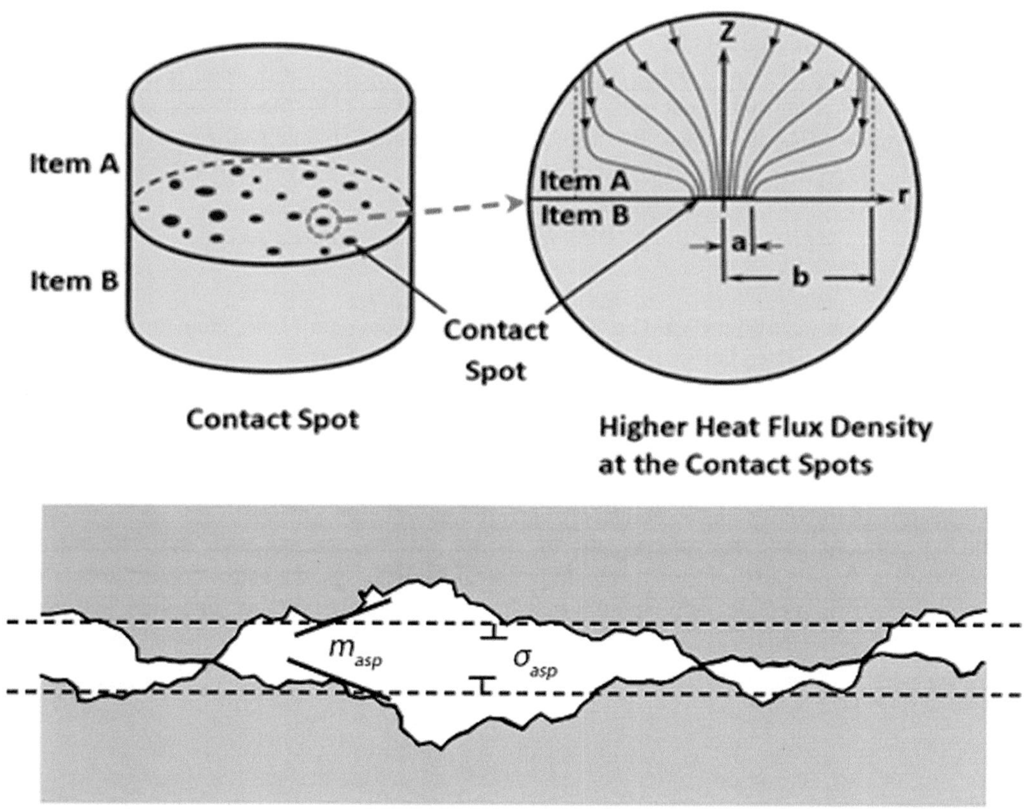

FIGURE 8.2 Microscopic view of a contact interface in a conforming rough surface.

8.2 THERMAL BOUNDARY CONDUCTANCE ESTIMATION AT CRYOGENIC TEMPERATURES

The number of energy carriers (phonons and electrons) in the heat flow is lower at low temperatures than at room temperature, resulting in a lower thermal boundary conductance. Multiple theoretical models, such as the acoustic mismatch model and diffuse mismatch model [1–3], have been developed to estimate the thermal boundary conductance

While these models are helpful for understanding the physical insights, no single one provides a generally applicable or practically accurate description. Particularly, no model offers a profound discussion on thermal boundary conductance at cryogenic temperatures. To address this limitation, a more empirical approach has been adopted to estimate thermal boundary conductance at low temperatures. After many years of pioneering work by Berman, Salerno, Kittel, Spivak, and others [4–7], an explicit empirical formula has been established as Equation (8.2):

$$\kappa_B = \alpha T^n \tag{8.2}$$

where

- n typically ranges from 0.75 to 2 at helium temperatures, depending on the materials and the condition of the interface.
- α is an asymptotic, near-linear function of the compression force normal to the interface (F_N, a temperature independent parameter).

NASA scientists measured thermal boundary conductance at temperatures between 1.6 K and 4.2 K, by applying compression forces from 22 N (4.9 lb) to 670 N (150 lb) across various metal interfaces and derived the values of n and α. Table 8.1 summarized representative values of these parameters for the most commonly used metals [5].

Figure 8.3 shows the experimental thermal boundary conductance of copper at 4.2 K measured by Berman et al. in the 1950s [4], along with the calculated values obtained using the parameters listed in Table 8.1, as function of applied forces. It is encouraging that these two results agree closely, despite the experimental data having been obtained 36 years before the parameters in Table 8.1 was established. Moreover, Equation 8.2 and Table 8.1 do not require detailed knowledge of the contact surface characteristics. This example demonstrates that Equation 8.2 provides a pragmatic tool for estimating thermal boundary conductance at the interface near liquid helium temperatures.

A deviation from the linear relationship was observed when larger force of 4,700 N was applied to two copper Cone–Socket specimens [8]. Because the taper can further amplify the normal force at the contact interface, the deviation indicates possible plastic deformation caused by high normal force at the interface.

Berman also measured the thermal boundary conductance at liquid nitrogen temperatures using the similar interface and the same range of applied forces as in the 4.2 K experiment. The experimental data obtained at both temperatures are shown in Table 8.2. Four conclusions can be drawn from Berman's measurements:

- At liquid nitrogen temperatures, the thermal boundary conductance is linearly proportional to the applied compressive force normal to the interface between two copper cylinders, up to approximately 250 lb.
- Equation 8.1 remains qualitatively valid at liquid nitrogen temperatures, but the parameters n and α are expected to differ from those at liquid helium temperatures.
- As shown in Table 8.2, the thermal boundary conductance at liquid nitrogen temperatures is approximately 30 times higher than that at liquid helium temperatures.
- The thermal boundary conductance at the metal contact surface was not sensitive to the "cycling history" of the contact surface.

TABLE 8.1
***n* and α at Liquid Helium Temperatures Under Different Applied Normal Force for Different Materials**

Force	Aluminum		Brass		Copper		Stainless Steel	
(N)	α (mW/K^n)	n	α (mW/K^n)	n	α (mW/K^n)	n	α (mW/K^n)	n
22	0.17 (±0.01)	1.05 (±0.02)	0.27 (±0.01)	1.22 (±0.02)	0.29 (±0.01)	1.60 (±0.02)	0.087 (±0.001)	0.77 (±0.007)
44	0.22 (±0.01)	1.07 (±0.03)	0.19 (±0.01)	1.56 (±0.02)	0.29 (±0.01)	1.66 (±0.03)	0.100 (±0.002)	0.78 (±0.01)
112	0.26 (±0.01)	1.31 (±0.03)	0.28 (±0.01)	1.53 (±0.02)	0.39 (±0.02)	1.67 (±0.03)	0.095 (±0.003)	0.84 (±0.01)
224	0.40 (±0.01)	1.36 (±0.03)	0.46 (±0.01)	1.43 (±0.02)	0.50 (±0.02)	1.79 (±0.04)	0.086 (±0.003)	1.00 (±0.01)
336	0.53 (±0.01)	1.37 (±0.03)	0.61 (±0.01)	1.38 (±0.02)	0.69 (±0.03)	1.76 (±0.04)	0.082 (±0.003)	1.10 (±0.01)
448	0.85 (±0.01)	1.21 (±0.03)	0.72 (±0.01)	1.38 (±0.02)	0.77 (±0.03)	1.85 (±0.05)	0.083 (±0.003)	1.17 (±0.02)
560	0.92 (±0.02)	1.28 (±0.03)	0.86 (±0.01)	1.36 (±0.02)	0.79 (±0.04)	1.97 (±0.06)	0.087 (±0.003)	1.21 (±0.02)
670	0.88 (±0.02)	1.40 (±0.04)	0.88 (±0.01)	1.41 (±0.02)	0.71 (±0.05)	2.15 (±0.06)	0.094 (±0.003)	1.23 (±0.02)

Source: Data from Kittel, P., Spivak, A.L. and Salerno, L.J.: *Thermal Conductance of Gold plated metallic Contacts at Liquid Helium Temperatures*, Advances in Cryogenic Engineering, 37, pt A, 241-248 (1992)] [5].

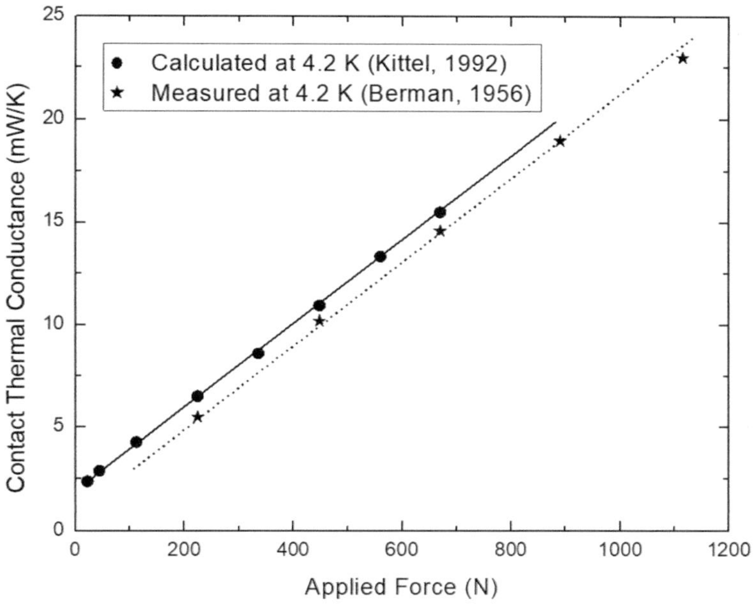

FIGURE 8.3 Plot of the measured thermal boundary conductance versus the estimated value between two copper flat contact surfaces at liquid helium temperatures.

TABLE 8.2
Measured Thermal Boundary Conductance at Liquid Nitrogen Temperatures

Applied Forces between Cu Rods		224 N 50 lbs.	448 N 100 lbs.	670 N 150 lbs.	890 N 200 lbs.
Measured Contact Thermal	At 77 K	160	325	490	660
Conductance (mW/K)	At 4 K	5.5	10.2	19	23
Conductance Ratio between 77 K and 4 K		29	31.8	33.5	34.7

Source: Data from Berman, R.: Some *Experiments on Thermal Contact at Low Temperatures*, Journal Applied Physics, 27, number 4 (1956) [4] and Berman, R. and Mate, C.F.: *Thermal Contact at Low Temperatures*, Letter to Nature, *Nature* 182, 1661–1663 (13 December 1958) [8].

We studied the thermal conductance at sub-Kelvin temperatures and found that Equation 8.2 was valid down to 300 mK. Details will be discussed in Section 8.4 *Examples of interface design and performance*.

In conclusion, Equation 8.2 can be used as an empirical formula to estimate the thermal boundary conductance. It is generally valid for most commonly used metals and over a wide temperature range from liquid nitrogen temperature down to sub-Kelvin region. The main limitations are that the applied force should remain below the plastic deformation threshold and the heat flow should be small—that is, the temperature difference (ΔT) across the interface should be small compared to the interface temperature.

8.3 THERMAL BOUNDARY CONDUCTANCE IMPROVEMENT

The thermal boundary conductance can be improved through the following approaches.

- **Improve the Surface Flatness:**

  A flat surface, rather than a smooth one, improves the thermal boundary conductance because improved flatness leads to a greater number of contact points between two surfaces (Figure 8.2). Figure 8.4 shows several common approaches used to improve surface conditions, among which flapping is the best method to reduce roughness. In contrast, a smooth surface is typically produced by buffing, which may cause the surface to become slightly curved, thereby reducing the contact area and consequently lowering thermal boundary conductance.

- **Material Selection and Proper Coating:**

  The interface between two gold-plated copper surfaces provides the highest thermal boundary conductance. Figure 8.5 lists commercial gold plating specifications, where Grade-C with Class 1 thickness is usually used to enhance thermal boundary conductance. However, thicker plating, such as Class 3, is preferred at millikelvin temperatures, where part of the gold acts as a filling material at the interface, increasing the effective contact area. For a similar reason, at liquid nitrogen temperatures and above, gold plating can be replaced by a thin, soft interstitial material, such as indium foil, to fill surface asperities.

- **Surface Treatment:**

  The interface surfaces should be thoroughly clean, degreased, and deoxidized to ensure good metal thermal contact.

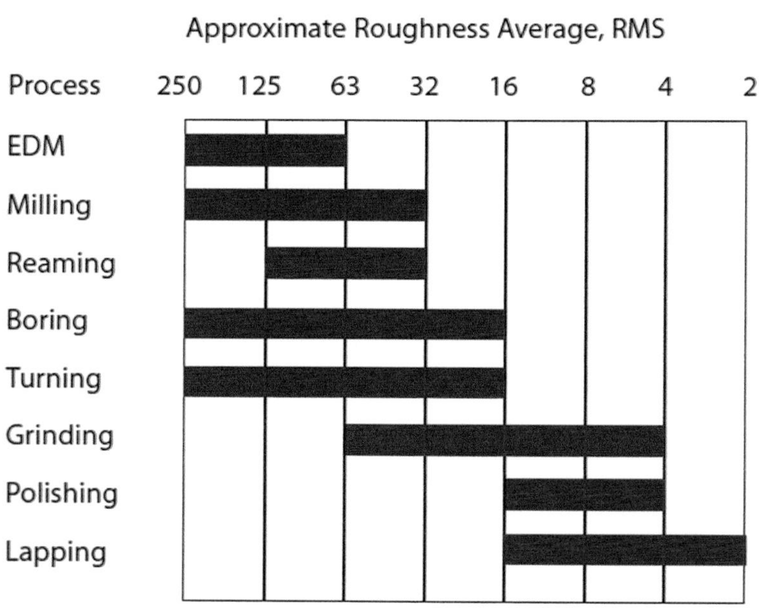

FIGURE 8.4 Flatten the contact surface.

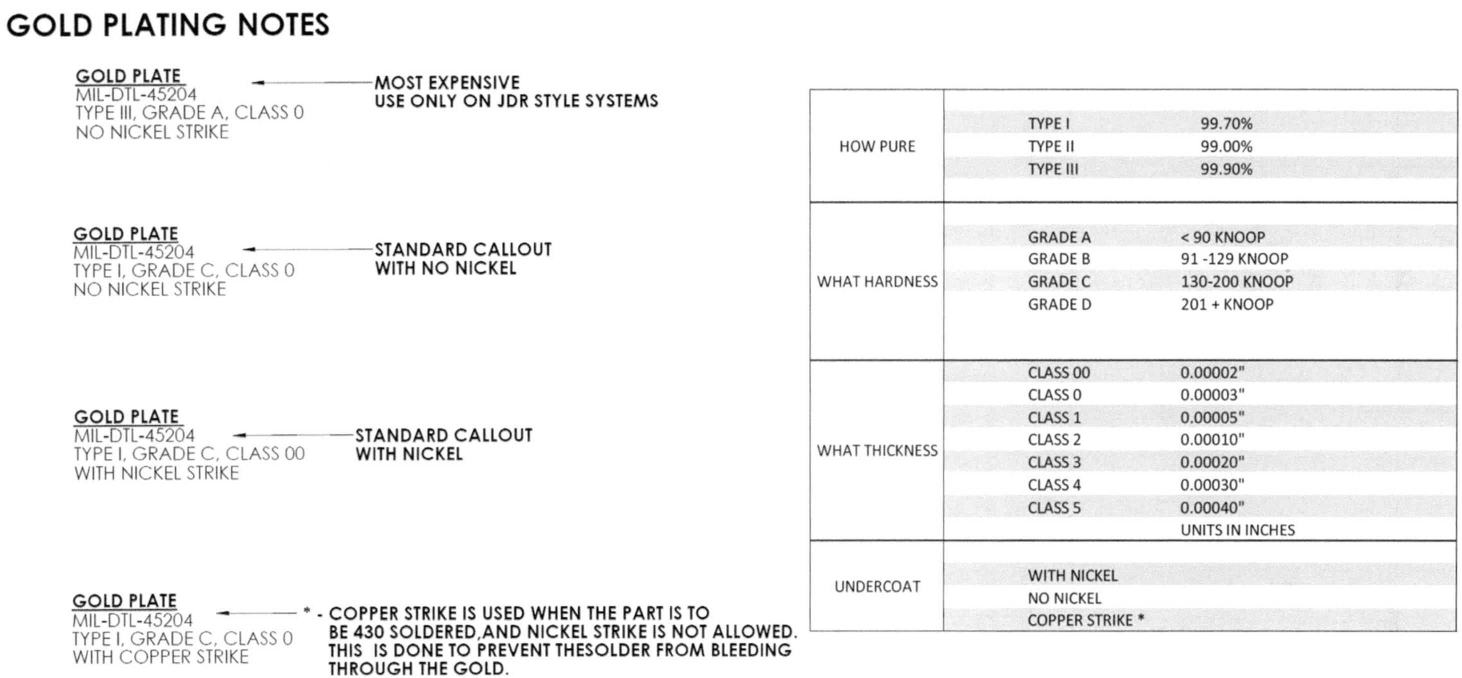

FIGURE 8.5 Gold plating specifications. (Projects accomplished by authors at Janis Research Company. Image courtesy of FormFactor, Inc.)

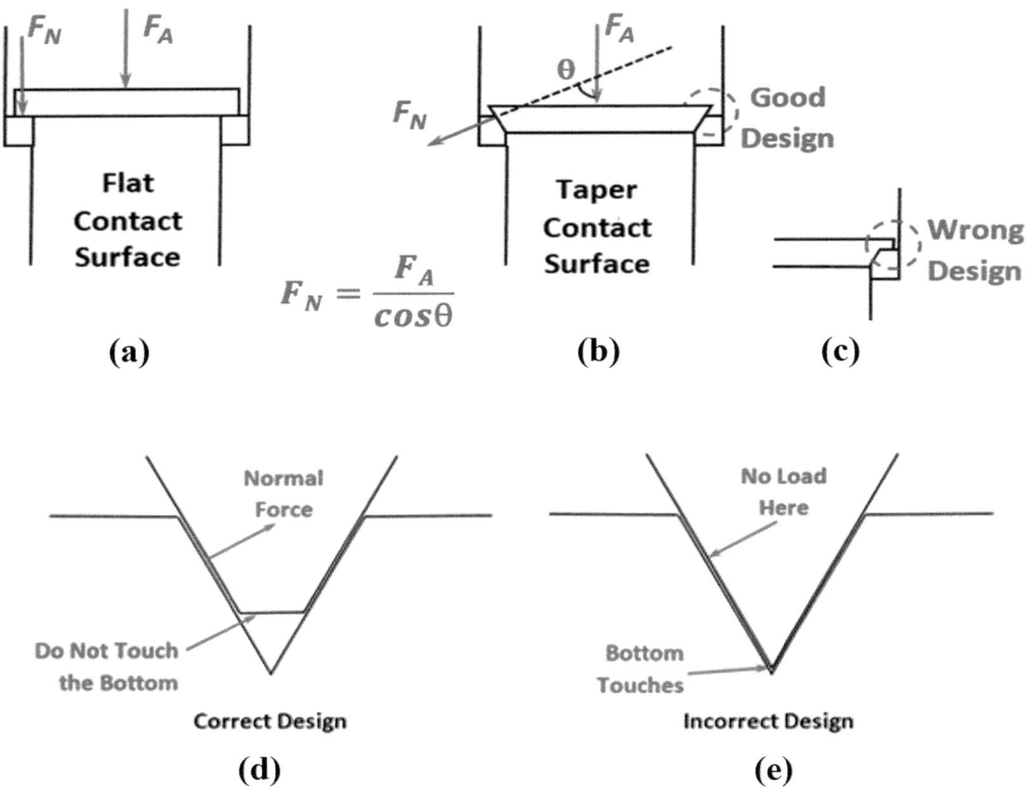

FIGURE 8.6 Contact interface designs. (a) The normal compression force at the interface equals the applied force; (b) The normal compression force is amplified to FN = FA/cos(θ); (c) Incorrect design: The added step cause θ become 0o, eliminating the force amplification; (d) Correct taper design; (e) Incorrect design: Contact at the bottom eliminates the force amplification.

- **Tapered Contact Interface (Figure 8.6):**
  A tapered (conical) contact interface can be incorporated where feasible to concentrate and amplify the normal force, thereby improving thermal boundary conductance. Figure 8.6(b) illustrates the typical design of a tapered interface, while Figure 8.6(c) and (e) shows some common mistakes in tapered interface design. Assuming the tapered interface has an angle θ relative to the applied force, as shown in Figure 8.6(b), the normal compression force on the taper interface is given by

$$F_N = \frac{F_A}{cos\,(\theta)} \tag{8.3}$$

where F_A is the applied force along Z-direction, and F_N is the force normal to the interface. Since $1/\cos(\theta) \geq 1$, the taper geometry functions as a mechanical force amplifier. The optimal value of the taper angle θ depends on various factors, such as the specific application, materials properties, and magnitude of the applied force. For example, an angle of 68° has been shown to be effective for gold-plated OFHC copper contact interfaces, providing a mechanical advantage by amplifying the applied force by a factor of approximately 2.7. However, as θ approaches 90°, $\cos(\theta)$ approaches zero, significantly increasing the risk of mechanical binding at the interface. Readers may refer to Reference [9] for a more detailed discussion on taper angle selection and its implications.

8.4 EXAMPLES OF INTERFACE DESIGN AND PERFORMANCE

8.4.1 Example #1: Thermal Boundary Conductance Study around 300 mK

While previous discussions focused on liquid helium temperatures, this example illustrates the contact design between two gold-plated surfaces operating at sub-Kelvin temperatures, using the empirical approach outlined in Equation 8.2 and Table 8.1. A top-loading sample probe, shown in Figure 8.7, includes five temperature stages, thermally anchored to corresponding thermal anchor stages on a wet He-3 cryostat. When the sample probe is fully inserted in the sample tube, these stages operate at temperatures of 150 K, 77 K, 4.2 K, 1.5 K, and 300 mK, respectively. Spring assemblies are installed between the stages along the sample probe to maintain contact and compliance. For easier discussion, the contact surface on the He-3 insert is referred to as the "**Anchor**", while the corresponding surface on the sample probe is referred to as the "**Disc**".

As the most important stage, ***Anchor #4*** is thermally bound to a He-3 pot at 300 mK. The heat flow from ***Disc #4*** on the probe to ***Anchor #4*** is approximately 1.89×10^{-5} W, according to the finite

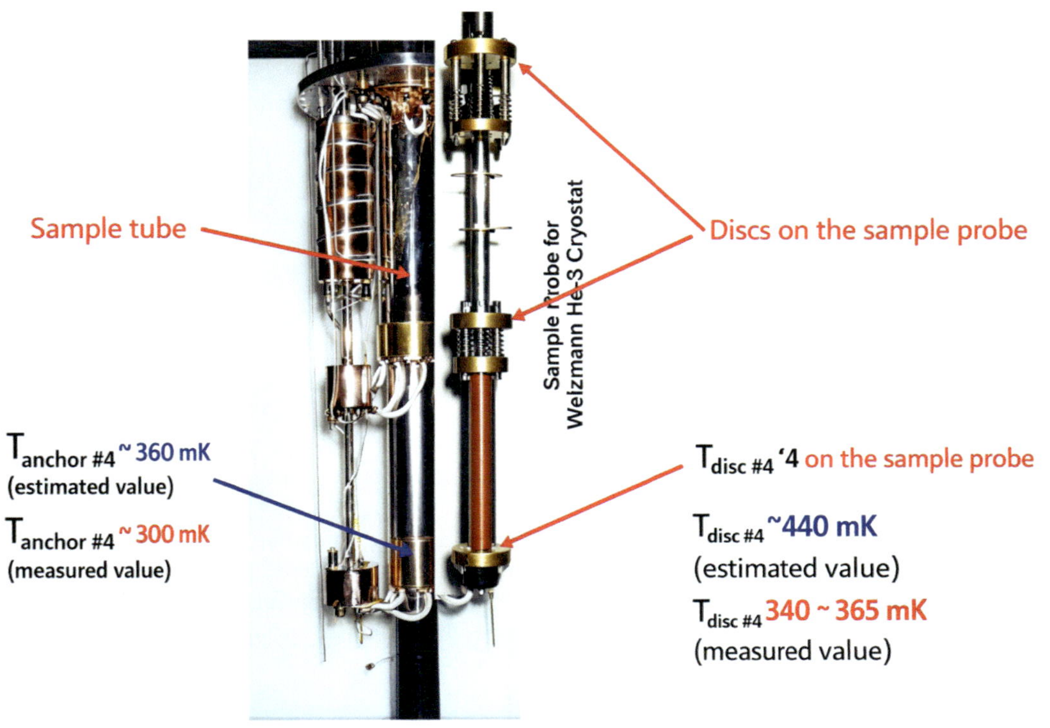

FIGURE 8.7 Thermal boundary conductance between sample tube and sample probe. (Reprint with permission from FormFactor, Inc.)

element analysis (FEA). The required temperature of ***Disc #4*** is below 0.4 K for the applications, which corresponds to a minimum thermal boundary conductance of 0.189 mW/K.

The empirical approach is used to estimate the normal force required to achieve sufficient thermal boundary conductance between gold-plated copper surfaces

1. Initial Parameters:
   Start with the parameters at 670 N in Table 8.1 (i.e., α =0.71, n = 2.15). Estimate the thermal boundary conductance for temperatures ranging from 300 mK to 400 mK.
2. Parameter Adjustment:
   Increase the parameters (α and n) to reach the required thermal boundary conductance.
3. Force Estimation:
   Treat the parameter $\boldsymbol{\alpha}$ as a linear function of the applied normal force, and calculate the normal force corresponding to the updated $\boldsymbol{\alpha}$.
   The author encourages the reader to complete these estimates as an exercise.
4. Final Design with Reasonable Safety Factor:
   - Anchor #4: concave, 68° taper.
   - Disc #4: is convex, 68° taper.
   - Applied force on Disc #4: 300 lbs (1,334 N).
   - Normal force applied on Anchor #4: 780 lbs (3,469 N).
     The thermal boundary conductance at the interface of two gold-plated copper surfaces at 0.3 K is estimated using:

$$\kappa_B = (0.00075F + 0.2075)T^{2.15} \quad (\text{mW/K}) \tag{8.4}$$

   where F is the normal force in newton, and T is the interface temperature in Kelvin.
5. Recorded Performance:
   - Anchor #4 temperature: ~300 mK
   - Disc #4 temperature: 340 mK–360 mK
   - Normal force on Anchor #4: ~3,470 N
   - Thermal boundary conductance: 0.315–0.47 mW/K
6. Conclusion and Discussion:
   - This empirical approach is applicable to millikelvin temperatures down to 300 mK.
   - Temperatures of Anchor #4 and Disc #4 are slightly different. The lower temperature should be used for conservative estimation during the design process.
   - Equation 8.2 is only valid when the temperature difference (ΔT) at the interface is small compared to the interface temperature.
   - The sample probe consists of five sections connected by four spring assemblies, which collectively need to travel longer distances.
   - Each disc must apply an appropriate force to the corresponding anchor. To achieve this, a hybrid spring assembly was developed (Figure 8.8), where conventional compression springs enable the travel while Belleville washers generate the required compression force. When a discs is stopped by the anchor stage, the combined spring constant shows discontinuity, as illustrated in Figure 8.9.

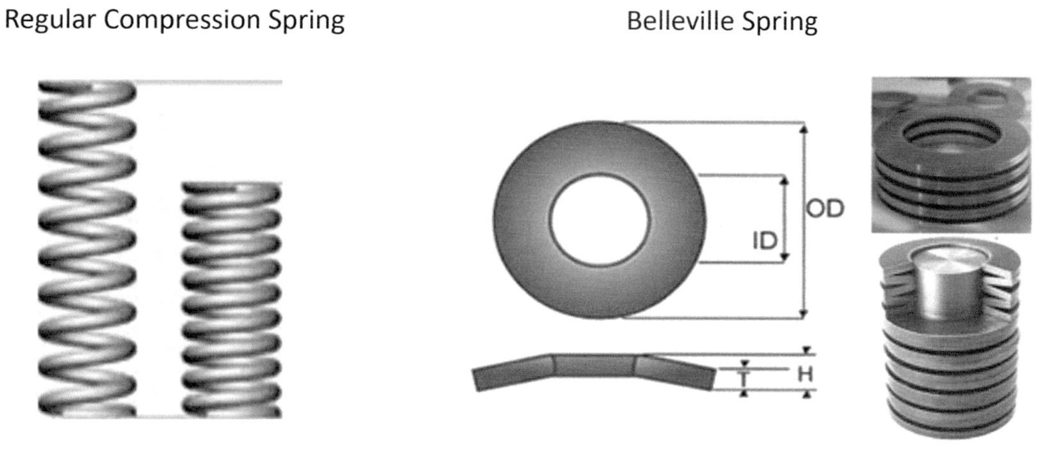

FIGURE 8.8 Hybrid spring assembly (Source of information from Lee Spring, www.leespring.com) [10].

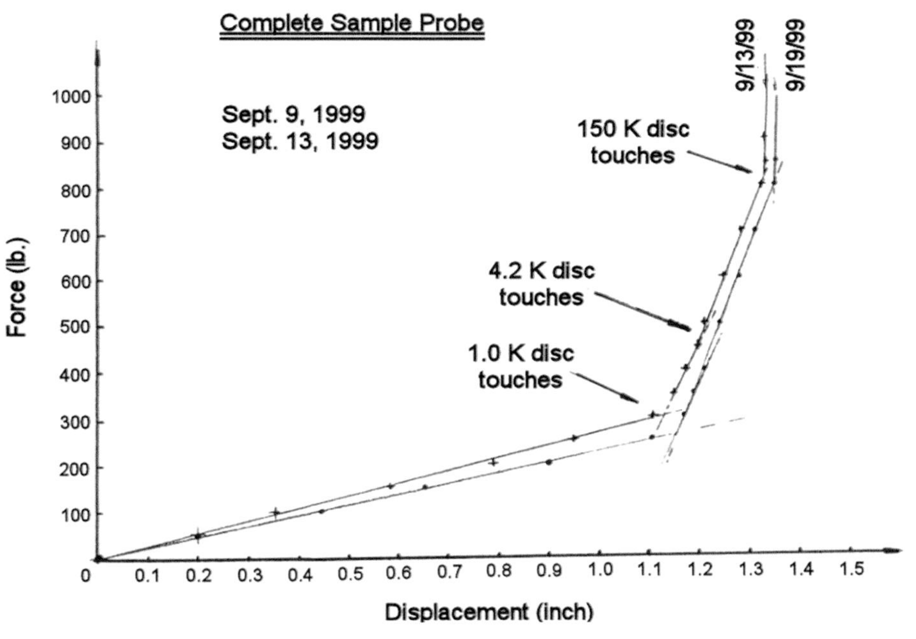

FIGURE 8.9 Performance of the hybrid spring assembly. (Projects accomplished by authors at Janis Research Company. Image courtesy of FormFactor, Inc.)

8.4.2 Example #2: Comparison between 300 mK He-3 Pot and Sample Chamber

This project aims to cool a sample chamber on a rigid sample probe by pressing it against a cold stage at 300 mK using mechanical force (Figure 8.10). The heat flow across the interface is approximately 1.06×10^{-6} W based on the analysis results. Similar to the approach in Example #1, the cold plate is concave with a 68° taper while the sample chamber is convex with the same taper. Both components are made of gold-plated copper. After vertical force was applied to the probe, the system reached the following performance:

- Cold plate temperature: 305 mK
- Sample chamber temperature: 330 mK
- Thermal boundary conductance: ~0.044 mW/K

8.4.3 Conclusions

- These examples demonstrated that the empirical approach as expressed by Equation 8.2 remains valid at sub-Kelvin down to 300 mK.
- It is reasonable to take the parameter $n = 2.15$ in Equation 8.2 and to treat the parameter α as a linear function of the applied normal force, i.e. the thermal boundary conductance can be considered linearly dependent on force at a fixed temperature. Different values of α can be

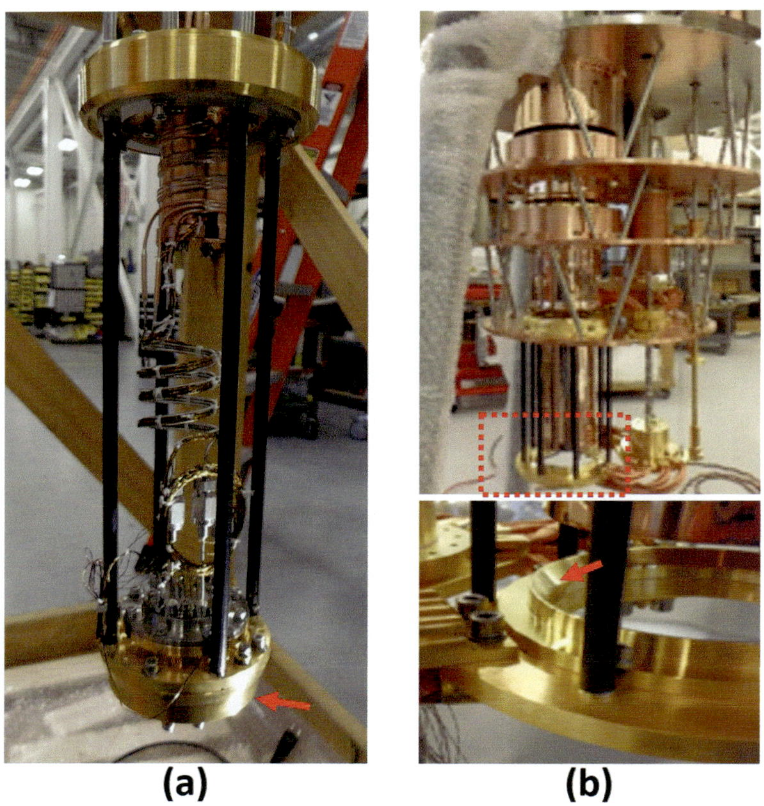

FIGURE 8.10 (a) Convex sample chamber and (b) concave cold stage. (Reprint with permission from FormFactor, Inc.)

determined experimentally for different normal forces until the estimated thermal boundary conductance meets the design requirement. A proper safety factor should be included in the final design.

- Using Equation 8.4 to estimate the normal force at the interface in Example #2 gives $F_N \approx 470$ N (105 lb). The corresponding applied force is approximately 39 lb, which is considered reasonable for this taper configuration.
- Although the parameter n in Table 8.1 appears applicable at 300 mK, the parameter α exhibits variation at sub-Kelvin temperatures, as the values listed in Table 8.1 are derived from measurements conducted at liquid helium temperatures.

REFERENCES

1. E.T. Swartz and R.O. Pohl, Thermal Boundary Resistance, *Review of Modern Physics*, 61(2), 605–668, (1989).
2. Pramod Reddy, Kenneth Castelino, and Arun Majumdar, Diffuse Mismatch Model of Thermal Boundary Conductance using Exact Phonon Dispersion, *Applied Physics Letters*, 87, 211908, (2005).
3. A.M. Clausing and B.T. Chao, Thermal Contact Resistance in a Vacuum Environment, *Journal of Heat Transfer*, 87(2), 243–251, (1965).
4. R. Berman, Some Experiments on Thermal Contact at Low Temperatures, *Journal Applied Physics*, 27(4), 318–323, (1956).
5. P. Kittel, A.L. Spivak, and L.J. Salerno, Thermal Conductance of Gold Plated Metallic Contacts at Liquid Helium Temperatures, *Advances in Cryogenic Engineering*, 37(pt A), 241–248, (1992).
6. L.J. Salerno and P. Kittel, Thermal Contact Conductance, National Aeronautics and Space Administration, Ames Research Center, Moffett Field, California, (February 1997).
7. L.J. Salerno, P. Kittel, and A.L. Spivak, *Thermal Conductance of Augmented Pressed Metallic Contacts at Liquid Helium Temperatures,* National Aeronautics and Space Administration, Ames Research Center, Moffett Field, California, USA, (August 1992).
8. R. Berman and C.F. Mate, Thermal Contact at Low Temperatures, Letter to Nature, *Nature*, 182, 1661–1663, (13 December 1958).
9. Jack W. Ekin, *Experimental Techniques for Low-Temperature Measurements*, Oxford University Press, Oxford, England, UK, (2006).
10. Lee Spring, 140 58th Street, Brooklyn, NY 11220, www.leespring.com

9 Thermometers at Cryogenic Temperatures

9.1 RUTHENIUM OXIDE (RUO)

The ruthenium oxide (RuO) sensor is a commercially available thick-film resistor with a negative temperature coefficient of resistance, meaning its resistance increases as temperature decreases [1]. It also exhibits positive magnetoresistance at magnetic fields above approximately 1.5 T, with the resistance increasing by about 10% at fields up to 18 T.

The RuO thermometer offers several advantages, including compact size, ease of mounting, good reproducibility under thermal cycling, and low cost. These qualities make it one of the most widely used sensors for cryogenic temperature measurement. It is particularly well-suitable for applications in high magnetic fields due to its relatively low magnetoresistance. Figure 9.1 shows an example of its magnetoresistance measured at 300 mK, where the resistance of a commercial RuO thermometer increases from about 5,000 Ω at zero magnetic field to nearly 5,300 Ω at 14 T—a change of roughly 6%.

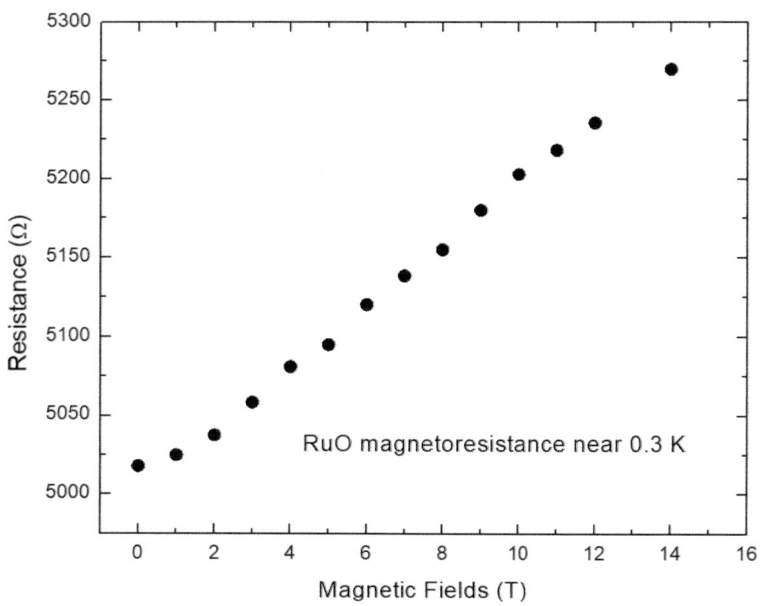

FIGURE 9.1 RuO sensor magnetoresistance up to 14 T.

DOI: 10.1201/9781042000050-9

However, the RuO thermometer has a limited operating temperature range: it loses sensitivity above approximately 20 K and becomes saturated below about 20 mK. It is also prone to self-heating effects below 0.3 K, which can reduce measurement accuracy at ultra-low temperatures.

9.2 CERNOX

The commercially available Cernox thermometer is fabricated by reactive sputtering of zirconium in a nitrogen–oxygen atmosphere [1]. The resulting thin film consists of conductive zirconium nitride (ZrN) embedded within a nonconductive zirconium oxide (ZrO_2) matrix. Similar to the RuO sensor, the Cernox thermometer exhibits a negative temperature coefficient of resistance. Cernox sensors offer good temperature sensitivity, excellent reproducibility through thermal cycling, and a wide operating temperature range—from 0.1 K to 420 K.

Cernox thermometers exhibit very small magnetoresistance above 4 K. For an example, resistance changes of less than 3% at 32 T for temperatures above 4.2 K have been reported [1]. This makes them one of the best choices for high-magnetic-field applications above 4 K.

However, below 4 K, Cernox sensors display a pronounced anomalous negative magnetoresistance, where the resistance decreases significantly in the presence of a strong magnetic field.

9.3 CERIUM MAGNESIUM NITRATE (CMN) THERMOMETER

9.3.1 Material

CMN thermometers are made of paramagnetic salt cerium magnesium nitrate with the molecular formula of $2Ce(NO_3)_3 \cdot 3Mg(NO_3)_2 \cdot 24(H_2O)$ [2–4]. The magnetic moments in CMN remain magnetically disordered in zero applied field due to extremely weak antiferromagnetic interactions, retaining a large spin entropy down to temperatures of only a few millikelvin (Figure 9.1). This property is one of the key factors that makes CMN an excellent material for temperature measurement in the millikelvin range.

The magnetic susceptibility of an ideal paramagnetic material follows Curie's law at sufficiently high temperatures, provided that there are no interactions between magnetic moments:

$$\chi = \frac{C}{T} \tag{9.1 a}$$

where T is the absolute temperature (K), and C is the material-specific Curie constant.

In CMN, however, weak interactions between neighboring magnetic moments lead to a deviation from the simple Curie behavior. As a result, the magnetic susceptibility is more accurately described by the Curie–Weiss law:

$$\chi = \frac{C}{T - \Delta} \tag{9.1 b}$$

where Δ is the Weiss constant, which reflects the effective molecular field arising from interactions between the magnetic moments. The measured value of Δ may also be influenced by crystal symmetry and, to a lesser extent, by demagnetizing effects associated with the sample geometry.

The Weiss constant of CMN is negative [4, 5] with typical value $|\Delta|$ below 2 mK. When the temperature approaches the value of $|\Delta|$, its impact on susceptibility becomes significant and cannot be neglected.

The Curie constant C is determined experimentally through calibration (see Section 9.3.3). Any temperature-independent background susceptibility is neglected in Equation (9.1b).

CMN contains crystallization water, which ensures a relatively large spacing (approximately 1 nm) between the magnetic ions. This large separation weakens magnetic interactions, thereby lowering the magnetic ordering temperature and enabling CMN thermometers to operate at temperatures down to a few millikelvin. However, CMN is sensitive to dehydration. When stored under vacuum at room temperature, it can lose up to two-thirds of its crystallization water within 250 h. To maintain proper functionality, CMN thermometer must be protected from dehydration.

CMN can be "diluted" by substituting some Ce+ ions with La+ ions, forming LCMN (La-diluted CMN). The molecular formula of LCMN is

$$(Ce_xLa_{1-x})_2\ Mg_3\ (NO_3)_{12}\bullet 24(H_2O)$$

where x denotes the faction of Ce^{3+} ions. A smaller x corresponds to a higher degree of dilution.

Because magnetic ordering is governed by the magnetic interaction strength between Ce^{3+} ions, reducing x decreases the number of magnetic moments and weakens their coupling, thereby lowering the magnetic ordering temperature. In practice, x typically ranges from 0.05 to 0.10 in functioning LCMN thermometers. We used an LCMN thermometer to measure a temperature of 0.88 mK in superfluid ^3He experiments. Paulson et al. reported a temperature of 0.41 mK using LCMN powder with $x = 0.05$ [6]. At significantly lower x, magnetic ordering is expected to vanish entirely.

It is worth noting that, although FAA is another widely used paramagnetic materials in ultra-low temperature (see Chapter 4), it is not employed as a thermometer. In contrast to CMN, which is used to measure entropy, FAA serves fundamentally different purposes and is utilized to extract entropy in ADR systems. In summary,

- The CMN thermometer, with a very low ordering temperature (~2 mK), allows the measurement of susceptibility as a function of temperature without significantly perturbing the system. To preserve this entropy and ensure accurate temperature readings, CMN thermometers should remain in zero magnetic field.
- FAA is used in ADRs as the refrigerant. Its function relies on applying a strong magnetic field to induce a large change in entropy, thereby enabling cooling with adiabatic demagnetization. FAA is chosen for its combination of relatively high entropy change and moderately low ordering temperature (~35 mK), which defines the practical low temperature limit for ADR operation.

9.3.2 Fabrication

CMN powder—sometimes mixed with silver powder to enhance thermalization—is packed into a cylindrical space situated between two concentric coils: the excitation coil and the pick-up coil. The excitation coil is typically wound with fine NbTi superconducting wire to eliminate ohmic heating when the exciting current passes through it at low temperatures. At room temperature, the excitation coil usually has a resistance of a few hundred ohms [7]. The pick-up coil, wounded with fine copper wire, functions as a voltage sensor rather than a current-carrying element and therefore does not generate ohmic heating during operation. Increasing the number of turns in the pick-up coil can enhance the induced voltage, thereby improving sensitivity. At room temperature, the pick-up coil resistance can typically reach several thousand ohms [7]. Finally, a niobium shield is typically installed around the coils to suppress the earth's magnetic field, as illustrated in Figure 9.2.

9.3.3 Operation and Calibration

When the CMN thermometer temperature is much higher than its magnetic ordering temperature, the temperature can be determined from the mutual inductance between the excitation and the pick-up coils, as expressed in Equation (9.2):

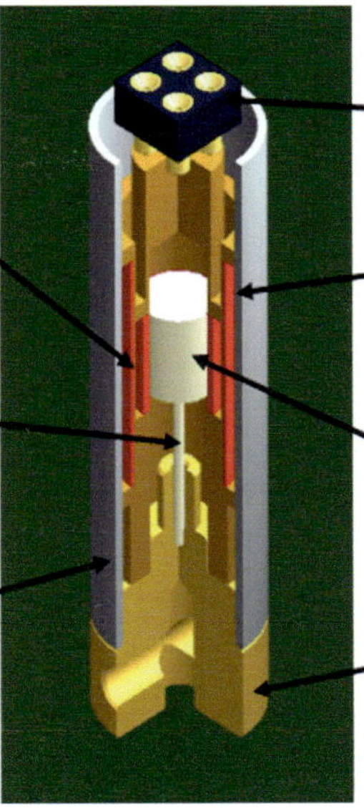

FIGURE 9.2 Schematic structure of CMN thermometer.

$$L_{12} = L_0 + \frac{L_1}{T} \tag{9.2}$$

where

- L_{12} is the mutual inductance between the excitation and pick-up coils
- L_0 and L_1 are parameters to be determined during thermometer calibration
- T is the temperature to be measured

The mutual inductance L_{12} can be measured using Lakeshore model 372S AC resistance bridge during the CMN thermometers calibration [8]. The measurement details are provided in Appendix.

Since the CMN thermometer is an inductive device, both calibration and operation must be performed in a zero (earth) magnetic field. The measured temperature should remain well above the magnetic ordering temperature but below 1 K, as the device exhibits poor sensitivity at higher temperatures.

Two primary methods are commonly used for CMN thermometer calibrations:

1. Calibration against a commercial thermometer
   In this method, the mutual inductance L_{12} of the CMN thermometer is measured as a function of temperature determined by a commercial calibrated Cernox or Germanium thermometer. A plot of L_{12} as the function 1/T yields a straight line, as illustrated in Figure 9.3. The interception and slope of this line correspond to the calibration parameters L_0 and L_1, respectively, in Equation (9.2).

   This approach is advantageous because both commercial thermometers provide nearly continuous and well-characterized calibrated data. However, caution is required when using a Cernox sensor for calibration. The lowest calibrated temperature for a Cernox sensor is usually around 0.1 K. To maintain accuracy, it is recommended to avoid data points near the end of the Cernox calibration range and use the temperature range from 0.12 K to 0.6 K for CMN

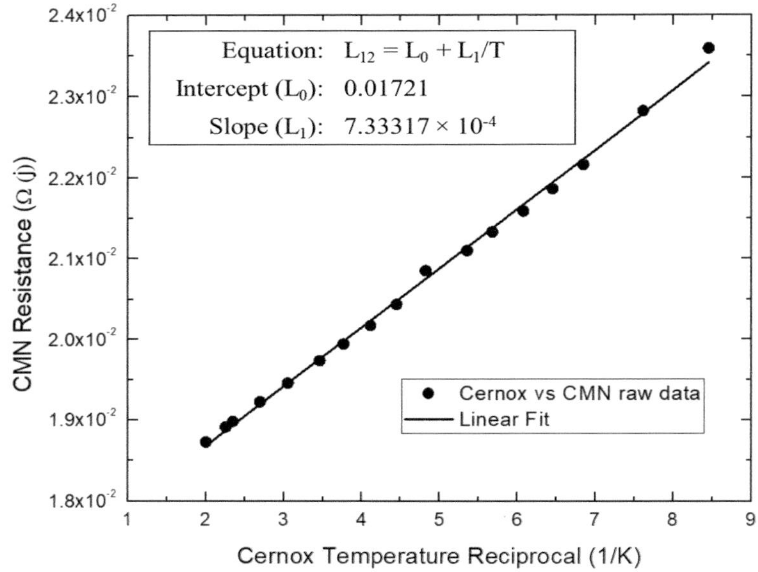

FIGURE 9.3 CMN calibration against Cernox thermometer.

thermometers calibration. If necessary, a linear extrapolation of the calibration curve can be applied to extend the calibration to temperatures below 10 mK.

2. Calibration using Fixed-Point Device (FPD) **[9]**

   FPD is another calibration tool. Similar to the CMN thermometer, the FPD also consists of an excitation coil and a pickup coil, as illustrated in Figure 9.4(a). Inside the coil assembly are several superconducting materials with distinct and well-defined transition temperatures. Table 9.1 lists the superconducting materials used in our previous projects.

   As the temperature changes and one of these samples enters the superconducting state, it expels magnetic flux via the Meissner effect. This expulsion alters the mutual inductance between the primary and secondary coils, resulting in a sharp voltage change at the secondary output, as shown in Figure 9.4(b).

Although FPD provides only a few calibration points, compared to Cernox or Germanium thermometers, it offers highly accurate and reproduced fixed -temperature reference, making it ideal for verifying or refining CMN calibration data.

To ensure the highest accuracy and reliability, it is recommended to use both methods—cross-calibrating the CMN thermometer against a commercial sensor and verifying it with fixed-point data. This combined approach minimizes systematic errors and provides a robust calibration in the millikelvin temperature range.

9.3.4 Sensitivity of CMN Thermometer

The CMN thermometer offers exceptional sensitivity at ultra-low temperatures, with a demonstrated capability to resolve temperature changes on the order of 10^{-7} K in the millikelvin range, as illustrated in Figure 9.5.

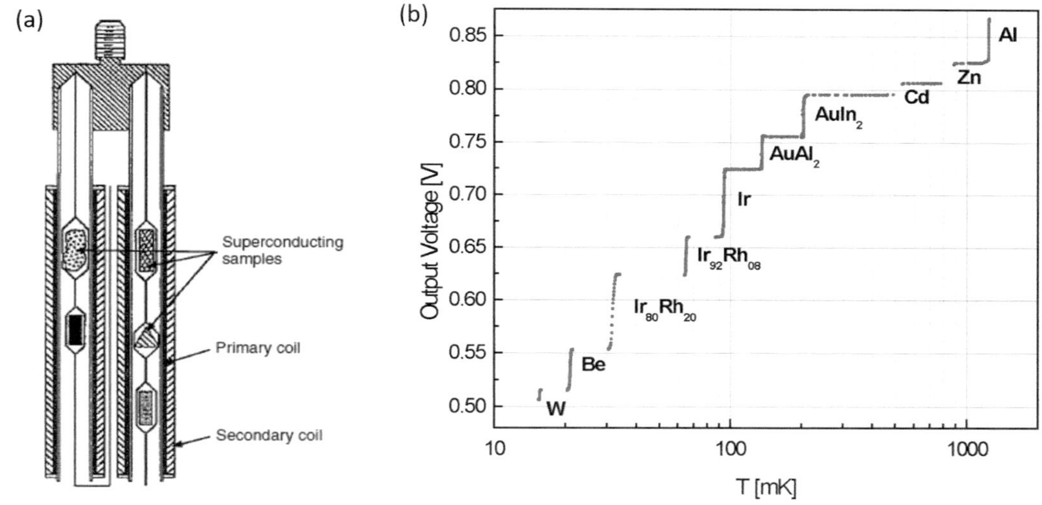

FIGURE 9.4 (a) Schematic illustration of a superconducting fixed-point device (FPD). (b) SC Transition temperature of FPD materials per SRD-1000. (Reprint with permission from Guglielmo Ventura and Lara Risegari, *The Art of Cryogenics; Low-Temperature Experimental Techniques,* Figure 8.5; page 187, First Edition 2008.*)* [10]

TABLE 9.1
FPD Material and Their SC Transition Temperatures

Material	In	Al	Zn	Cd	$AuIn_2$	$AuAl_2$	Ir	W
SRD1000 Superconducting Transition Temperature (K) at 0 T	3.400	1.182	0.850	0.5195	0.2080	0.145	0.0998	0.015

Source: Data from Guglielmo Ventura and Lara Risegari, *The Art of Cryogenics; Low-Temperature Experimental Techniques,* Table 8.10; page 187, First Edition 2008.

9.3.5 Summary of CMN Thermometer

The CMN thermometer offers excellent sensitivity and stability for ultra-low-temperature measurements. It is capable of resolving temperature changes as small as a few tenths of a microkelvin in the millikelvin range, with negligible power dissipation and high resolution below 1 K. Due to its low heat capacity and short thermal time constant, the CMN thermometer responds rapidly to temperature changes and is straightforward to operate. However, it must be used in a zero magnetic field, as its performance is severely affected by external magnetic fields. In addition, the thermometer exhibits poor sensitivity above 1 K, and its calibration must be verified during the first cool-down to ensure measurement accuracy.

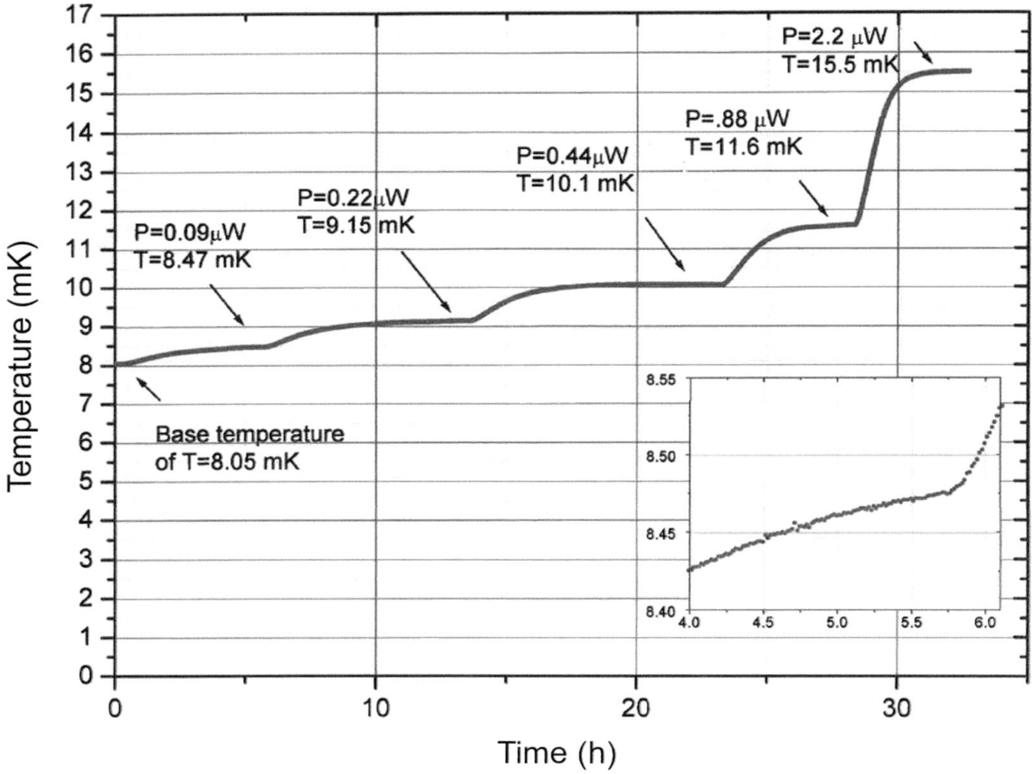

FIGURE 9.5 Resolution and noise level of CMN thermometer. (Projects accomplished by authors at Janis Research Company. Image courtesy of FormFactor, Inc.)

9.4 MELTING CURVE THERMOMETER (MCT)

9.4.1 He-3 Melting Curve

Figure 9.6 illustrates the phase diagram of He-3 in zero magnetic field, showing the solid, liquid, and gaseous phases as functions of pressure and temperature [11, 12]. At low temperatures, the solid and liquid phases coexist along the melting curve, commonly referred to as the mixture. In this section, we focus on the behavior of the He-3 melting curve below 1 K.

The He-3 melting curve exhibits "anomalous" behavior below approximately 315 mK, where the melting pressure decreases with increasing temperature (dP/dT < 0), contrary to the behavior observed for most materials. According to the Clausius–Clapeyron relation, the melting curve is described by Equation (9.3).

$$\frac{dP}{dT} = \frac{S_L - S_S}{V_L - V_S} \tag{9.3}$$

where

- S_L is the molar entropy of liquid He-3
- S_S is the molar entropy of solid He-3
- V_L is the molar volume of liquid He-3
- V_S the molar volume of solid He-3

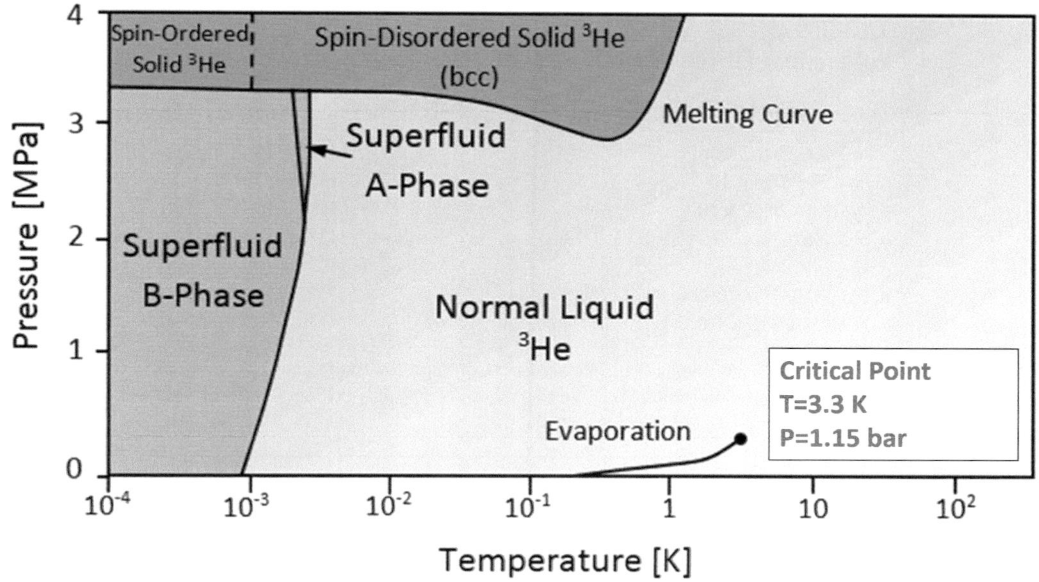

FIGURE 9.6 He-3 phase diagram at low temperatures. (Reprint with permission from Taylor & Francis Group.) [13]

The sign of the melting-curve slope is determined by the entropy difference between the liquid and solid phases, since the molar volume difference remains positive and nearly constant over this temperature range.

Below 315 mK, the molar entropy of the liquid is lower than that of the solid ($S_L < S_S$), while the liquid's molar volume remains larger than that of the solid ($V_L > V_S$). As a result, the slope dP/dT remains negative in this temperature range. A pressure minimum occurs at approximately 29.3 bar and 315 mK, where dP/dT = 0, indicating that the entropies of liquid and solid He-3 are equal at this point.

Between 0.9 mK and 1.0 K, the pressure along the He-3 melting curve can be represented as a polynomial function of temperature, as given in Equation (9.4):

$$p = \sum_{i=-3}^{9} \alpha_i T^i \tag{9.4}$$

In the equation, P is the pressure in megapascal (MPa), and T is the temperature in kelvin on the PLTS-2000 scale. The polynomial coefficients a_i, corresponding to the PLTS-2000 scale [14–16], are listed in Table 9.2. Four fixed points on the melting curve can be used for thermometry. These include the curve minimum T_{min}, the transitions into the superfluid phases A (T_A) and B ($T_{A\text{-}B}$), and the transition into antiferromagnetic ordering of the nuclear spins (T_N), listed in Table 9.3.

9.4.2 He-3 MCT Structure

A He-3 MCT contains a fixed amount of He-3 in a sealed volume at an appropriate pressure. Within the temperature range of interest—typically below 1 K—the He-3 remains on its melting curve, although the liquid-to-solid ratio varies with temperature. The temperature is determined from the measured pressure using the polynomial expression, as given by Equation (9.4).

The structure of a typical MCT is illustrated in Figure 9.7. The sensing element consists of a parallel-plate capacitor, where one plate is rigidly mounted and the other is exposed to the pressure

TABLE 9.2
Polynomial Fitting of Helium-3 Melting Curve

Temperature Scale PLTS-2000	Coefficient a_i (bar/Ki)
$a_{-3} = -1.3855442 \times 10^{-12}$	$a_4 = 7.1499125 \times 10^1$
$a_{-2} = 4.5557026 \times 10^{-9}$	$a_5 = -1.0414379 \times 10^2$
$a_{-1} = -6.4430869 \times 10^{-6}$	$a_6 = 1.0518538 \times 10^2$
$a_0 = 3.4467434 \times 10^0$	$a_7 = -6.9443767 \times 10^1$
$a_1 = -4.4176438 \times 10^0$	$a_8 = 2.6833087 \times 10^1$
$a_2 = 1.5417437 \times 10^1$	$a_9 = -4.5875709 \times 10^0$
$a_3 = -3.5789853 \times 10^1$	

Source: Data from Jiaojie Yan, Construction of ultra-low temperature environment and transport measurements, Ph.D. thesis, School of Physics, Peking University (May, 2022)] [14].

TABLE 9.3
Fixed point of the Provisional Low Temperature Scale PLTS-2000

Fixed Points	Pressure (bar)	Temperature (mK)
T_{min}	29.3113	315.24
T_A	34.3407	2.444
T_B	34.3609	1.896
T_N	34.3934	0.902

Source: Data from Jiaojie Yan, Construction of ultra-low temperature environment and transport measurements, Ph.D. thesis, School of Physics, Peking University (May, 2022)] [12].

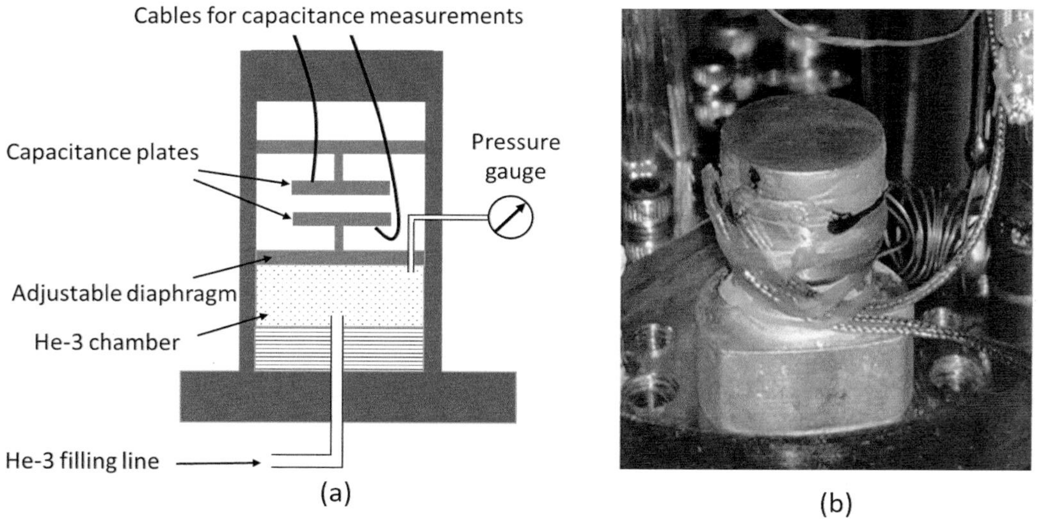

FIGURE 9.7 (a) Sketch of He-3 melting-curve thermometer. (b) Image of the actual MCT. (Reprint with permission from Jiaojie Yan, Construction of ultra-low temperature environment and transport measurements, Ph.D. thesis, School of Physics, Peking University (May, 2022).) [14]

of the He-3 mixture. As the pressure changes with temperature along the melting curve, the movable plates deflect, producing a corresponding change in capacitance. Electrodes connected to both plates enable capacitance measurements, which typically falls within the picofarad range.

9.4.3 Operation

9.4.3.1 Calibration

Accurate operation of an MCT relies on proper calibration of its internal pressure against capacitance. For an example, pure He-3 gas is introduced into the MCT at 4.2 K, and the pressure–capacitance relationship is measured over a range of 25–35 bar. This span fully encompasses the operating range of the MCT, ensuring reliable pressure–capacitance mapping throughout subsequent low-temperature measurements. The resulting calibration curve is later used in conjunction with Equation 9.6 and Table 9.3 to determine temperature from measured capacitance during operation.

9.4.3.2 Confinement of He-3

Once calibration is completed, the He-3 can be confined using the "blocked-capillary method".

The MCT is filled with He-3 through a thin CuNi capillary and pressurized to 35 bar (508 psi). As the MCT cools from 4.2 K, the internal pressure decreases, requiring additional He-3 to maintain the pressure at 35 bar—particularly when He-3 begins to liquefy. At approximately 0.8 K (highlighted by a circle in Figure 9.8), solid He-3 forms inside the capillary, creating a blockage that traps a fixed quantity of the He-3 mixture within the MCT for the rest of the cooldown. Below this temperature, the internal pressure evolves solely in response to temperature changes along the He-3 melting curve, allowing the MCT to operate as a primary thermometer.

9.4.3.3 Interpretation of Pressure Readings

As shown in Table 9.3, the He-3 melting curve exhibits a minimum temperature of approximately 0.315 K at a pressure of 29.3 bar. For pressure above this value, two corresponding temperatures are

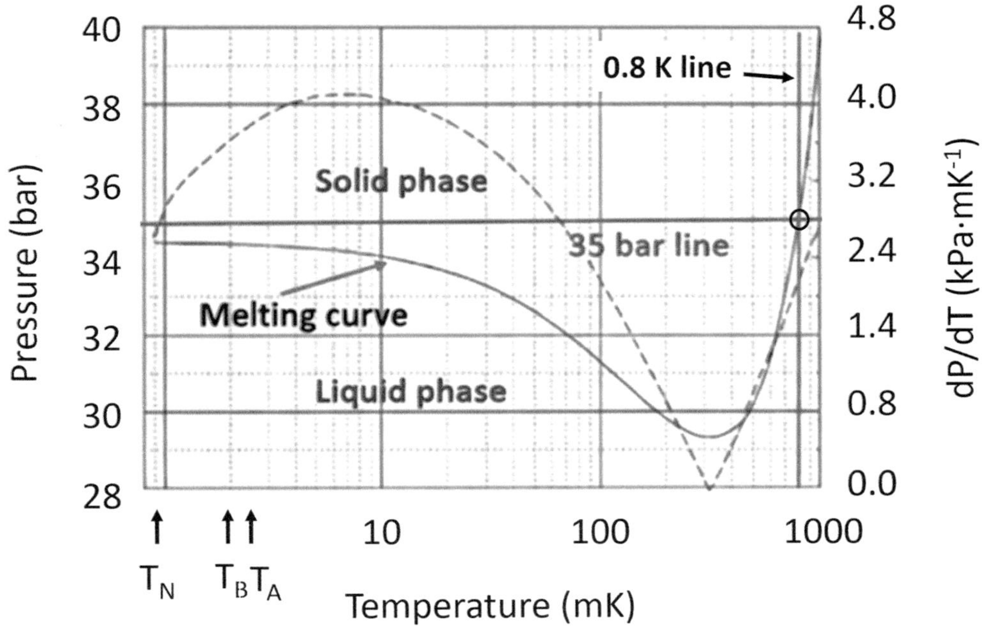

FIGURE 9.8 Confining He-3 in MCT (figure not strictly to scale). The solid line is He-3 melting pressure curve in MPa per PLTS-2000. The dashed line is its slope *dP/dT* in kPa/mK [14].

possible—one above and one below 0.315 K. To determine the correct value, the pressure trend is monitored during cooling the MCT: if the pressure increases with time, the thermometer is below 0.315 K, whereas an decreasing pressure indicates operation above the melting curve minimum. This procedure ensures accurate temperature determination throughout the millikelvin range.

9.4.4 Summary of Key Features

The melting-curve thermometer offers several notable advantages for millikelvin thermometry. It can reach temperatures of a few millikelvin and provides exceptionally high sensitivity, with a typical temperature resolution of about 1 μK. The instrument is highly reproducible, exhibits zero power dissipation, and is inherently resistant to radio-frequency interference. Its short thermal time constant enables rapid response to temperature changes. The main drawback of the MCT is its relatively complex and relatively large setup compared with other cryogenic thermometers.

Interested readers may refer to reference [14, 17], which presents a detailed description of the MCT's working principle, design, and construction, as well as its operation and performance.

REFERENCES

1. Frank Pobell, *Matter and Methods at Low Temperatures*, Third Edition, Springer Berlin, Heidelberg, Berlin, Germany, (2007).
2. O.V. Lounasmaa, *Experimental Principles and Methods Below 1 K*, Academic, London (1974) Guglielmo Venturra and Lara Risegari, The Art of Cryogenics, Low-Temperature Experimental Techniques, 2008 Elsevier Ltd.
3. Robert C. Richardson, *Low Temperature Technique*, Cornell University, Ithaca, New York, USA, (1982).
4. F. Pobell, *Matter and Methods at Low Temperatures*, Third Edition, Springer, Berlin, (2007).
5. D.S. Greywall and P.A. Busch, Fast cerium magnesium nitrate (CMN) thermometer for the low millikelvin temperature range, *Review of Scientific Instruments*, 60(3), 471–473, (1 March, 1989).
6. D.N. Paulson, M. Krusius, J.C. Wheatley, R.S. Safrata, M. Koláč, T. Těthal, K. Svec, and J. Matas, Magnetic Thermometry to Below One Millikelvin with Lanthanum-Diluted Cerium Magnesium Nitrate, *Journal of Low Temperature Physics*, 34, 63–82, (1979).
7. Small Coil LLC 13 Overlook Drive, Atkinson, NH, 03811, USA Phone: 603-489-9878 Email: info@smallcoil.com
8. User's Manual: Model 372AC Resistance Bridge andTemperature Controller Lake Shore Cryotronics, Inc. 480 Olde Worthington Rd, Westerville, OH, USA Website: www.lakeshore.com
9. Christian, Enss, and Siegfried, Hunklinger, Low-Temperature Physics, Springer Berlin, Heidelberg, Berlin, Germany, (2005).
10. Ventura, Guglielmo, and Lara Risegari, *The Art of Cryogenics; Low-Temperature Experimental Techniques*, First Edition, Elsevier Science, Amsterdam, Netherlands, (2008).
11. Robert C. Richardson, *Nobel Lecture*, Cornell University of Physics, Ithaca, New York 14853, USA, (1996).
12. David M. Lee, The Extraordinary Phases of Liquid 3He, *Reviews of Modern Physics*, 69(3), 645–666, (1997).
13. Zuyu Zhao and Chao Wang (Eds.), *Cryogenic Engineering and Technologies: Principles and Applications of Cryogen-Free Systems*, CRC Press (Taylor & Francis Group), Boca Raton, FL, USA, (2020).
14. Jiaojie Yan, Construction of Ultra-Low Temperature Environment and Transport Measurements, Ph.D. thesis, School of Physics, Peking University (May, 2022).
15. R.L. Rusby, M. Durieux, A.L. Reesink, et al. The Provisional Low Temperature Scale from 0.9 mK, *Journal of Low Temperature Physics*, 126, 633–642, (2002).
16. Report of the proceedings of the89th meeting of Comité international des poids et mesures (CIPM), Bureau International des Poids et Mesures Time: October 2000 Location: Sèvres, France
17. Xi Lin, *Introduction of Low Temperature Experiments*, Peking University Press, Beijing, China, (2005).

10 Impact of High-Magnetic Fields on Millikelvin Temperatures

An increasing number of studies in fundamental physics and novel material development are being conducted under extreme conditions—such as ultra-low temperatures, high-magnetic fields, high pressures, and ultra-fast laser excitation. Among these, the combination of ultra-low temperatures and high-magnetic fields has become one of the most widely used experimental environments. Notable examples include scanning tunneling microscopy (STM) at millikelvin temperatures in high-magnetic fields [1], as well as microkelvin nuclear-adiabatic demagnetization refrigerators (NADR) [2, 3]. This chapter examines the challenges associated with this combination and proposes practical solutions.

10.1 GENERAL CHALLENGES

10.1.1 Eddy Current Heating

Low temperatures and high-magnetic fields are often regarded as two *competing* parameters, since variations in magnetic field induce eddy-current heating in conductive materials. For a circular conductive disc in a uniform magnetic field, the heating power generated by the eddy currents can be expressed as in Equation (10.1):

$$Q = \frac{VR^2(dB/dt)^2}{8\rho} \tag{10.1}$$

where

V: volume of the object exposed to eddy current heating
R: radius of the circular disc
dB/dt: change rate of the magnetic field
ρ: resistivity of the disc material

As an example, a typical circular OFHC mixing chamber cold plate in DR has a radius of 5.08 cm and thickness of 1.27 cm, and the electrical resistivity of copper is 1.68×10^{-8} Ω·m. If the magnet is ramped from 0 to 15 T within 30 min, the induced eddy current in the copper plate generates an estimated dissipative heating power of approximately 0.137 mW.

DOI: 10.1201/9781042000050-10

10.1.2 Limited Cooling Power

A DR has very limited cooling power near its base temperature. For example, a DR-based millikelvin scanning tunneling microscope typically operate between 10 mK and 20 mK, and the fridges typically provide approximate 10 μW of cooling power at 20 mK (see Figure 10.1) [1]. The eddy current heat load, as large as 0.137 mW, warms the microscope significantly above its target operating temperature during magnet ramping.

This challenge is even more pronounced for a nuclear adiabatic demagnetization refrigerator (NADR), which is typically a single-shot system with significantly lower cooling power during demagnetization, making it particularly susceptible to eddy current heating.

10.1.3 Vibration-Induced Eddy Current

Vibration must be minimized in any ultra-low temperature system operating in a magnetic field because mechanical motion produces fluctuations in magnetic flux through conductive materials, generating eddy-current heating. Although this type of eddy-current heat load is usually in the range of micro- or even pico-watts, it can still perturb the system's base temperatures, particularly in sub-millikelvin NADR systems.

10.1.4 Long Recovery Time

A millikelvin system typically requires a long time to cool down and stabilize at its base temperatures. Once it is warmed up during a magnetic field change, the temperature recovery can take several hours. In a NADR system, the situation is considerably more challenging. If the nuclear demagnetization stage—containing dozens of moles of high-purity copper—is heated by eddy currents, it may require several days to re-cool the copper to the appropriate initial temperature for demagnetization. This prolonged recovery time arises from the large nuclear heat capacity of copper in that temperature range [2, 4, 5].

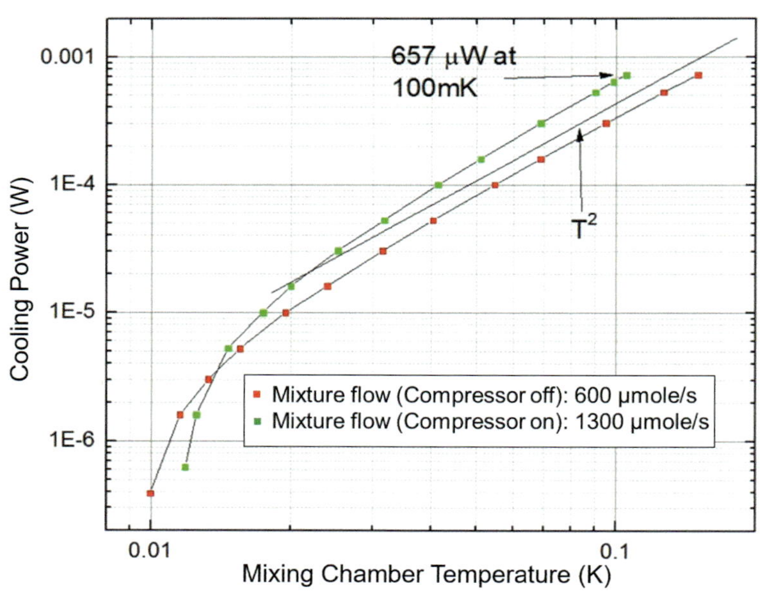

FIGURE 10.1 Typical cooling power of a "high power" DR. (Projects accomplished by authors at Janis Research Company. Image courtesy of FormFactor, Inc.)

10.2 SOLUTIONS

10.2.1 Eddy Current Reduction Design

Proper cryostat design should be considered to minimize eddy currents. Taking NADR as an example, the demagnetization stages in the early developed NADR systems were constructed from bundles of Formvar-insulated, annealed AWG 46 copper wires. Each bundle contained approximately 10,000 wires, with an overall diameter of about 0.2 in and a length of 20 in. [5, 6]. Based on measured cross-sectional areas of copper relative to the total bundle area, the copper volume fraction in the bundle is estimated to be approximately 67%. A similar NADR stage was built into the cryostat used for our study on superfluid He-3.

Eddy current losses in these stages are negligible in the insulated copper wires due to the small cross-sectional areas. However, this benefit comes at the cost of increased construction complexity. In particularly, an efficient heat exchanger is required between the NADR stage and the sample to maintain adequate spin-phonon thermalization.

Most contemporary NADR stages are made of a long copper cylinder with slices (Figure 10.2). This design concept can be applied to conductors with different shapes and configurations, such as circular or rectangular plate or rod. Compared to the copper wire bundles, the sliced design offers several advantages. It provides proven thermal conduction due to higher occupancy of copper; For example, the configuration illustrated in Figure 10.2 achieves a copper occupancy of 74% [7]. It also significantly simplified the fabrication, including machining, annealing, assembling, and handling, and it also eliminates parasitic heat contributions associated with the wire insulations.

10.2.2 Vibration Reduction

To minimize vibration-induce eddy currents, several effective approaches can be implemented, such as installing gimbals to intercept vibrations transmitted from the environment to the system (Figure 10.3) and seating the cryostat on a vibration-isolated optical table (Figure 10.4). Further

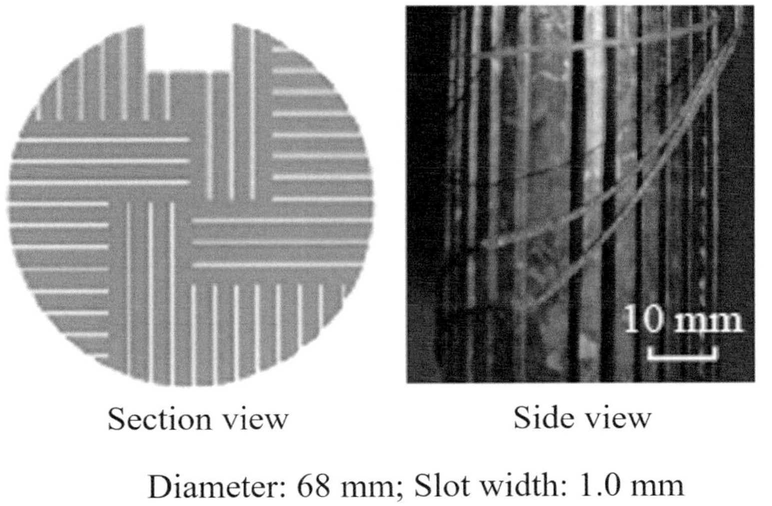

FIGURE 10.2 Slices on NADR stage (68 mm in diameter; Slot width: 1.0 mm) to prevent eddy current. (Reprint with permission from Jiaojie Yan, Realization and transportation research of ultra-low temperature system, Ph.D. thesis, School of Physics, Peking University (May, 2022).) [2]

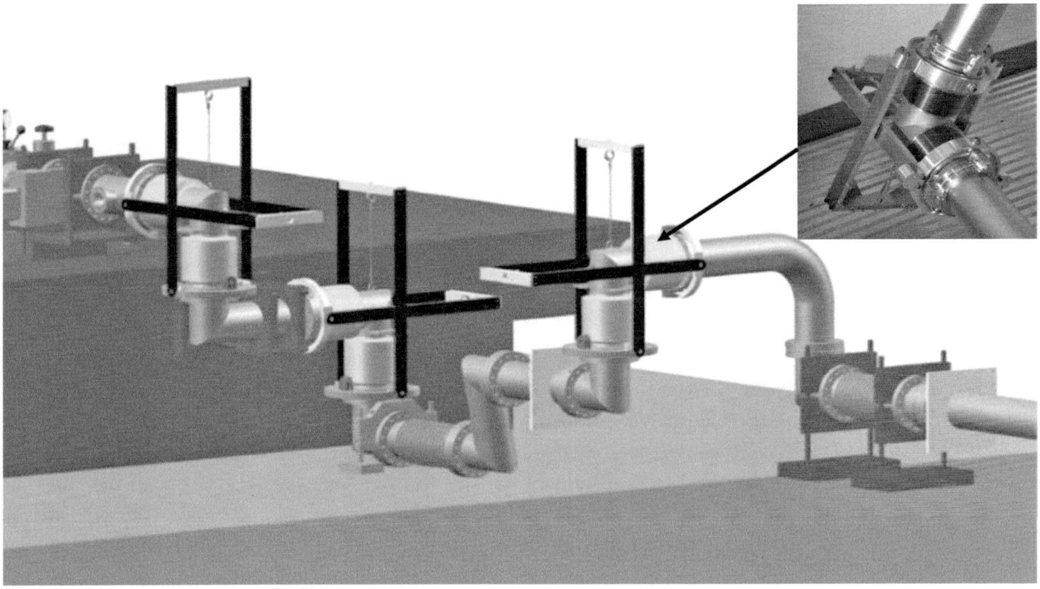

FIGURE 10.3 Gimbals [8] and that installed on ULT STM systems. [Reprint with permission from the author, Young Jae Song, Alexander F. Otte, Vladimir Shvarts, Zuyu Zhao, Young Kuk, Steven R. Blankenship, Alan Band, Frank M. Hess, and Joseph A. Stroscio Joe, *Invited Review Article: A 10 mK scanning probe microscopy facility,* Review of Scientific Instruments 81, 121101 (2010).] [1]

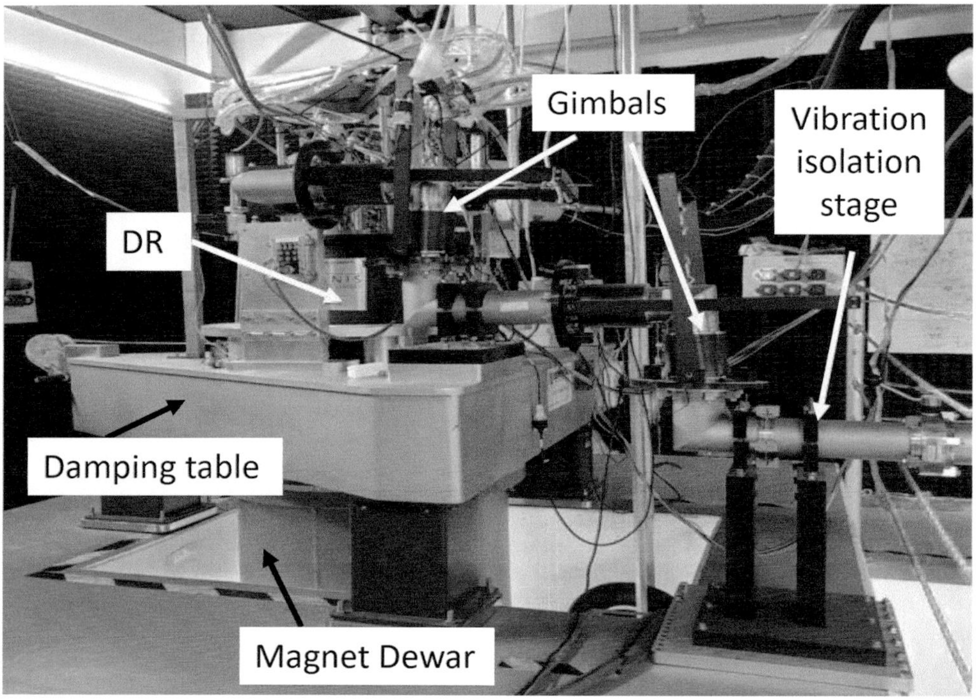

FIGURE 10.4 NADR system seated on an optical table. [Reprint with permission from Jiaojie Yan, Realization and transportation research of ultra-low temperature system, Ph.D. thesis, School of Physics, Peking University (May, 2022).] [2]

discussions on vibration reduction are presented in Chapter 16: Vibration Measurement and Reduction.

10.2.3 Compensation Magnet

Install a compensating magnet alongside the main magnet to create a low-field region around a localized volume where eddy current heating is undesirable. This low-field region is typically a cylindrical volume with a peak field of approximately 100–200 Gauss. A common practice is to place the compensating magnet around DR mixing chamber.

10.2.4 Sample Thermalization Improvement

When components such as thermal links or sample plates are cooled in a magnetic field, the cool-down time is strongly impacted by their nuclear spin heat capacity, where shorter cool-down time is always desired. Figure 10.5 illustrates the nuclear spin heat capacity of Ag and Cu. The nuclear heat capacity follows a $(B/T)^2$ dependence, within certain experimental errors, when the condition $\mu B \gg k_B T$ is satisfied, where μ and k_B is the nuclear magnetic moment and Boltzmann constant, respectively. This quadratic dependence breaks down as μB approaches $k_B T$, and becomes invalid at very low temperatures due to internal nuclear-nuclear interactions. At millikelvin temperatures, silver has a nuclear spin of ½ with a nuclear magnetic moment $\mu(^{109}Ag) = 0.13\ \mu_N$, where $\mu_N = 5.05 \times 10^{-27}$ A·m² is the nuclear magneton. The nuclear spin heat capacity of silver is roughly an order of magnitude smaller than that of copper within the relevant temperature and magnetic field range—typically at 10–20 mK and 9–18 tesla—resulting in significantly shorter cool-down times. Therefore, components such as thermal links and sample plates (e.g., STM cradles, see Figure 10.6) are often

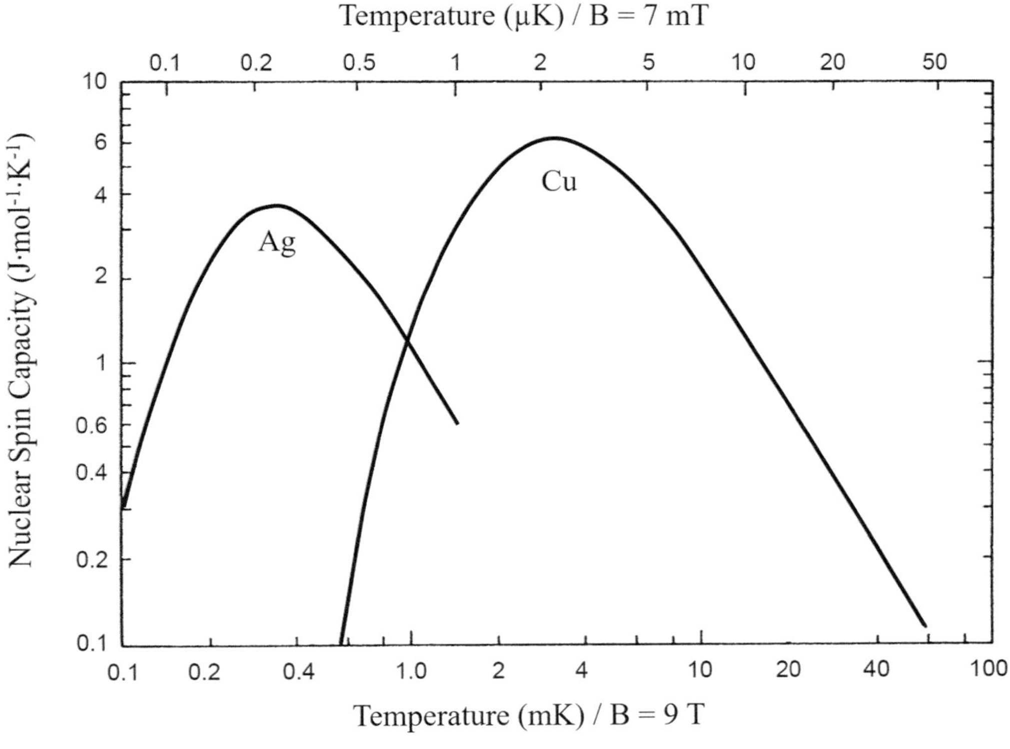

FIGURE 10.5 Nuclear spin heat capacity of Ag and Cu [5].

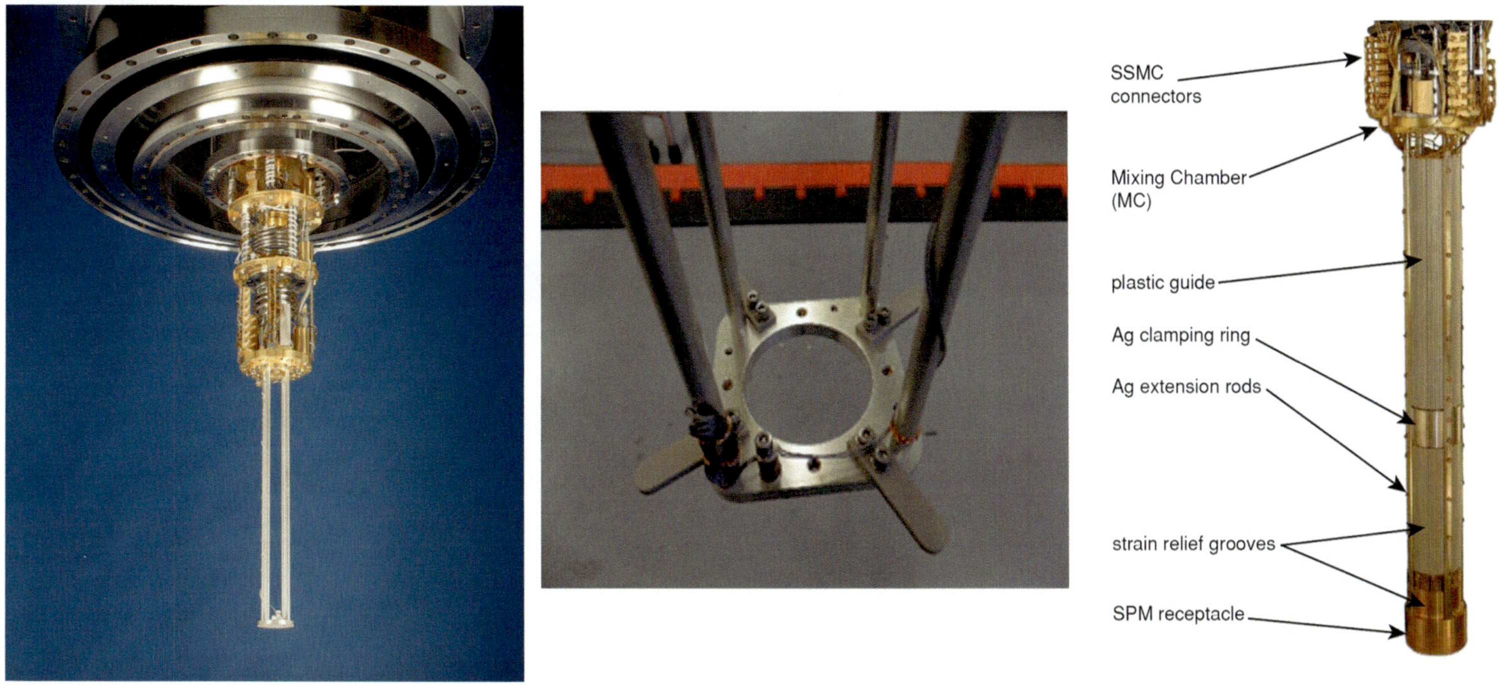

FIGURE 10.6 ULT STM thermal link and STM cradle made of 5x9 Ag. (Projects accomplished by authors at Janis Research Company. Image courtesy of FormFactor, Inc.)

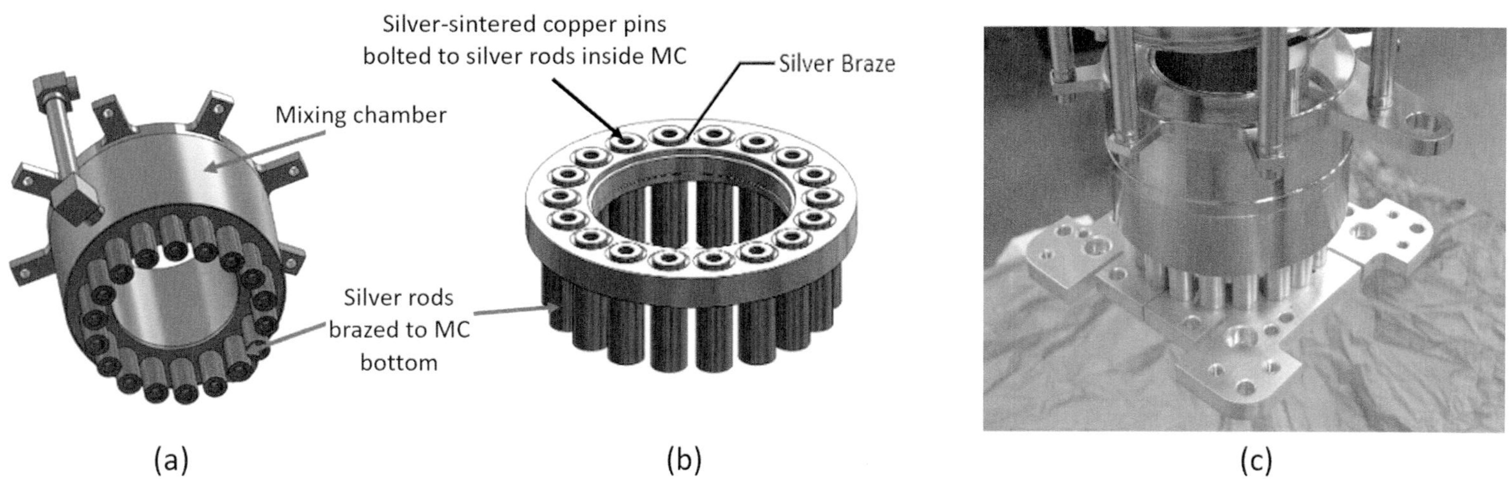

FIGURE 10.7 Customized low eddy current MC. (Projects accomplished by authors at Janis Research Company. Image courtesy of FormFactor, Inc.)

fabricated from silver rather than copper [9]. To maintain the component rigidity, nonannealed 99.95% silver is preferred over annealed pure silver. In case that annealing is required, silver can be processed in open air, whereas copper requires annealing under vacuum.

It is worth noting that 10–16 mK is a common starting temperature for nuclear demagnetization cooling for the copper NADR stage after precooling from higher temperature in a constant magnetic field. Because the nuclear spin heat capacity increases rapidly during cool-down—proportional to $(B/T)^2$ in that temperature range (Figure 10.5), the precooling time for a copper NADR stage can easily reach 2 days or longer [1, 2].

10.2.5 Special Mixing Chamber Design

Since stainless steel has a much higher electrical resistivity than copper, it can be used to make the bottom plate of a mixing chamber to reduce eddy current heating within the plate. To maintain good thermal conduction between the mixture and the sample, silver rods are brazed to the stainless steel mixing chamber bottom plate with non-magnetic AgCuNi braze alloy, with the top of the silver rods extending into the mixing chamber and making direct contact with the mixture. Silver-sintered copper rods are then bolted onto the silver rod tops from inside the mixing chamber, as shown in Figure 10.7(a) and 10.7(b). In this configuration, the silver rods remain isothermal with the mixture while experiencing minimal eddy-current heating. Customized low–eddy-current cold plates or sample holders can then be bolted onto the silver rods, as illustrated in Figure 10.7(c). This approach has been successfully implemented in several ultra-low-temperature scanning tunneling microscope systems [10].

Another solution is to use plastic mixing chamber or sample chamber as discussed in Chapter 18.

10.3 ULT Thermometers in High Magnetic Fields

User-friendly ULT thermometers capable of measure millikelvin or lower temperatures in high-magnetic fields is long time expected by the community. Existing options, such as Pt NMR thermometers, He-3 melting curve thermometers, and noise thermometers, can operate in this regime, but they are typically expensive or complex to set up and use [11, 12]. The Coulomb blockade thermometer presented by Hahtela *et al.* [13] offers be a promising alternative, although it still requires further development.

References

1. Young Jae Song, Alexander F. Otte, Vladimir Shvarts, Zuyu Zhao, Young Kuk, Steven R. Blankenship, Alan Band, Frank M. Hess, and Joseph A. Stroscio Joe, Invited Review Article: A 10 mK Scanning Probe Microscopy Facility, *Review Of Scientific Instruments*, 81, 121101, (2010).
2. Jiaojie Yan, Jianing Yao, Vladimir Shvarts, Rui-Rui Du, and Xi Lin, Cryogen-Free One Hundred Microkelvin Refrigerator, *Review of Scientific Instruments*, 92(2), 025120, (2021).
3. H. Zu, W. Dai, and A.T.A.M. de Waele, Development of Dilution Refrigerators—A Review, *Cryogenics*, 121, 103390, (2022).
4. Satoshi Abe and Koichi Matsumoto, Nuclear Demagnetization for Ultra-Low Temperatures, *Cryogenics*, 62, 213–220, (2014).
5. Frank Pobell, *Matter and Methods at Low Temperatures*, Third Edition, Springer-Verlag, Berlin Heidelberg, (2007).
6. E.B. Osgood and J.M. Goodkind, *Physical Review Letters*, 18, 894, (1987).
7. Jiaojie Yan, private communication, Max Planck Institute for Solid State Research, Heisenbergstraße 1, 70569 Stuttgart, Germany, email: yanjaojie95@gmail.com
8. W.P. Kirk and M. Twerdochlib, Improved Method for Minimizing Vibration Motion Transmitted by Pumping Lines, *Review of Scientific Instruments*, 49, 765–769, (1978).
9. Olli V. Lounasmaa, Nuclear Magnetism in Copper, Silver, and Rhodium Metals at Positive and Negative SpinTemperatures in the Nano- and Picokelvin Regimes, *Reviews of Modern Physics*, 69, 1–136, (1997).

10. Shuheng H. Pan, private communication, Institute of Physics, Chinese Academy of Sciences, Beijing, China (span@iphy.ac.cn).
11. Christian Enss and Siegfried Hunklinger, *Low-Temperature Physics*, Springer-Verlag, Berlin Heidelberg, (2005).
12. A. Fleischmann, A. Reiser, and C. Enss, Noise Thermometry for Ultralow Temperatures, *Journal of Low Temperature Physics*, 201, 803–824, (2020).
13. O.M. Hahtela, A. Kemppinen, J. Lehtinen, A.J. Manninen, E. Mykkänen, M. Prunnila, N. Yurttagül, F. Blanchet, M. Gramich, B. Karimi, E.T. Mannila, J. Muhojoki, J.T. Peltonen, and J.P. Pekola, Coulomb Blockade Thermometry on a Wide Temperature Range, Applied Physics (physics.app-ph), 2020 *Conference on Precision Electromagnetic Measurements (CPEM)*, Denver (Aurora), CO, USA, (2020).

11 Sample Probe for High-Voltage Operations at Millikelvins

11.1 SAMPLE PROBE PLATFORM

The objective of this project is to develop a top loader (sample probe) for neutron scattering. The probe needs to be operated in a cryogen-free He-3 cryostat with high voltage capabilities (10 kV DC) at He-3 temperatures [1]. A prototype, developed and tested on a commercial cryogen-free He-3 platform, is illustrated in Figure 11.1. Details of the cryostat precooled by the pulse tube refrigerator are provided in Chapter 1.

A load lock assembly is an essential component for a top-loading sample probe, which allows users to change samples while keeping cryostats at cryogenic temperatures. Details of the load-lock assembly can be found in Chapter 2.

To minimize the heat load to the sample at He-3 temperatures, multiple thermalizing stages are installed on the sample probe to intercept the parasitic heat load. These thermal stages are referred to as 40K stage, 10K stage, 3K stage, and 0.3K stage. The first three stages are thermalized to the first stage, the second stage regenerator, and the second stage of the pulse tube refrigerator, respectively. The 0.3K stage is thermalized to the He-3 pot cold plate, which could reach 0.3 K during regular operations.

A hermetically sealed four-channel model SHV-10kV [2] type feedthrough, rated to 10 kV DC, is installed on the top of the sample probe to provide a reliable high voltage DC signal into the cryostat vacuum space. The four channels are secured with a custom-machined G-10 retainer which is mechanically anchored to the feedthrough flange (Figure 11.2).

11.2 HIGH ELECTRIC VOLTAGE WIRES

The wire assembly needs to meet the following requirements:

- The bare probe must withstand at least 10 kV DC between the wires, the feedthrough, and the test cryostat when the lower ends of the wires left floating during testing on a cryogen-free He-3 system.
- The leakage current between the probe and the test cryostat should be less than 0.01 mA across the full operating range from room temperature to the lowest He-3 temperatures.
- The bare probe should impose minimal heat load so that the He-3 cryostat can reach the base temperature of 0.4 K or lower.

DOI: 10.1201/9781042000050-11

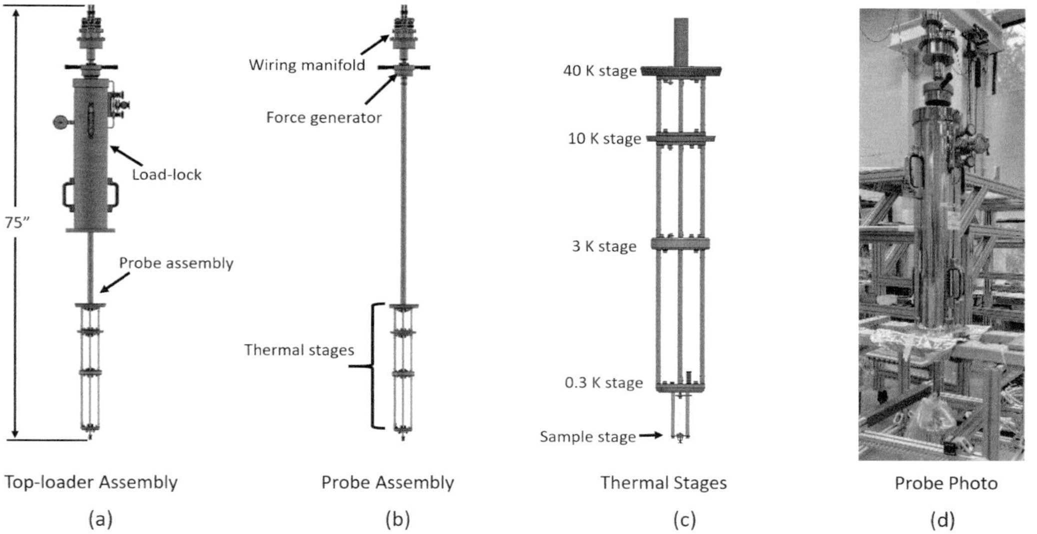

FIGURE 11.1 (a) Top loader assembly with load-lock, (b) probe assembly, (c) thermal stages at different temperatures on the sample probe, (d) photo of top loader assembly with load-lock (Projects accomplished by authors at Janis Research Company. Image courtesy of FormFactor, Inc.).

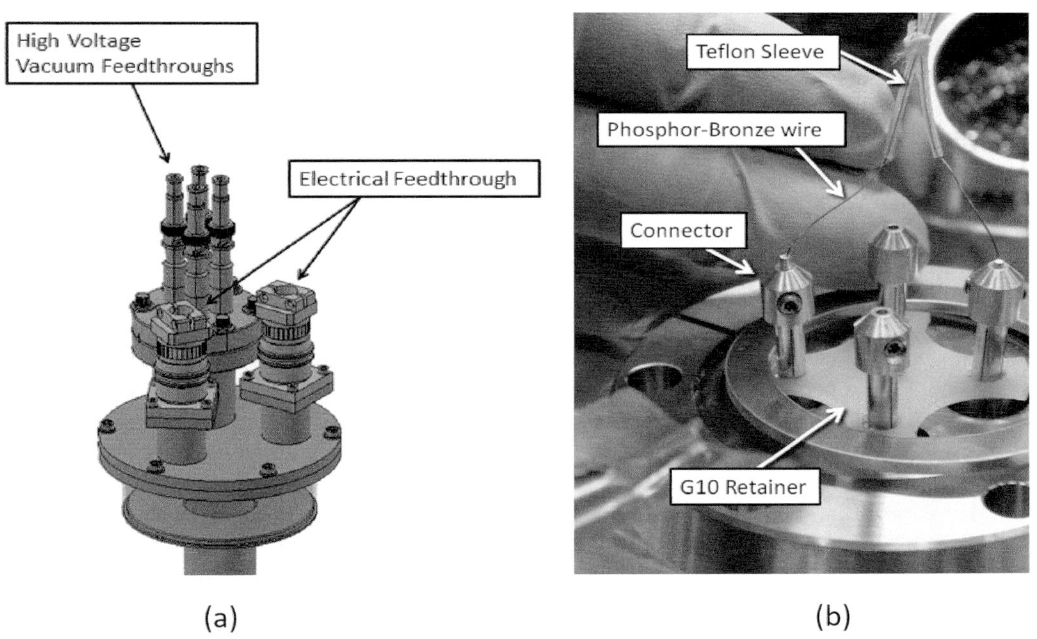

FIGURE 11.2 (a) High voltage electrical feedthrough (SHV-10kV type) on 2.75" CF Flange [2], (b) vacuum-side wire connection and G-10 retainer. (Projects accomplished by authors at Janis Research Company. Image courtesy of FormFactor, Inc.).

TABLE 11.1
The Resistivities of PhBr in μOhm.cm at Different Temperatures

Temperature (K)	10	20	50	77	100	150	200	250	295
PhBr Electrical Resistivity (μΩ·cm)	8.58	8.58	8.69	8.89	9.07	9.48	9.89	10.3	10.7

Source: Data from Jack W. Ekin, Experimental Techniques for Low-Temperature Measurements, Oxford, 2006 [5].

TABLE 11.2
Thermal Conductivity of Sapphire in $W{\cdot}m^{-1}K^{-1}$ at Different Temperatures

Temperature (K)	4	10	20	40	77	100	150	200	300
Sapphire Thermal Conductivity $W{\cdot}m^{-1}K^{-1}$	230	2,900	15,700	12,000	1,100	450	150	82	47

Source: Data from Jack W. Ekin, Experimental Techniques for Low-Temperature Measurements, Oxford, 2006 [5].

0.008" bare phosphorous bronze (PhBr) alloy wires with 0.001" thick polyimide insulation were selected. The typical dielectric strength of the insulation layer is approximately 3.5 KV, which generally increases at cryogenic temperatures [3]. Teflon sleeve is installed around the wires to improve the electrical insulation due to its high dielectric strength (approximately 16 kV/mm) [4]. This wire also provides reasonably good electrical conductance as listed in Table 11.1 for the project. At the same time, its lower thermal conductivity conducts low enough heat and enables the He-3 cryostat to reach the desired base temperature.

11.3 ELECTRIC WIRE THERMAL ANCHOR

Teflon sleeve needs to be removed and replaced by an alternative material at the at the thermal anchor stages for a better "direct" thermal contact. Among available dielectric materials, sapphire (single-crystal Al_2O_3) provides promising thermal conductivity at cryogenic temperatures as well as excellent electrical insulation. Listed in Table 11.2 are the thermal conductivity of sapphire at different temperatures [5]. It is notable that the thermal conductivity of sapphire is of the same order of magnitude as that of OFHC copper at 4 K, which is approximately 630 $W{\cdot}m^{-1}K^{-1}$.

Sapphire also offers high dielectric strength. Its breakdown voltage is approximately 1.2 kV per mil at room temperature [6], implying a 9-mil-thick sapphire plate can theoretically withstand more than 10 kV voltage difference. Furthermore, its breakdown voltage is also expected to increase at cryogenic temperatures. In our experiment, we considered a safety factor of seven (7) on the breaking voltage and used 0.062" thick Sapphire plate. The PhBr wires are clamped between two

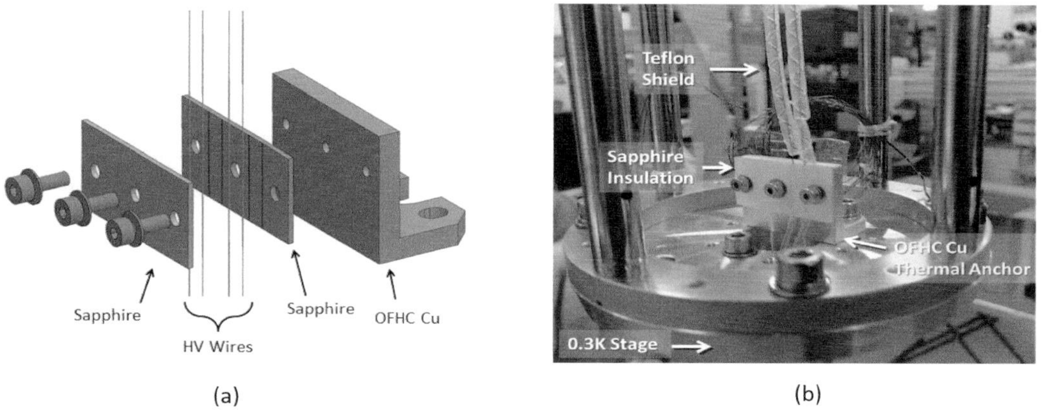

FIGURE 11.3 (a) HV Thermal anchor assembly schematic and (b) HV thermal anchor assembly on He-3 plate of Sample probe (Projects accomplished by authors at Janis Research Company. Image courtesy of FormFactor, Inc.)

sapphire plates at each thermal anchor stage (Figure 11.3), with one plate thermally connected to a gold-plated OFHC copper adapter mounted on the sample probe cold plate.

11.4 PERFORMANCE TEST

Performance testing was carried out using a TDK-Lambda PHV12.5R50 high-voltage DC power supply [7], which can provide up to 12,500 V and 50 mA. Tests were performed at two different temperatures: 3 K and 400 mK.

At 3 K, the He-3 cryostat and He-3 pot were precooled by the PTR, after which the voltage on the sample probe was gradually increased from 0 to 10,000 V over 55 minutes with the power supply current limited to 2 mA. No electrical breakdown occurred, the temperatures at all cryostat stages remained stable, and no leakage current was detected within the 0.01 mA resolution. The 10 kV bias was held for 30 minutes before being ramped down.

In the 400 mK test, the cryostat was cooled to its base temperature, as shown in Figure 11.4. Using the same voltage-ramping procedure, the applied bias was increased to 10 kV and maintained at this level for 15 minutes. Throughout the test, no leakage current exceeding 0.01 mA was detected, and the cryostat temperatures remained stable, indicating robust high-voltage performance under low-temperature operating conditions.

11.5 CONCLUSION

A custom-designed top-loading sample probe was developed for operation in a cryogen-free He-3 cryostat. The probe demonstrated reliable performance down to 400 mK while sustaining a maximum applied bias of 10 kV DC. High-voltage operation did not produce any measurable impact on the cryostat temperatures, and no leakage current above 0.01 mA was observed throughout testing.

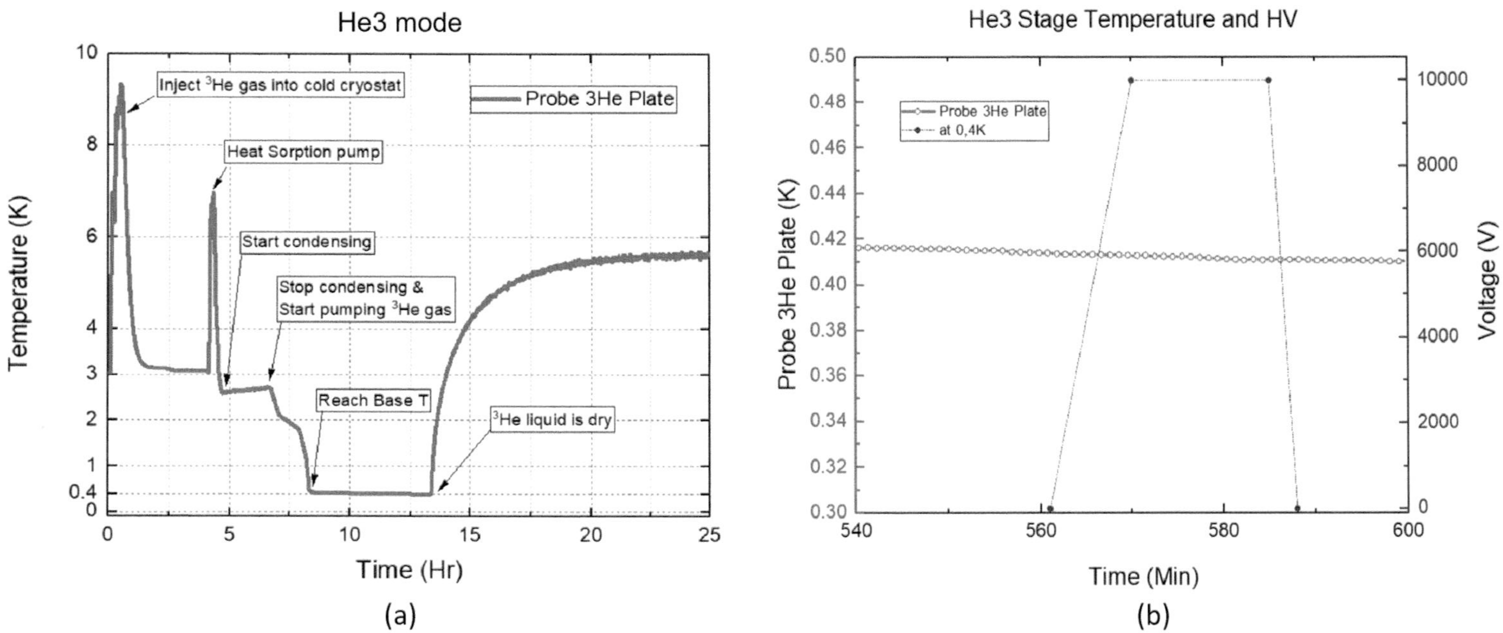

FIGURE 11.4 (a) Cooldown to Base Temperature 0.4 K and (b) high voltage test to 10 kV while sample stage is set at base temperature 0.4 K. (Projects accomplished by authors at Janis Research Company. Image courtesy of FormFactor, Inc.)

REFERENCES

1. Final Tech Report; Principal Investigator: Jeonghoon Ha, *Ultra-Low Temperature Cryostat for Application of High Electric Field in a Neutron Beam*, DOE SBIR Final Technical Report, Woburn, MA, USA (January 29, 2020).
2. MDC Precision, LLC. 30962 Santana Street, Hayward, CA, USA. www.mdcprecision.com
3. NEMA MW 1000 (National Electrical Manufacturers Association – Magnet Wire Standard).
4. John H. Moore, Christopher C. Davis, Michael A. Coplan, with a chapter by Sandra C. Greer, *Building Scientific Apparatus*, Cambridge University Press, Cambridge, England,, UK (2009).
5. Jack W. Ekin, *Experimental Techniques for Low-Temperature Measurements*, Oxford University Press, Oxford, England, UK, (2006).
6. Rayotek Scientific, Inc. 11499 Sorrento Valley Rd, San Diego, CA 9, USA. www.rayotek.com. Rayotek Scientific, Inc. provides technical sapphire material data sheets covering optical, thermal, and mechanical properties for engineering and scientific applications.
7. TDK-Lambda Americas Inc. 405 Essex Road, Neptune, NJ, USA. www.us.lambda.tdk.com

12 Ultra-High Vacuum Technologies

Ultra-high vacuum (UHV) technologies are playing an increasingly important role in industry and scientific research, including the semiconductor industry, surface science, materials science, and related fields [1–6]. UHV is usually defined as vacuum pressure in the range of 10^{-10} torr or lower at room temperature, where the pressure is primarily determined by the outgassing status of components under vacuum rather than by the pumping capacity. To achieve UHV conditions, several factors need to be considered, such as the correct design, appropriate material selection and treatment, proper cleaning and bakeout procedures, and the right pumping station, among others.

12.1 DESIGN PRINCIPLE

Preventing virtual leakage is the most important principle to be followed in UHV system design. Any geometric feature that can trap gas—often referred to as a "pocket" must be avoided. Virtual leakage usually occurs inside joints, crevices, or blind tapped holes. Figure 12.1 illustrates examples of both proper and improper welding joint designs for UHV systems. For blind holes, vented screws or bolts should be used to allow trapped air to escape during UHV pumping, thereby preventing virtual leaks.

12.2 MATERIAL SELECTION AND TREATMENT

UHV systems are typical characterized by two main factors: vacuum level and cleanliness. The achievable vacuum level is determined primarily by the outgassing rate or vapor pressure of the materials, while cleanliness refers to the absence of impurities or contaminants on the surface.

A UHV-compatible material should exhibit a very low outgassing rate, and its vapor pressure should be in the range of 10^{-10} torr or lower at room temperature. In addition to low outgassing, suitable materials should exhibit high temperature resistant during bakeout and low porosity. When a material is used to fabricate the vacuum chamber, it must also provide high mechanical strength and meet the necessary machining, welding, and mounting requirements. Finally, proper cleaning is essential to remove surface impurities before the components are used in UHV systems.

12.2.1 Metallic Material

Austenitic stainless steel (commonly referred to as stainless steel) and OFHC copper are the two most commonly used metals in UHV systems for both room temperature and cryogenic applications.

DOI: 10.1201/9781042000050-12

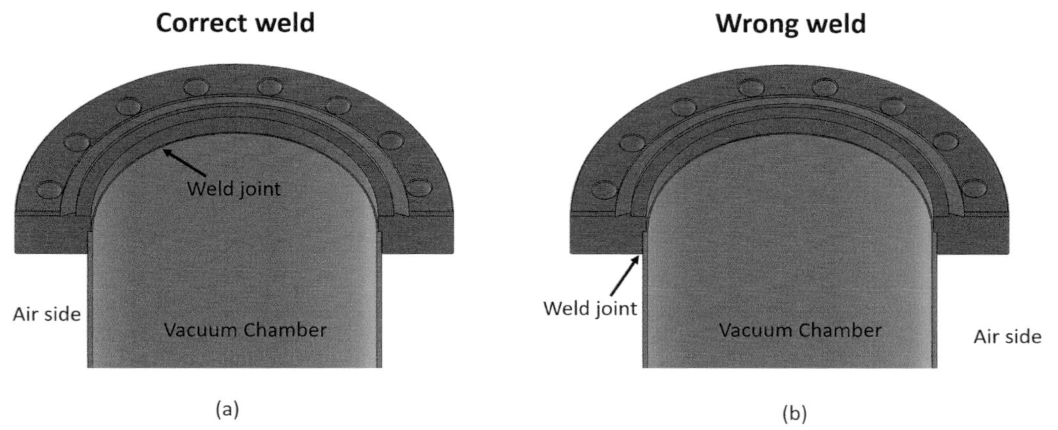

FIGURE 12.1 Welding joint design for UHV systems.

TABLE 12.1
Saturated Vapor Pressures of Metals at Different Temperatures

Metal		Melting Point	Temperature [°C] at vapor pressure [mbar]			
Symbol	**Name**	**[°C]**	**10^{-7}**	**10^{-5}**	**10^{-3}**	**10^{-1}**
Ag	Silver	961	618	742	907	1135
Al	Aluminum	660	745	892	1087	1362
Au	Gold	1064	871	1034	1252	1558
Cu	Copper	1084	773	921	1117	1390
Fe	Iron	1538	946	1110	1324	1618
In	Indium	156	525	644	806	1037
Mo	Molybdenum	2622	1670	1926	2258	2709
Ni	Nickel	1455	992	1157	1371	1662
Pd	Palladium	1554	915	1084	1308	1621
Pt	Platinum	1768	1377	1601	1896	2301
Sn	Tin	231	732	883	1088	1380
Ti	Titanium	1670	1134	1325	1576	1921
W	Tungsten	3414	2240	2560	2973	3527
Zn	Zinc	419	147	207	287	398

Source: Data from Jack W. Ekin, Experimental Techniques for Low-Temperature Measurements, Oxford, 2006 [7].

Titanium and aluminum are also good candidates for many UHV components. For high-temperature UHV applications, tungsten, titanium, and tantalum are particularly suitable due to their exceptionally low outgassing rates even at elevated temperatures. Table 12.1 lists the vapor pressures of several commonly used metals.

12.2.1.1 Copper

Copper components should be gold-plated, with the exception of certain commercial parts such as copper gaskets for ConFlat flanges (see Figure 12.2(a) and (b)). Regardless of the material and coating, all components must be thoroughly cleaned to UHV standards. A common mistake in UHV

(a) (b) (c)

FIGURE 12.2 (a) Gold-plated copper radiation shield, (b) Bare ConFlat flanges with copper gaskets, and (c) Electropolished stainless steel OVC of large dilution refrigerator. (Projects accomplished by authors at Janis Research Company. Image courtesy of FormFactor, Inc.)

fabrication is buffing copper components before gold plating to improve appearance. This process can introduce buffing compounds that are extremely difficult to remove through cleaning or bakeout. Once embedded in the copper, these tiny particles may migrate through the thin gold layer during the later applications. Although the particles may not affect the achievable vacuum level, they can potentially contaminate the UHV environment and potentially compromise sensitive samples.

12.2.1.2 Stainless Steel

Stainless steel components should be electropolished (EP) to smooth microscopic surface roughness and remove potential contaminants embedded during machining. Generally, commercial stainless steel ConFlat components do not require EP since they are manufactured to UHV standard. However, if a ConFlat flange must undergo EP, the knife edge should be carefully protected, typically with tape, to prevent it from being etched. Excessively long EP durations should be avoided. Regardless of the procedure, the knife edge should be inspected thoroughly after the electropolish.

Cryogenic UHV systems, such as low-temperature STM systems, often require use of low- or nonmagnetic materials. For these applications, selected stainless steels, such as 316L and 321L, or 316LN and 321LN, are appropriate choice. Additionally, all stainless steel components should be properly annealed to further reduce the residual magnetism. More details regarding stainless steels of different grades are presented in Section 12.6.1.

12.2.1.3 Titanium and Aluminum

Titanium is another popular material for UHV applications, as illustrated in Figure 12.3. The naturally formed surface oxide layers exhibit very low outgassing rates after proper UHV bakeout, making it well suited for UHV environments. In addition, electropolishing is generally not required when titanium is used in the fabrication of UHV components.

Similarly, after proper cleaning and bakeout, the naturally formed surface oxide layers on aluminum also have very low outgassing rates, make aluminum suitable for UHV applications. For example, Aluminum 6061 was used in our projects to replace expensive gold-plated copper radiation shields in large UHV cryostats. When using aluminum in UHV fabrication, several guidelines are recommended:

FIGURE 12.3 Titanium chamber for ARPES cryostat shield. (Projects accomplished by authors at Janis Research Company. Image courtesy of FormFactor, Inc.)

- Typically, high-purity aluminum alloys are preferred for UHV applications. However, scratches should be avoided during handling because they increase the effective surface area and can increase outgassing rates.
- The bakeout temperature for aluminum components is usually limited to approximately 150 °C, above which the mechanical strength of the material may decrease.
- Silver plating is sometimes applied to aluminum parts, which not only improves the appearance but also makes the surface smoother to reduce outgassing.

12.2.1.4 Brass

As a zinc alloy, brass is widely used in many applications due to its low cost and favorable mechanical properties, including good corrosion resistance and excellent machinability. Brass also offers intermediate electrical and thermal conductivities, making it suitable for certain specialized applications, such as vapor-cooled high-current leads, discussed in Chapter 13. However, zinc has a vapor pressure of approximately 10^{-8} torr at 123 °C (see Table 12.1), which is a typical UHV bakeout temperature. Therefore, brass is considered not UHV-compatible, even though its vapor pressure is very low at room temperature. Consequently, brass components should be avoided in UHV environments.

12.2.2 Nonmetallic Materials

The following nonmetallic materials are generally considered in UHV applications:

- Ceramic: UHV compatible; widely used in high-temperature applications.
- Teflon (PTFE, FEP): Generally UHV compatible, though some users prefer to avoid it in their critical UHV environment.
- Kapton (Polyimide): UHV compatible.
- Kapton Tape (Silicone Adhesive): Acceptable for some UHV applications but should be avoided in critical UHV systems. Kapton tape with acrylic adhesive is generally prohibited.
- PEEK (Polyether Ether Ketone): Generally UHV compatible, except for the glass-filled PEEK, which tends to outgas.

- Vespel SP-1: Generally UHV compatible, but bulk Vespel rods often degas excessively and should be avoided.
- Optical fibers and fiberglass: Acceptable for UHV applications.

12.2.3 Wires, RF cables, and Feedthrough

It is crucial to select proper materials and reliable vendors when purchasing commercial UHV items such as vent screws for blind tapped holes, phosphor-bronze helicoils, electrical feedthroughs, connectors, wires, RF cables with PTFE dielectric, thermometers, heaters, epoxies, and other specialized items. Among these, DC wires and RF cables are often critical for the success of experiments in UHV systems. Although vendors rarely label their products as "UHV Certified," many commercially available wires and cables have been successfully used in UHV-compatible setups. Detailed information on UHV-compatible wires, RF cables and connectors, and recommended vendors is provided in Chapter 17.

12.3 GUIDELINE FOR CLEANING UHV ITEMS

- All welding and machining burns should be removed.
- After EP and gold plating, clean items in a UHV ultrasonic cleaner for at least 10 min using a neutral-pH, residue-free detergent such as Alconox® or Liquinox® [8], to remove chemical residues.
- Smaller items, such as wires, connectors, and screws, can be placed in a beaker filled with acetone, which is then placed in an ultrasonic cleaner for cleaning. For additional cleaning, 2-propanol is recommended after acetone. Each ultrasonic cleaning cycle should last about 10 min.
- Solder flux can be removed by ultrasonic cleaning in warm water, followed by alcohol and acetone.
- Rinse items with deionized (DI) water, regardless of the cleaning method used.
- Dry items immediately after rinsing to prevent water marks.
- Store clean and dry items in plastic bags/containers or wrap them with UHV-compatible aluminum foil.

12.4 BAKEOUT

Bakeout is an essential step in UHV preparation to reduce the surface desorption of impurities and remove the adsorbed gas molecules. UHV systems are typically baked at 150–250 °C for days, although the exact temperature and duration depend on the specific system and the desired ultimate pressure. It is notable that RF cables require particular attention and thorough bakeout to remove water that can be trapped in the dielectric material between the outer and inner conductors of coaxial cables.

12.5 MACHINING TOOLS AND COOLANT

For the fabrication of UHV components, it is highly recommended to maintain a dedicated set of clean machining and cutting tools to avoid cross-contamination from non-UHV work. Even trace residues from oils, greases, or shop debris can significantly increase outgassing and compromise vacuum performance.

Listed below are two commonly used cutting coolant options suitable for machining UHV parts:

a) Hocut® 795 B
   Heavy-duty machining & grinding fluid for all metals
   Supplier: DTS Industrial Supply
   266 Karl Street Berea, Ohio USA
   https://dtsindustrial.com/hocut-795-b.html
b) Kool Mist Formula #77-Mist Spray Coolant
   Recommended mixture: 4.0 fl. oz. concentrate per 1.0 gallon of water.
   Supplier: Kool Mist
   218 Norwalk Blvd, Santa Fe Springs, CA 90670, USA
   https://koolmist.com/

12.6 UHV AND CRYOGENICS

Since the late 20th century, the scientific community has demanded UHV systems operated at millikelvin temperatures for advanced researches [9, 10]. Examples include STM with atomic resolution operated at low millikelvin temperatures and ARPES to study materials with transition temperatures below 1 K.

12.6.1 Challenges

Cryogenics is a double-edged sword for vacuum systems. On one hand, cold surfaces provide effective cryopumping, naturally adsorbing residual gases and improving vacuum level. On the other hand, the sample is one of the coldest items in the system and therefore has the highest absorptivity for residual gases and impurities. This makes sample particularly vulnerable to contamination. Additionally, impurities released during a warm sample loading may be readsorbed by the sample afterward.

Material selection for cryogenic UHV environments is also more demanding. Although the general UHV design and fabrication remain the similar to those used for room temperature systems, material selection and handling need more cautious. Materials that perform well at room temperature may lose those characteristics at low temperature, for example, becoming brittle. Additionally, most of cryogenic UHV systems require low- or nonmagnetic components, therefore 316L and 321L stainless steel, rather than standard 316 and 321 grades, are commonly used for these systems [11].

- Grade 316L stainless steel is the low carbon version of 316 and is immune to grain boundary carbide precipitation (sensitization). It is the most popular stainless steel for millikelvin UHV systems.
- Grade 321L stainless steel is the low carbon version of 321, offering excellent resistance to intergranular corrosion though is used less frequently than 316L.
- Grade 316LN stainless steel, enriched with nitrogen, exhibits even lower magnetic permeability and is often used for components located near the sample, such as the IVC can.

12.6.2 Examples of milli-Kelvin UHV systems

Over the past several decades, we have applied the UHV techniques described above to develop multiple millikelvin UHV systems for various scientific applications [12, 13]. All of these systems achieved the vacuum levels on the order of 10^{-10} torr at room temperature. They were baked at different temperatures for varying durations and evacuated using different pumping configurations. Representative examples and key parameters are summarized in Table 12.2 for reference.

TABLE 12.2
Summary of Cryogenic UHV Systems with the Sample Temperatures, the Bakeout Temperatures, the Bakeout Time, the Achieved Pressures at Room Temperature, and the Pump Configurations.

System	Sample Temperature (mK)	Bakeout Temperature (°C)	Bakeout Time (h)	Achieved Pressure at RT (torr)	Pump Configuration
Sub-K angle-resolved photoelectron spectroscopy (ARPES)	640	90	408	2.3×10^{-10}	2× Pfeiffer TMU 521 2× Pfeiffer Hipace 80
Top-loading He-3 cryostat for STM application	300	150	48	1.0×10^{-10}	1× Pfeiffer TMH 260 2× Varian Vaclon 1× 150 diode ion pump
Bottom-loading He-3 cryostat (1) for STM application	300	< 70	300	1.0×10^{-10}	3× Boc Edwards STP 301 turbo 3× Gamma Vacuum 500L/s ion pump
Bottom-loading He-3 cryostat (2) for STM application	300	120	24	5.0×10^{-10}	1× Varian V250 1× Varian V70 1× Varian Starcell 125 ion pump 1× Lesker titanium sublimation pump
Bottom-loading dilution refrigerator for STM application	12	N/A	N/A	In the range of 10^{-10}	1× Pfeiffer HiPace 80 1× Vacuubrand MV10 1× Leybold Turbovac 340M

Source: Data from communications between authors and users.

REFERENCES

1. P.A. Redhead, J.P. Hobson, and E.V. Kornelsen, *The Physical Basis of Ultrahigh Vacuum*, American Institute of Physics, Melville, NY, USA, (1993).
2. J.M. Lafferty (Ed.), *Foundations of Vacuum Science and Technology*, John Wiley & Sons, New York, NY, USA, (1998).
3. L.N. Rozanov (Ed.), *Vacuum Technique*, CRC Press (Taylor & Francis Group), Boca Raton, FL, USA, (2002).
4. H.F. Dylla, The Development of Ultrahigh and Extreme High Vacuum Technology for Physics Research, Thomas Jefferson National Accelerator Laboratory, Newport News, Virginia 23606, USA, *Journal of Vacuum Science & Technology A*, 21, S25–S33, (2003).
5. A. Roth, *Vacuum Technology*, North Holland (Elsevier), Amsterdam, Netherlands, (2012).
6. Zuyu Zhao and Chao Wang (Eds.), *Cryogenic Engineering and Technologies: Principles and Applications of Cryogen-Free Systems*, CRC Press (Taylor & Francis GroupBoca Raton, FL, USA, (2020).
7. Jack W. Ekin, *Experimental Techniques for Low-Temperature Measurements*, Oxford University Press, Oxford, United Kingdom, (2006).
8. Alconox, LLC. Address: 30 Glenn St., Suite 309, White Plains, NY 10603, USA. Tel: (914) 948-4040. Email: cleaning@alconox.com Website: www.alconox.com
9. Oswald Gröbner, *Vacuum and Cryopumping*, CERN (European Organization for Nuclear Research), Geneva, Switzerland, 396–409, (2004).

10. Oswald Gröbner, *Vacuum System*, CERN (European Organization for Nuclear Research), Geneva, Switzerland, 331–342, (2006).
11. Outokumpu Oyj, *Handbook of Stainless Steel*, Outokumpu, Oyj, Espoo, Finland, (2013).
12. Andreas Mandelis, Focus on Vacuum and Cryogenics, *Physics Today*, 67(10), 60–62, (2014).
13. O. Korneta, C. Diep, V. Luppov, and Z. Zhao, Breakthrough Instruments and Products Janis Research Company, LLC Cryogen Free Sub-Kelvin ARPES System, *Review of Scientific Instruments*. 91, 059501, (2020).

13 Vapor-Cooled Current Leads for High-Field Magnets

In modern scientific researches, high-field magnets are commonly built from low-Tc superconductors operated in liquid helium. The current leads that supply electrical power to these magnets play a critical role in overall system performance. Beyond simply delivering current, they must be carefully designed to minimize the heat load to liquid helium, as excessive heat transfer increases cryogen consumption and reduces operating efficiency. A high cryogen boil-off rate can also introduce vibration to the magnet, posing additional risk of magnet quench, especially when operating at high field.

Current leads introduce heat into the helium bath through two primary mechanisms: conductive heat load from room temperature, and Joule heating produced by the electrical current flowing through the leads. To mitigate these heat loads, vapor-cooled brass tubes are commonly used as current leads. Brass provides a practical balance between electrical resistivity and thermal conductivity, enabling effective current transport while limiting parasitic heat flow to the cryogen.

This chapter presents the designs of vapor-cooled brass current leads that have been widely implemented in our cryogenic products. These leads are designed to operate at currents ranging from 25 A to 150 A, making them suitable for most laboratory applications.

13.1 BACKGROUND

13.1.1 Brass Thermal Conductivity

Brass is an alloy primarily composed of copper and zinc. As shown in Table 13.1, the average thermal conductivity of brass from room temperature to liquid helium temperatures is only about 12% of that of pure copper, which is the most commonly used materials for electrical conductors. This substantially lower thermal conductivity indicates that, for components of identical dimensions, conductive heat transfer from room temperature to liquid helium is much lower in brass than in copper.

13.1.2 Vapor Cooling

The evaporated cold helium vapor flows through the brass tube, intercepting part of the heat from the current leads. Effective vapor cooling is crucial to minimizing helium loss. The enthalpy changes of helium gas from 4 K to 300 K at 1 atm is approximately 1,541.83 J/g, as listed in Table 13.2. This substantial enthalpy change allows the helium vapor to absorb a significant amount of heat from the current leads, thereby reducing the thermal load on the liquid helium reservoir. For example, when one mole (4.0 g) of liquid helium—approximately 40 ml.—boils off and the resulting vapor warms from 4.2 K to room temperature within one hour, it provides approximately 1.71 W of cooling power.

DOI: 10.1201/9781042000050-13

TABLE 13.1
Average Thermal Conductivity of Brass and Copper in W·m^{-1}·K^{-1}

Material	Copper		Brass	
Temperature range (K)	300–77	300–4	300–77	300–4
Average thermal conductivity (W/cm·K)	4.1	5.7	0.81	0.67

Source: Data from author's note.

TABLE 13.2
Enthalpy Change of Helium Vapor from 4 K to 300 K Under 1.0 Atmospheric Pressure

1,542 J·g^{-1}	6,168 J·mol^{-1}

Source: Data from Jack W. Ekin, *Experiment Techniques for Low Temperature Measurements,* Oxford University Press, 2006 [1].

From these considerations, several conclusions can be drawn:

- Helium vapor provides substantial cooling capacity. As it warms up from cryogenic to room temperatures, it effectively absorbs heat from the current leads.
- Vapor cooling operates as a convective heat exchange process. The brass tube is designed with optimized geometry to maximize surface area for convection. To further enhance convective efficiency, a tube made of Teflon or nylon can be inserted inside the brass tube. This inner tube, with a smaller diameter than the brass tube's inner diameter and sealed at the lower end, forces the cold helium vapor to flow through the narrow annular gap between the two tubes, improving convective heat exchange.
- Brass tubes offer additional benefits. They are cost-effective and nonmagnetic, and therefore suitable for use in high magnetic field environments.
- Vapor cooling effectively intercepts Joule heating in the current leads. This resistive heating occurs only during the magnet energizing and deenergizing phases and is absent when the magnet operates in the persistent mode.

13.2 BRASS LEAD DESIGN

13.2.1 General Guide Model

Figure 13.1 illustrates the steady-state thermal balance between helium vapor and magnet current leads in a vapor-cooled system. This image shows how evaporated helium gas ascends through the brass current lead, absorbing heat by convective exchange from both conductive heat load and Joule heating.

Equation (13.1) represents this steady-state thermal balance, incorporating conductive heat transfer, ohmic heating, and convective cooling by helium vapor. The design of current leads is based on this relationship, which models heat transfer to the liquid helium under the assumption of ideal convective cooling—that is, the vapor temperature is assumed to be equal to the lead temperature at every point along the lead.

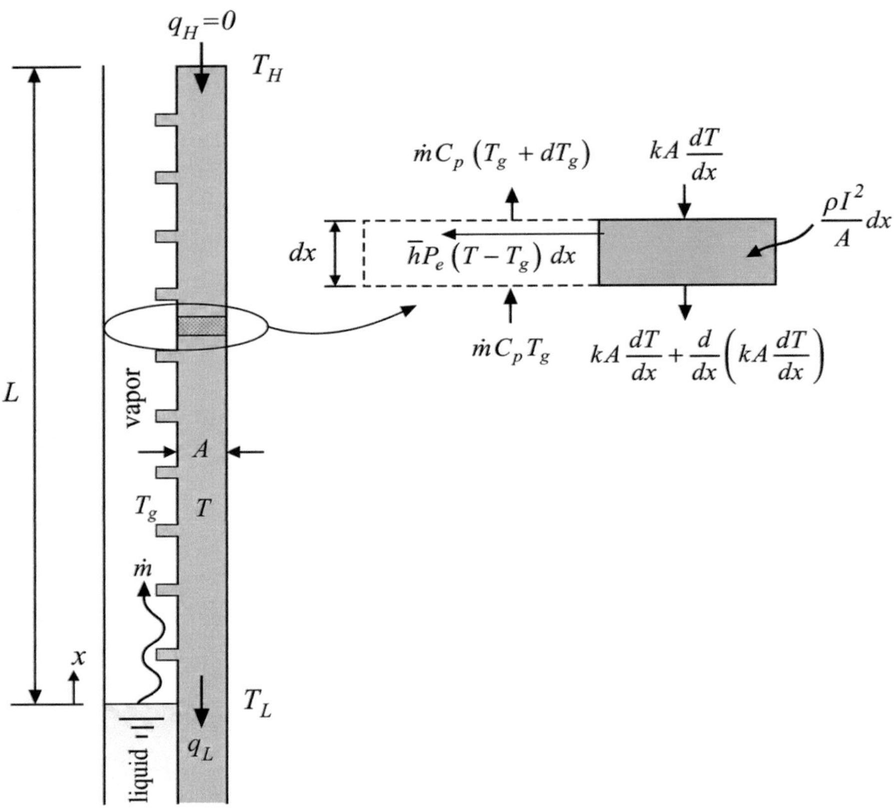

FIGURE 13.1 Energy balance in a vapor-cooled current lead. (Reprint with permission from Ho-Myung Chang, Jung Joo Byun, Hong-Beom Jin, *Effect of convection heat transfer on the design of vapor-cooled current leads,* Cryogenics 46 (2006) 324–332.) [2]

$$\frac{d}{dx}(A\kappa(T)\frac{dT}{dx})-\dot{m}C_p(T)\frac{dT}{dx}+\frac{\rho(T)I^2}{A}=0 \tag{13.1}$$

where

- T: Temperature of the current leads (as a function of position)
- dx: Differential length element
- A: Cross-sectional area of the current lead
- κ(T): Temperature-dependent thermal conductivity of the current lead
- $\dot{m}$: Helium mass flow rate
- I: Current in the leads
- ρ(T): Temperature-dependent electrical resistivity of the current lead
- C_p (T): Heat capacity of cold vapor. Helium vapor is treated as an ideal gas, which has a constant heat capacity of 5.19 $J \cdot g^{-1} \cdot K^{-1}$ under 0.1 MPa.

The first term in Equation (13.1) represents conductive heat transfer from ambient temperature down to the liquid helium. The second term accounts for heat removal by the cold helium vapor flow, and the third term corresponds to Joule heating generated by the current in the lead. Under steady-state operating conditions, where the heat removed by the mass flow equals the combined conductive and Joule-heating inputs, the total energy balance is zero. This equation serves as the foundation for the design of vapor-cooled current leads.

13.2.2 Brass Tube Dimensions Estimation

A rough preliminary estimation of the brass tube dimensions is necessary to initiate the design. The total heat load to liquid helium is given by Equation (13.2)

$$Q = \kappa \frac{A}{L} \Delta T + \rho \left(\frac{A}{L}\right)^{-1} I^2 \tag{13.2}$$

where:

- κ (W•cm^{-1}•K^{-1}): average thermal conductivity of the brass leads over the relevant temperature range,
- I (A): electric current in the leads,
- A (cm^2): cross-sectional area of the current leads,
- L (cm): distance between the inner surface of the dewar's top flange and the surface of the liquid helium,
- ρ (Ω•cm): average electrical resistivity of the brass leads over the relevant temperature range (Table 13.3).
- ΔT: temperature difference along the brass tube (see the note below)

Note that, in the liquid helium dewar filled with cold vapor, the top of the current lead usually is much colder than that room temperature, therefore ΔT is much smaller than 296 K. The actual conductive heat load should be estimated from the actual temperature gradient along the vapor-cooled leads.

The minimum heat load from the brass tube is obtained by setting ∂Q/∂(A/L) = 0, yielding

$$\frac{A}{L} = I \left(\frac{\rho}{\kappa \Delta T}\right)^{\frac{1}{2}} \tag{13.3}$$

Using the thermal conductivity and electric resistivity values in the range of 4–300 K listed in Tables 13.1 and 13.3, the resulting ratio A/L is typically in the range of 10^{-3} to 10^{-4} cm times the maximum operating current.

This estimation approach provides a guideline for the initial design only and does not decide final and highly accurate optimized parameters. Detailed thermal and electrical modeling, including the real temperature profile of the vapor-cooled brass tubes, is required for final design. Furthermore, the practical design options are also constrained by the availability of specific commercial brass tubing.

TABLE 13.3
Electrical Resistivity of the Brass at Different Temperatures

Temperature (K)	296	77	4.2*	300 ~ 4**
Brass electrical resistivity (μΩ·cm)	7.2	4.7	1.0	3.6 (in average)

Source: Data from Jack W. Ekin, *Experiment Techniques for Low Temperature Measurements,* Oxford University Press, 2006 [1].
*This value varies depending on the alloy's purity and exact composition.
**Measured value by author on the brass tube for this project

Based on the commercial brass tubing availability and the typical helium dewar dimensions, we selected the following tube to design our first pair of vapor-cooled brass current leads, rated up to 60 A.

- Outer diameter: 0.25 in (0.635 cm)
- Wall thickness: 0.032 in (0.081 cm)
- Cross-sectional area per tube: 0.162 cm^2 (total for the pair: 0.324 cm^2)
- Total length: 90 cm
- Typical length above the liquid helium (not immersed): 76 cm (30.0 in)

Using the above-liquid length L = 76 cm (the section where vapor-cooling is effective) gives the A/L of 4.26×10^{-3} cm, which is close to that estimated using Equation (13.3).

13.3 VAPOR-COOLED BRASS CURRENT LEAD EXAMPLE

13.3.1 Brass Leads with Superconducting Wires

The vapor-cooled brass high-current leads typically begin at room temperature and terminate approximately 1.5 in above the top of the magnet (Figure 13.2(a)). Each lead has five drilled holes that facilitate the entry and exit of cold helium vapor. Three of these holes (designated as *Hole A*) are located near the liquid helium surface, while the remaining two holes (*Hole B*) are positioned approximately 3.0 in below the upper limit of the brass tube assembly inside the dewar. These holes have an approximate diameter of 0.1 inches and are spaced 1.0 in apart, as shown in Figure 13.2(a).

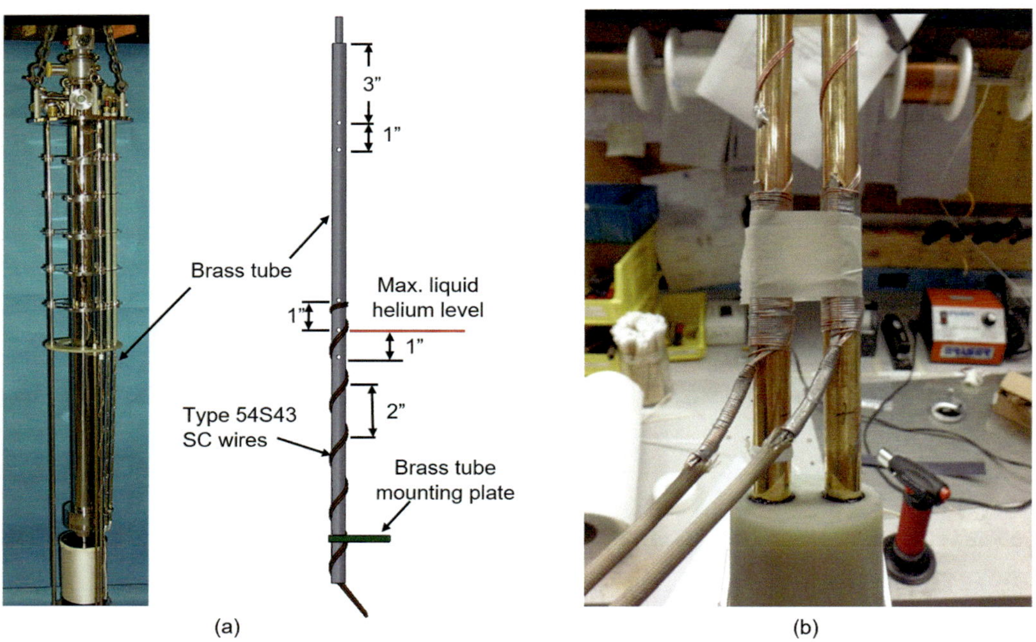

(a) (b)

FIGURE 13.2 (a) Vapor cooled brass magnet high current leads design, and (b) connection between the brass leads and the magnet leads. (Projects accomplished by authors at Janis Research Company. Image courtesy of FormFactor, Inc.)

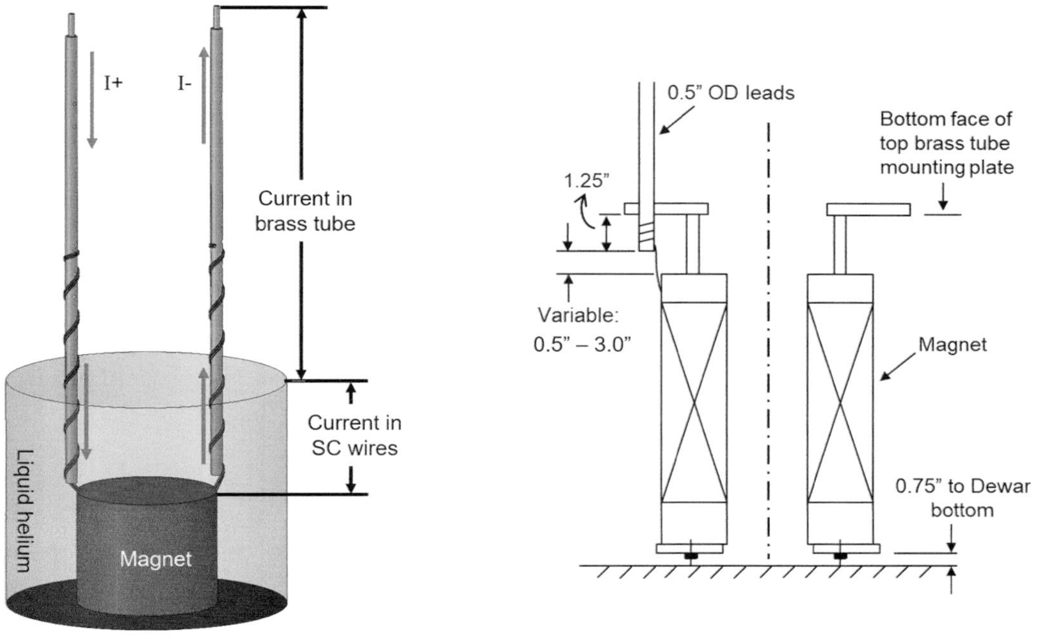

FIGURE 13.3 Current flow in the vapor cooled brass current leads.

In operations, cold helium vapor ideally enters the brass tube through Hole A, exchanges heat with the current lead through convection cooling, and exits through Hole B.

A copper rod is soldered to the top of each brass tube to provide an electrical interface for connecting the power supply cables. This rod also seals the brass tube, preventing helium vapor from escaping into the ambient environment.

Cu/SC type 54S43 bare multifilament superconducting wires with approximate 0.76 mm diameter [3] are used to connect the brass leads to the superconducting magnet leads extending from the end of the magnet winding. This wire is composed of 54 NbTi superconducting filaments embedded in a copper matrix. Each filament has an approximate diameter of 0.060 mm. The copper-to-superconducting material ratio is 1.3:1. The connection wires originate from the middle Hole A and extend several inches below the bottom of the brass tube. Wires spiral around the tube with a pitch of approximately one turn every two inches and are spot-soldered to the brass using 60/40 solder [4]. At the lower end of the brass tubes, the connecting wires overlap with the magnet leads for a few inches and are soldered together using the same 60/40 solder, as illustrated in Figure 13.2(b).

Type 54S43 wires remain superconducting when immersed in liquid helium, allowing them to bypass the brass tube and carry current without generating Joule heating. However, when the liquid helium level drops and the wires become exposed to helium vapor, part of the wires transition to the normal (resistive) state. The current would be shared by the normal 54S43 wire and the brass tubes, resulting in some Joule heating (Figure 13.3).

Since the critical current of Type 54S43 superconducting wire depends on the local magnetic field (refer to Table 13.4), multiple wires are typically bundled together to mitigate the risk of quenching in the magnet's fringe field. A greater number of Type 54S43 wires is required for higher operating currents. The exact number of wires soldered onto the brass tube is determined by the maximum operation current of the magnet.

TABLE 13.4
Critical Field of Type 54S43 Wire with 0.076 mm in Diameter

Magnetic Fields (T)	Critical Current of Type 54S43 Wire with 0.076 mm in Diameter (A)
3	710
5	435
7	280
9	90

Source: Data from Supercon Inc. www.supercon-wire.com/content/fine-sc-wires [3].

13.3.2 Performance Test

Brass current leads for the Test:

- Outer diameter: 0.25 in (0.635 cm)
- Wall thickness: 0.032 in (0.082 cm)
- Total length inside the dewar: 35.5 in (90 cm)

Test dewar specifications

- Dewar neck length: 17.5 in (44.5 cm)
- Helium reservoir height: 24 in (61.0 cm)
- Liquid helium level during the test: approximately 40% of reservoir height (9.6 in/24.4 cm)
- Length of brass current leads exposed to helium vapor: ~31.9 in (81 cm)

The experimental setup is shown in Figure 13.4. The lower ends of the brass tubes were connected using Type 54S43 superconducting wire, forming a superconducting shunt between the two vapor-cooled leads. The background heat load, conductive heat load from brass leads, and Joule heating are measured with the following steps, and the test results are summarized in Table 13.5.

Step 1: Background heat load measurement

- Raise the brass leads so that the bottoms of the brass tubes are above the liquid helium level.
- Measure the helium boil-off rate resulting from the parasitic heat load of the test setup.
- The measured liquid helium boil-off rate was 240 mL/h, corresponding to a background heat load of approximate 0.169 W on the liquid helium.

Step 2: Brass tube conductive heat load measurement

- Lower the brass tubes into liquid helium as illustrated in Figure 13.4, while no electrical current was applied.
- The helium boil-off rate remained unchanged, indicating that the conductive heat load from the vapor-cooled brass tubes was very small and is intercept by the cold helium vapor (see the discussion below).

Step 3: Brass tube Joule heating measurement

- When electric current was applied to the brass leads, the liquid helium boil-off rate increased.
- The helium boil-off rate increased monotonically with increasing the applied current.

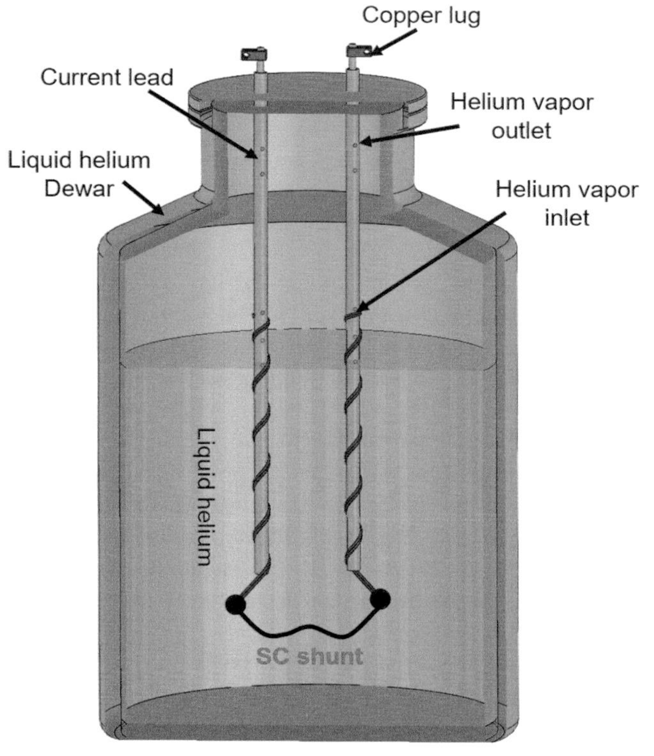

FIGURE 13.4 Set up of vapor cooled brass current leads test.

TABLE 13.5
Test Results

Liquid Helium Level	Current lead Length above the Helium Liquid	Current Applied to the Brass Tubes (I)	Additional Helium boil-off Rate above to the Background Boil-off Rate	Un-intercepted Joule Heating that Caused Additional Boil-off (Q_A)	Voltage across the Pair of Brass Tubes (V)	Total Electrical Power Applied to the Brass Tubes (Q_E = IV)
(%)	(inch/cm)	(A)	(c.c./hour)	(W)	(V)	(W)
41.30%	31.59/80.23	0	0	0	0	0
40.60%	31.76/80.67	25	48	0.035	0.18	4.5
39.80%	31.95/81.15	50	120	0.087	0.19	9.5
39.00%	32.14/81.63	80	280	0.203	0.20	16.0

Note that one (1) watt heat load boils approximately 1,380 c.c. liquid helium per hour.

Source: Data from author's note.

13.3.3 Discussions and Conclusions

13.3.3.1 Conductive Heat Load Interception

The fact that the helium boil-off rate does not change when the brass tube lower end is submerged in liquid helium indicates that the enthalpy carried away by the boiled-off helium vapor generated from the background heat load (0.169 watt) is sufficient to intercept all conductive heat load through the brass tubes. It is noted that the helium vapor vented from the dewar is typically much colder than ambient temperature, indicating that the vented vapor retains surplus enthalpy that can also be utilized for vapor cooling.

A significantly higher helium boil-off rate is expected while ramping the current. During this process, users should direct as much of the vented cold helium vapor as possible through the current leads. This practice maximizes convective heat interception, reduces the thermal load on the cryogenic system, and minimizes the risk of overheating the brass leads.

13.3.3.2 Voltage Behavior During Current Ramp

No significant voltage increase was observed across the brass leads as the current increased. This suggests that the effective electrical resistance of the lead assembly decreases when the helium boil-off rate increases. Two mechanisms contribute to this effect: a greater length of the Type 54S43 wires becoming superconducting, and lower brass electrical resistance in higher and colder vapor flows. Both factors contribute to a relatively stable voltage across the brass leads as the current increases.

13.3.3.3 Under- or Over-Design of Brass Current Leads

A significantly higher-than-expected voltage across the brass current leads indicates an under-designed lead, where the brass A/L ratio is too small and the electrical resistance is too high. Excessive Joule heating not only increases the helium boil-off rate significantly but also poses a risk of quenching the superconducting wires. To mitigate this, brass tubes with a larger A/L ratio should be used.

Conversely, a much lower voltage suggests an over-designed lead, where the A/L of the brass is unnecessarily large. This leads to an excessive conductive heat load to the cryogenic system. For improved efficiency, the A/L of the brass tubes can be reduced to the optimized range.

13.3.3.4 Pros and Cons of 54S43 Superconducting Wire

The use of Type 54S43 multifilament superconducting wire as the connecting conductor between the brass tubes and the magnet leads offers several advantages as well as notable limitations. These wires are easy to install, cost-effective, and allow flexible adaptation to different operating-current requirements. However, their performance is sensitive to mechanical vibration, particularly when the magnetic field is changing, which can trigger a quench either in the wire itself or in the magnet. In addition, the relatively low critical field of Type 54S43 makes it unsuitable for higher-field magnet systems. Therefore, although Type 54S43 wires are convenient and practical for many applications, their use must be evaluated carefully when operating in environments with strong or variable magnetic fields.

13.3.3.5 Commercial Applications

As listed in Table 13.6, our commercial vapor-cooled brass current leads are classified into four categories defined by the maximum current rating. The parameters for each category are selected to cover a specific range of currents for commercial applications. Since these current leads are intended for reliable field use, their designs are intentionally conservative, providing proper safety margins against overheating and quenching. For operating currents above 100 A, an alternative vapor-cooled brass current-lead design has been developed in which the Type 54S43 superconducting wires are replaced by a superconducting bus bar, offering improved stability and higher current-carrying capability.

TABLE 13.6
Tube Dimensions, Current, and Number of SC Wires

Maximum Current	Tube Outer Diameter	Tube Wall Thickness	# of SC Wires
60A or lower	0.25"	0.032"	1
61A to 80A	0.50"	0.032"	2
81A to 100A	0.50"	0.032"	3
Above 100A	Refer to Section 13.4		

Source: Data from author's note.

13.4 BRASS LEADS WITH SUPERCONDUCTING BUS BAR

13.4.1 Superconducting Bus Bar

A superconducting bus bar is an electrical conductor composed of a superconducting material that is thermally and electrically bonded along its entire length to a highly conductive metal, such as copper or aluminum [5, 6]. The conductive metal serves as a stabilizer, providing an alternate current path if the superconductor transitions to the normal (resistive) state. This stabilization prevents localized overheating or quenching, thereby enhancing safety and reliability. Superconducting bus bars allows the transmission of large electrical currents in cold helium vapor environments without quenching and are commonly used in our systems with superconducting magnets operating above 12 T and carrying currents above 100 A.

Although a superconducting bus bar is similar to superconducting magnet leads, a key distinction lies between these two items. Magnet leads are a direct extension of the superconducting magnet winding and are made from the same superconducting material—such as niobium-titanium (NbTi) or niobium-tin (Nb_3Sn). These leads must remain immersed in liquid helium during high-current operation to prevent quenching, whereas a superconducting bus bar is a separate stabilized conductor designed for operation in cold helium vapor.

13.4.2 Design and Fabrication

In this design, Type 54S43 superconducting wires are replaced by a superconducting bus bar. The fabrication procedures for the vapor-cooled brass current lead with a superconducting bus bar are outlined below (also see Figures 13.5–13.8):

- Terminate the brass tubes several inches below the top of the helium reservoir belly.
- Pre-tin the bottom ends of the brass tubes and the OFHC copper adapter (see Figure 13.5) using 430 solder [7].
- Solder the OFHC copper adapter to the pre-tinned end of the brass tubes using 430 solder.
- Pre-tin the flat surfaces of both the superconducting bus bar and the OFHC copper adapter using 60/40 solder.
- Solder the flat surface of the superconducting bus bar to the OFHC copper adapter.
- Wrap the superconducting bus bar and the OFHC copper adapter together with bare copper wire (see Figure 13.6).
- Solder the superconducting bus bar, copper adapter, and bare copper wire together using 60/40 solder to ensure a secure, low-resistance joint.
- Thoroughly clean all flux residues from the assembly.

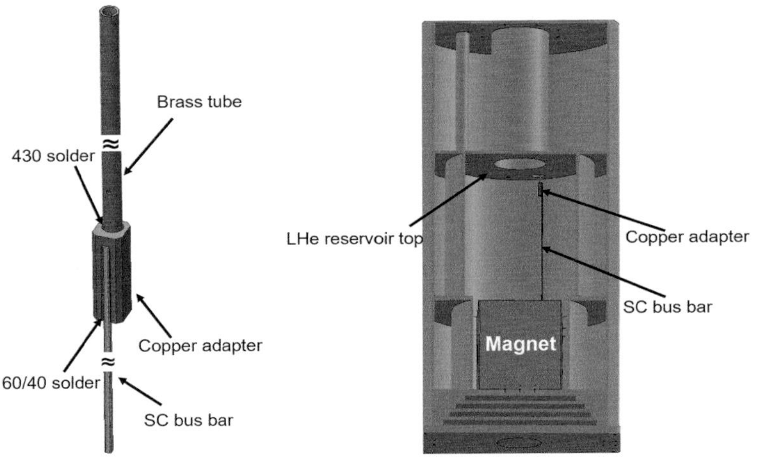

FIGURE 13.5 Vapor cooled brass high current lead with SC buss bar.

FIGURE 13.6 Vapor cooled brass magnet high current lead with SC buss bar. (Projects accomplished by authors at Janis Research Company. Image courtesy of FormFactor, Inc.)

- Wrap insulation tape around the joint between the superconducting bus bar and the copper adapter (see Figure 13.7).
- Terminate the superconducting bus bar approximately 3-4 inches below the top of the magnet to allow sufficient overlap with the magnet leads.
- Wrap the bus bar-magnet lead overlapping section with bare copper wire, then solder the assembly together using 60/40 solder (see Figure 13.8).
- Wrap insulation tape around the joint between the superconducting bus bar and the bare copper wire.

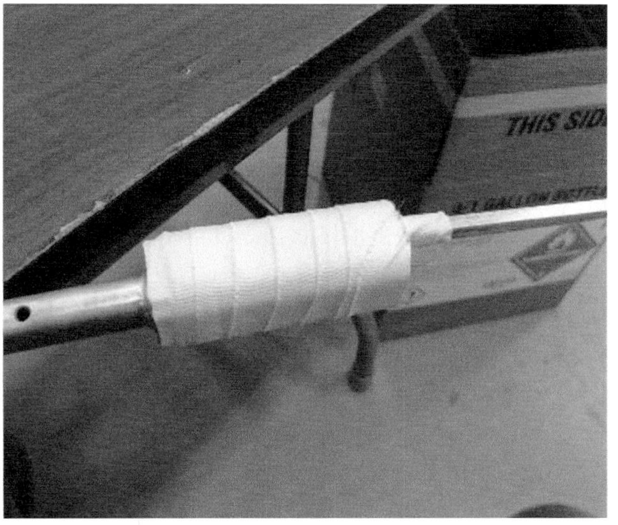

FIGURE 13.7 Cover the joint with insulation tapes (Projects accomplished by authors at Janis Research Company. Image courtesy of FormFactor, Inc.)

FIGURE 13.8 Connect bus bar to the magnet lead. (Projects accomplished by authors at Janis Research Company. Image courtesy of FormFactor, Inc.)

REFERENCES

1. Jack W. Ekin, *Experiment Techniques for Low Temperature Measurements*, Oxford University Press, Oxford, England, UK, (2006).
2. Ho-Myung Chang, Jung Joo Byun, and Hong-Beom Jin, Effect of Convection Heat Transfer on the Design of Vapor-Cooled Current Leads, *Cryogenics*, 46, 324–332, (2006).
3. Supercon, Inc 830 Boston Turnpike, Shrewsbury, MA, USA Email: sales@supercon-wire.com Website: www.supercon-wire.com/content/fine-sc-wires
4. 60/40 solder is a near-eutectic tin–lead (Sn–Pb) alloy (60% Sn, 40% Pb) with a melting range of 361°F–374°F (183°C–190°C). It is widely used for electronics, wiring, and general soft soldering due to its good wetting and ease of use.

5. L. Belova, M. Genet, J.-L. Perinet-Marquet, P. Ivanov, and C. Urpin, Design and Manufacture of the Superconducting Bus-Bars for the LHC Main Magnets, *IEEE Transactions on Applied Superconductivity*, 12(1), 1305–1309, (2002).
6. Product: Superconducting bus bar pair Manufacturer: Cryomagnetics, Inc. Length: 24 in Current class: 50–150 A systems Function: Low-loss cryogenic interconnect between current leads and superconducting magnet Website: cryomagnetics.com/products/50-150amp-superconducting-buss-bar-24-pair.
7. 430 solder is a high-temperature 96% Sn–4% Ag alloy with a melting temperature of 430°F (221°C). It is commonly used for superconducting joints, vacuum systems, and other high-reliability applications. Typical uses include copper-to-copper and stainless-steel-to-copper joints (with suitable flux).

14 Copper Thermal Link Fabrication

14.1 COPPER THERMAL LINKS

Copper is a widely used material in cryogenic engineering due to its good thermal conductivity at cryogenic temperatures. Because of the reduced electron–phonon scattering at cryogenic temperatures, the thermal conductivity of copper exhibits a pronounced increase, peaking near 20 K, as illustrated in Figure 14.1.

Braided copper is one of the most commonly used flexible thermal links [1]. Figure 14.2 shows typical thermal straps made of highly flexible fine copper wires by Technology Applications, Inc. (TAI) [2]. As an example, the CS-64 thermal strap from TAI consists of two rows, each containing 19 cables. Each cable is fabricated from stranded OFHC copper wires with an overall diameter of 0.10 inch (2.54 mm), and the distance between the mounting lugs is 3.94 in (100 mm). When the thermal boundary resistance at the interface is taken into account, the effective thermal conductance of this strap are approximately 0.44 W/K at 300 K, 1.17 W/K at 40 K, 1.1 W/K at 10 K, respectively [2]. For practical designs, users should use the appropriate thermal conductance values corresponding to the operating temperature range. These flexible thermal straps have been widely used in many fields, including aerospace systems, cryogenic systems, and vibration-sensitive scientific research.

When commercial thermal links do not meet the requirements or cannot accommodate spatial constraints, custom-made thermal links using highly flexible copper "ropes" may be employed [3], as illustrated in Figure 14.3. Each rope contains approximately 10,374 annealed OFHC copper fine wires (AWG 46; 0.00157 in in diameter). Although a 0.25" diameter rope has a "gross" cross-sectional area of 0.317 cm^2, the actual cross-sectional area of the copper wires is approximately 0.13 cm^2, corresponding to a "filling factor" of approximately 41%.

14.2 EXAMPLES OF THERMAL LINK DESIGN

14.2.1 Thermal Link Design for a large "zero-loss" Liquid Helium Dewar

14.2.1.1 Objective

The objective of this project to develop a "zero loss" liquid helium cryostat using a Model PT415RM pulse tube refrigerator (PTR). The concept is to surround the liquid helium reservoir with multiple radiation shields, referred to as the 50K shield, 10K shield, and 4K shield, which are thermally anchored to the first stage, the second-stage regenerator, and the second stage of the PTR, respectively, as illustrated in Figure 14.4. When the 4K radiation shield reaches near the liquid helium temperature, the heat load to the helium reservoir becomes negligible, resulting in "zero" helium boil-off.

DOI: 10.1201/9781042000050-14

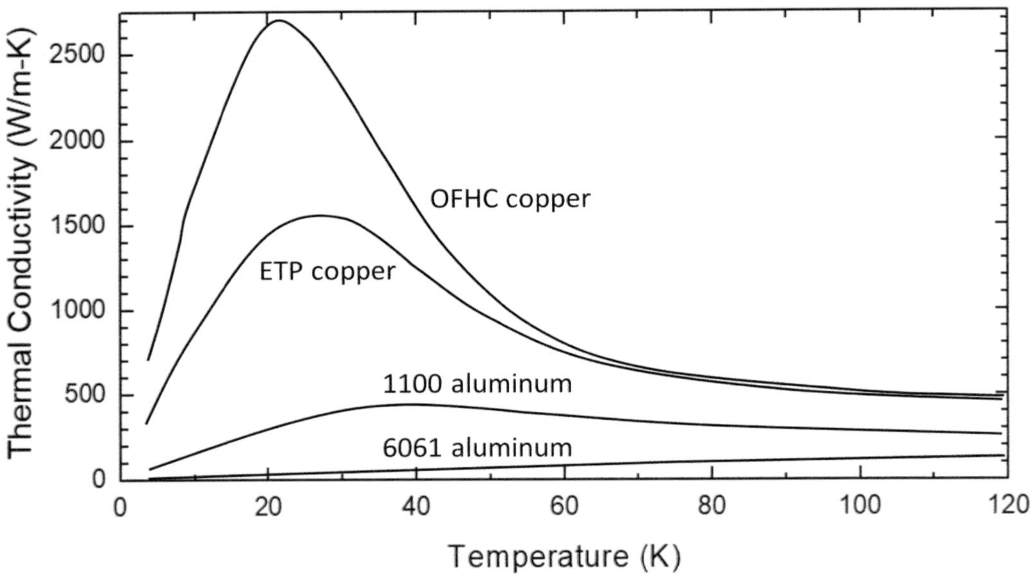

FIGURE 14.1 Thermal conductivity versus temperature of copper and aluminum.

Since the experiment requires minimal vibration, all thermal links between the PTR and the radiation shields are implemented using the flexible copper ropes, as described in the previous section. These copper ropes must provide sufficiently high enough thermal conductance while maintaining minimal temperature gradient along their length.

14.2.1.2 Thermal Link Design

1. **Parameters for Design Analysis**
   - An effective thermal conductivity of 3.5 $W{\cdot}cm^{-1}{\cdot}K^{-1}$ is assumed for the copper rope over the entire cryogenic temperature range in this project.
   - Each copper rope is assumed to have a cross-sectional area of 0.13 cm^2.
   - The cooling performance is evaluated with reference to the load map of model PT415RM PTR (Figure 14.5)
2. **Performance**
   Thermal link #1: Between the PTR first stage and the 50K cold plate
   The total heat load from the room temperature OVC to the 50K radiation shield without vapor cooling is estimated to be 31 W [5]. According to the load map of the model PT415RM PTR, the operating temperature of the 1st stage under this load is taken as 45 K.
   Thermal link design:
   - Total number of copper ropes: 64
   - Length of copper braids: 12.7 cm

   Table 14.1 presents a comparison between the designed and experimentally measured temperatures at the PTR first stage, thermal links, and 50K cold plate.

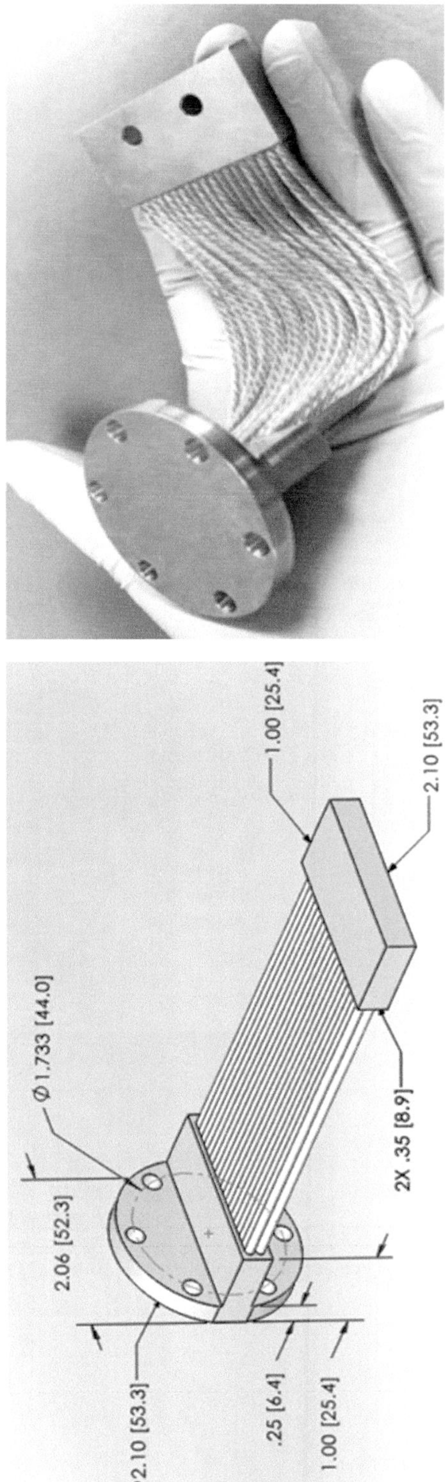

FIGURE 14.2 TAI's thermal strap made of flexible pure OFHC copper strand-woven rope braid and the end mounting-lugs. (Reprint with permission from Technology Applications, Inc. www.techapps.com/contact-us) [2].

FIGURE 14.3 Home-made flexible thermal link made of OFHC copper rope. (Projects accomplished by authors at Janis Research Company. Image courtesy of FormFactor, Inc.)

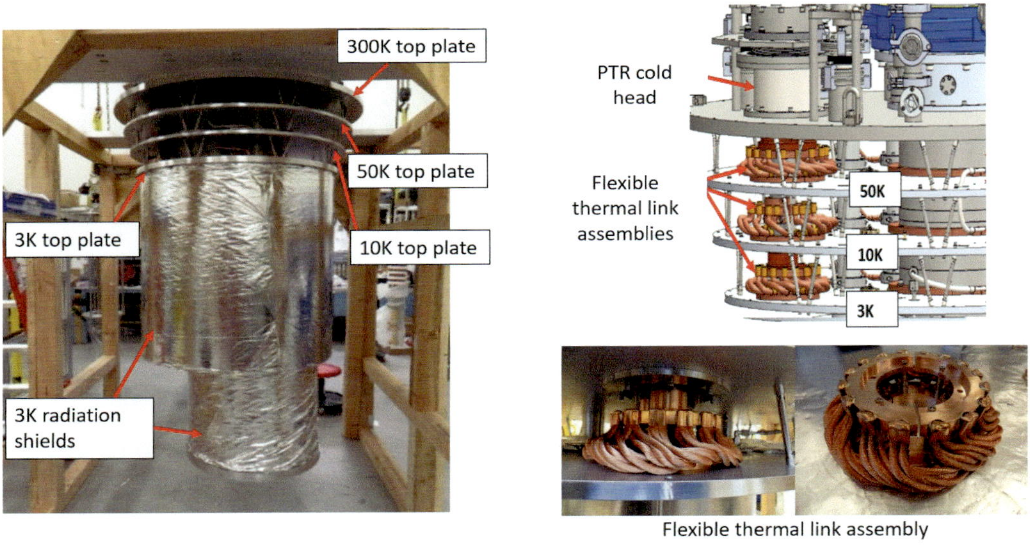

FIGURE 14.4 Zero loss liquid helium dewar. (Projects accomplished by authors at Janis Research Company. Image courtesy of FormFactor, Inc.)

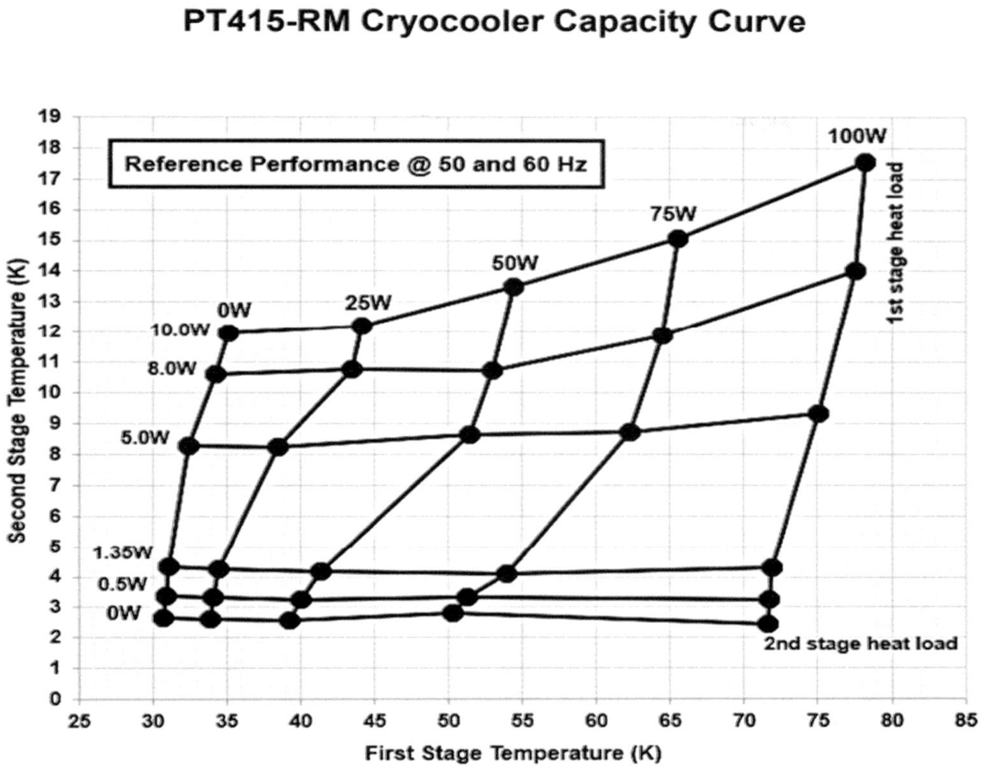

FIGURE 14.5 Load map of model PT415RM PTR. (Reprint with permission from Courtesy from Bluefors Cryocooler Technologies Inc.; https://bluefors.com/contact/bluefors-syracuse) [4].

TABLE 14.1
The Performance of Thermal Link #1 between the PTR First Stage and the 50K Cold Plate (Projects Accomplished by Authors at Janis Research Company. Courtesy of FormFactor, Inc.)

	Performance (K)	Design (K)
PTR first stage	52.5	45
Thermal link lower end	64.7	59
T gradient along thermal link	12.2	14
50K cold plate	76	59 (Expected)

Thermal link #2: Between the PTR second-stage regenerator and the 10K cold plate

The total heat load from the 50K radiation shield to the 10K radiation shield without vapor cooling is estimated to be 1.66 W [5]. According to our analysis based on the PTR load map and Reference [5], the temperature of the second-stage regenerator is taken as 25 K to start with.

Thermal link design:

- Total number of copper ropes: 48
- Length of copper ropes: 12.7 cm

Table 14.2 shows a comparison between the designed and experimentally measured temperatures at the PTR second-stage regenerator, thermal links, and 10K cold plate.

Thermal link #3: Between the PTR second stage and the 4K cold plate

The total heat load from the 10K radiation shield to the 4K radiation shield without vapor cooling is estimated to be 0.133 W [5]. According to the PTR load map, the temperature of the second stage is expected to be 5 K or slightly higher under this load.

Thermal link design:

- Total number of copper ropes: 64
- Length of copper ropes: 12.7 cm

Table 14.3 shows a comparison between the designed and experimentally measured temperatures at the PTR second stage, thermal links, and 4K cold plate.

TABLE 14.2
The Performance of Thermal link #2 between the PTR Second Stage Regenerator and the 10K Cold Plate (Projects accomplished by authors at Janis Research Company. Courtesy of FormFactor, Inc.)

	Performance (K)	Design (K)
PTR second stage regenerator	22.4	25
Thermal link lower end	23.3	26
T gradient along thermal link	1.1	1.0
10K cold plate	26	26 (Expected)

TABLE 14.3
The Performance of Thermal link #3 between the PTR Second stage and the 4K Cold Plate (Projects Accomplished by Authors at Janis Research Company. Courtesy of FormFactor, Inc.)

	Performance (K)	Design (K)
PTR second stage	4.29	5
Thermal link lower end	4.49	5.1
T gradient along thermal link	0.2	0.06
4K cold plate	6.8	5.1 (Expected)

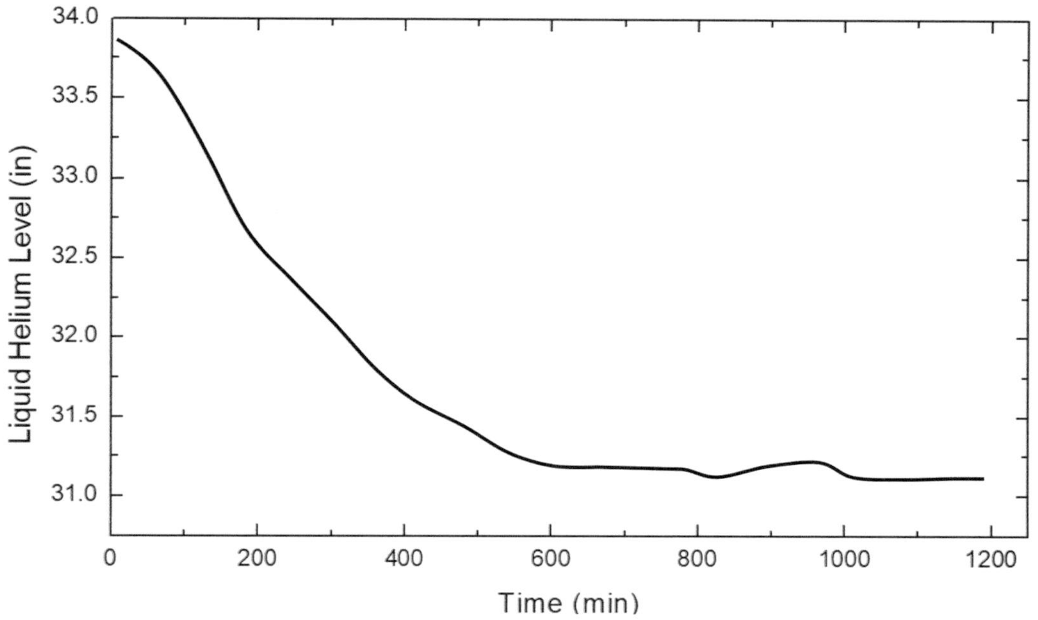

FIGURE 14.6 Helium loss rate. (Projects accomplished by authors at Janis Research Company. Image courtesy of FormFactor, Inc.)

14.2.1.3 Discussion

Temperatures of all cold plates were measured under "zero" helium loss conditions, i.e. with no vapor cooling. Under these conditions, all the heat loads are adsorbed by the PTR. The cold plate temperatures are expected to be lower when vapor cooling is present.

This example demonstrates that OFHC copper rope is a good candidate for a flexible thermal link. Although the discrepancies between the designed values and the measured results are within acceptable limits, more accurate predictions could be achieved using finite element analysis (FEA) or more advanced thermal design software.

The measured temperature difference between the thermal link and the 4K cold plate is much larger than expected, indicating that the thermal boundary conductance at helium temperature requires careful consideration in future designs.

Although the helium boil-off rate is reduced to nearly zero using the above thermal link design, a model HeRL20 helium reliquefier was installed to achieve true zero-loss operation, as illustrated in Figure 14.6.

When flexible thermal links are not required, alternative designs using rigid thermal links with full copper filling can be considered, especially when very high thermal conductance is needed for rapid cool-down of large systems with high heat capacities. An example of a rigid thermal link consists of two sets of annealed OFHC copper strips (Figure 14.7). Each set contains nine strips, with each strip measuring approximately 15.4 cm in length, 3.81 cm in width, and 0.081 cm in thickness, resulting in a cross-sectional area of 0.31 cm^2 per strip. All strips are soldered to a copper block and stacked to fit within the compact space inside the cryostat. The annealing process not only improves thermal conductivity but also provide a degree of mechanical flexibility.

FIGURE 14.7 Semi-rigid thermal link made of annealed copper strips. (Projects accomplished by authors at Janis Research Company. Image courtesy of FormFactor, Inc.)

14.2.2 Thermal Link Design below 1 K

14.2.2.1 Objective

This project aims to cool the rotating sample holder in a sub-kelvin ARPES cryostat (Figure 14.8(a)) to a temperature below 1 K. The holder is thermally anchored to the Yoke, a cooling stage in the cryostat, through a flexible thermal link, while still allowing a minimum rotational tilt of ±30°.

14.2.2.2 Challenge and Solution

The primary challenge is the development of a suitable flexible thermal link that fits within a limited space and accommodates an irregular geometry, which must be addressed through three key design requirements:

- The flexible thermal link must be sufficiently long to allow the sample holder to rotate by ±30°.
- As part of the cold finger, the Yoke reaches a temperature of 640 mK. This requirement limits the allowable temperature gradient along the thermal link between the Yoke and the sample holder be 360 mK or less.
- The contact thermal resistance at both ends of the thermal link must be minimized.

To meet these requirements, a customized flexible OFHC copper strand was fabricated by loosening individual copper wires from an unraveled commercial copper rope, as shown in Figure 14.8(b). To characterize the thermal performance, a four-wire electrical resistance measurement was performed on the copper wires at liquid nitrogen temperature, as illustrated in Figure 14.9. The test configuration ensures the contact thermal resistance comparable to that of the thermal joint when installed on the Yoke.

Based on the measurement, the effective thermal conductivity of the copper strands was determined to be 1.39 $W \cdot cm^{-1} \cdot K^{-1}$ using the Wiedemann–Franz law. This value was adopted for the design, although it is significantly lower than that of bulk OFHC copper at liquid nitrogen temperature. The heat load from the ARPES experiment to the sample is estimated to be 12 mW, as

FIGURE 14.8 (a) Flexible thermal link between the rotating sample holder and the cold finger and (b) flexible thermal link made of flexible copper braids. (Projects accomplished by authors at Janis Research Company. Image courtesy of FormFactor, Inc.)

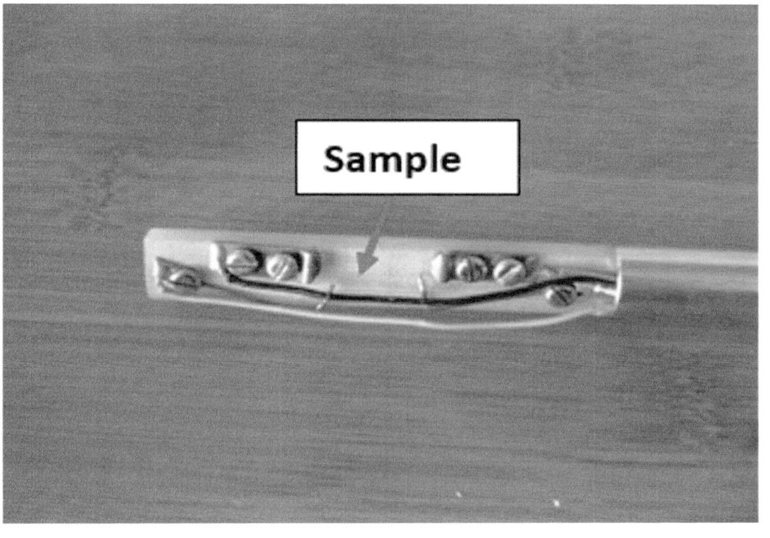

FIGURE 14.9 Four wire electrical conductivity measurement on a sample copper wire. (Projects accomplished by authors at Janis Research Company. Image courtesy of FormFactor, Inc.)

discussed in Chapter 1. The maximum allowable temperature gradient along the thermal link is specified as 360 mK.

In the final design, the thermal link is constructed from 18 flexible OFHC copper strands, each with a length of 1.50 in. The number of strands (18) represents the maximum that can be accommodated within the limited space of the sample holder, while the length of 1.5 in is the minimum required for the ±30° rotation.

The total cross-sectional area of 18 strands is 0.123 cm^2. The calculated temperature gradient along the thermal link is 267 mK. Under these conditions, the sample temperature is expected to be 907 mK. Experimentally, the temperature of the rotating sample mount was measured to be 840 mK, as illustrated in Figure 14.10.

14.2.3 Summary

Commercial OFHC copper rope or braid is a good candidate for a flexible thermal link, although it has limited copper filling fraction. The key factor in successful thermal link design is the use appropriate effective thermal conductivity values for calculations, including the impact of the contact thermal resistance. Proper the mounting plate design is essential to minimize the thermal boundary resistance. An effective home-made mounting plate design is presented in the next section.

Based on our experience, an effective thermal conductivity in the range of 2.5–3.5 W·cm^{-1}·K^{-1} can be considered a reasonably safe estimate for OFHC copper rope with mounting plates, as illustrated in Figure 14.11. The application temperature is another critical factor in design of flexible copper thermal links, as the thermal conductivity of copper is strongly temperature dependent. For example,

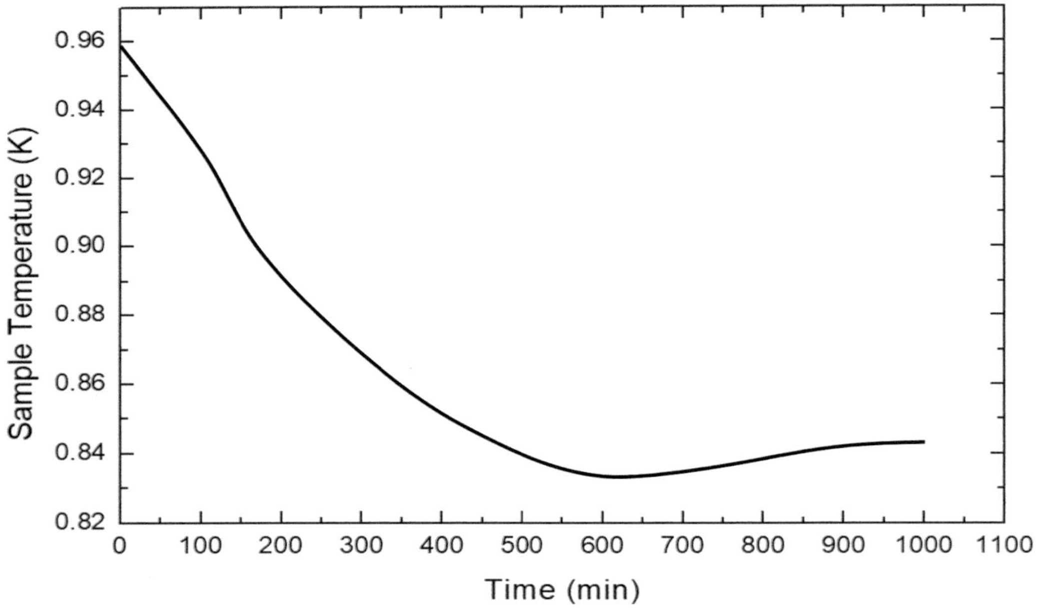

FIGURE 14.10 Measured rotating sample holder temperature. (Projects accomplished by authors at Janis Research Company. Image courtesy of FormFactor, Inc.)

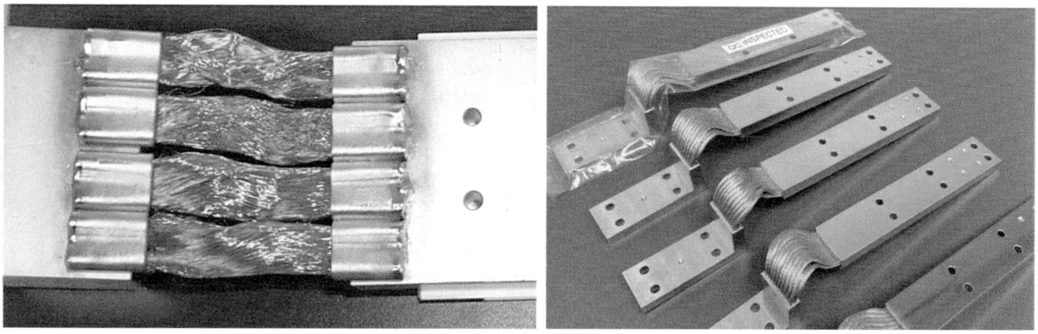

(a) (b)

FIGURE 14.11 (a) Home-made flexible thermal link with mounting plate (Projects accomplished by authors at Janis Research Company. Image courtesy of FormFactor, Inc.), (b) Commercial flexible thermal link with mounting plate. (Reprint with permission from Technology Applications, Inc. www.techapps.com/contact-us) [2].

the thermal conductivity near liquid helium temperature is nearly twice that at 300 K. When direct thermal measurements are unavailable, the Wiedemann–Franz law offers a useful method for estimating effective thermal conductivity.

Finally, in application where a specific cool-down time or very high thermal conductivity is required and mechanical flexibility is not critical, alternative design—such as rigid thermal links fabricated from annealed OFHC copper strips—may be preferable to copper braids.

14.3 HOME-MADE FLEXIBLE THERMAL LINK WITH MOUNTING PLATE

When developing a customized flexible thermal link, establishing a reliable thermal connection between the link and the mounting block is crucial. This section presents an effective method for attaching flexible thermal links to a custom-made mounting block. The fabrication process is illustrated in Figure 14.12.

Step 1: Copper Mounting Block Preparation (Figure 14.12(a))
A copper mounting block is prepared, and a shallow groove is machined to match the diameter of the copper tube. A copper tube with an inner diameter of 0.5 in is silver-soldered onto a gold-plated OFHC copper plate.

Step 2: Copper Braids Installation (Figure 14.12(b))
Four copper ropes are inserted into the copper tube. Each rope consists of approximately 10,374 annealed OFHC copper fine wires (AWG 46).

Step 3: Thermal Contact Enhancement (Figure 14.12(c))
A compressive force of approximately 400 pounds is applied to crush the copper tube, improving thermal contact between the tube and the copper wires.

Step 4: Final Assembly (Figures 14.12(d) and 14.13)
The copper wires and tube are fused together, and the joint is subsequently gold-plated to complete the fabrication.

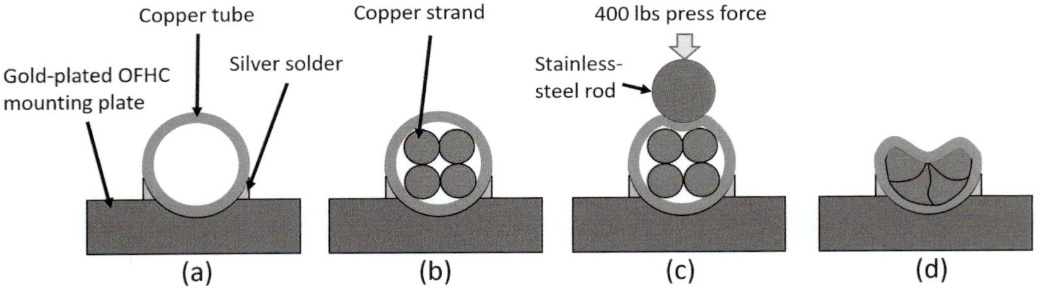

FIGURE 14.12 Thermal link and mounting plate fabrication: (a) preparation of the copper mounting block; (b) insertion of copper braids into the copper tube; (c) pressing of the copper tube with the copper braids inside; and (d) completed assembly.

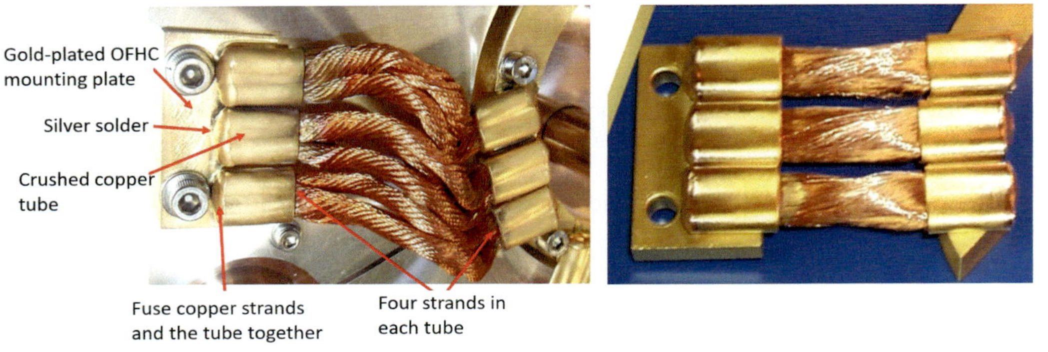

FIGURE 14.13 Home-made thermal link with mounting plate. (Projects accomplished by authors at Janis Research Company. Image courtesy of FormFactor, Inc.)

REFERENCES

1. Zuyu Zhao and Chao Wang (Eds.), *Cryogenic Engineering and Technologies: Principles and Applications of Cryogen-Free Systems*, CRC Press (Taylor & Francis Group) Boca Raton, FL, USA, (2020).
2. Technology Applications, Inc (TAI). 5303 Spine Rd, Boulder, CO, USA Website: www.techapps.com TAI specializes in thermal management solutions for high-performance and extreme-environment applications.
3. New England Wire Technologies 130 North Main Street, Lisbon, NH, USA Website: www.newenglandwire.com The company is a custom wire and cable manufacturer specializing in engineered electrical interconnect solutions.
4. Bluefors Cryocooler Technologies Inc. 6682 Moore Rd, Syracuse, NY, USA Website: www.bluefors.com U.S. manufacturing and support center of Bluefors, focused on cryocooler production and service.
5. Wang, C. Extracting Cooling From Pulse Tube and Regenerator in a 4K Pulse Tube Cryocooler, *15th International Cryocooler Conference* (2009).

15 Multilayer Superinsulation

15.1 INTRODUCTION

Multilayer superinsulation (MLI) is a widely used and highly effective method for reducing blackbody radiation in cryogenic system. In this chapter, we share our practical experiences with the design, application, and installation of MLI, highlighting best practices and lessons learned from our custom projects.

15.1.1 Radiation and Emissivity

A blackbody is a hypothetical object that absorbs all incident electromagnetic radiation, regardless of wavelength or direction, which neither reflects nor transmits radiation. Therefore, all the radiation it emits originates from the body's own thermal energy. This emitted radiation is known as blackbody radiation. Despite its name, a blackbody is not necessarily black in appearance. For example, the Sun closely approximates a blackbody at a temperature of approximately 5778 K and appears bright rather than black. A blackbody is, however, an idealized concept. Real materials do not absorb all incident radiation at every wavelength and are therefore not perfect emitters.

The emissivity (ε) of a material is a dimensionless property that characterizes its ability to emit thermal radiation. It is defined as the ratio of the radiant energy emitted by a real surface to that emitted by a blackbody at the same temperature. Emissivity generally refers to the total hemispherical emissivity, which accounts for radiation over all wavelengths and emission directions. It depends primarily on the material properties and conditions, rather than on its visible color or appearance.

By definition, the emissivity of a perfect blackbody is unity ($\varepsilon = 1$), whereas that of all real materials is less than one. At a given temperature, materials with lower emissivity emit lower radiant energy density. Although emissivity can be defined as a wavelength-dependent parameter, this topic is not discussed in this chapter.

In practice, emissivity is often determined indirectly by measuring absorptivity, defined as the fraction of incident radiation absorbed by a surface (Refer to Appendix). According to Kirchhoff's law of thermal radiation, a material's emissivity equals its absorptivity at thermal equilibrium.

The emissivity of a material depends on temperature and surface conditions, such as surface finish or oxidation conditions. As a result, emissivity values reported in the literature often vary significantly. Appendix A3 provides emissivity values for commonly used laboratory materials at

DOI: 10.1201/9781042000050-15

cryogenic temperatures, compiled from technical handbooks and peer-reviewed publications. It also includes the values successfully applied in our cryostat designs.

15.1.2 Graybody Thermal Radiation

Figure 15.1 illustrates a simple model of radiative heat exchange between two parallel plates without the involvement of a third surface. The surface areas, temperatures, and emissivities of Plate 1 and Plate 2 are denoted by A_1 and A_2, T_1 and T_2, and ε_1 and ε_2, respectively. The distance between these two plates, d, is sufficiently small that the geometry correction can be neglected when estimating the radiative heat transfer.

The total thermal radiation emitted by Plates 1 and 2 is given by

$$E_1 = A_1 \varepsilon_1 \sigma T_1^4$$

and

$$E_2 = A_2 \varepsilon_2 \sigma T_2^4$$

where $\sigma = 5.670374419 \times 10^{-8}$ W·m^{-2}·K^{-4} is the Stefan–Boltzmann constant.

If both plates have the same surface area ($A_1 = A_2 = A$), the radiative heat-transfer analysis is simplified because the geometric contribution of each surface is identical. Consequently, area-related effects cancel out, and the net radiative heat exchange depends only on the surface temperatures (T_1, T_2), emissivities (ε_1, ε_2), and the separation distance (d).

The net radiant heat exchange between these two plates, E_{12}, is given by

$$E_{12} = A \frac{\varepsilon_1 \varepsilon_2}{\varepsilon_1 + \varepsilon_2 - \varepsilon_1 \varepsilon_2} \sigma (T_1^4 - T_2^4) \tag{15.1}$$

In many practical situations, $T_1 \gg T_2$ and $\varepsilon_1 \gg \varepsilon_2$. Under these conditions, Equation (15.1) can be simplified further as

$$E_{12} \approx A \varepsilon_2 \sigma T_1^4 \tag{15.2}$$

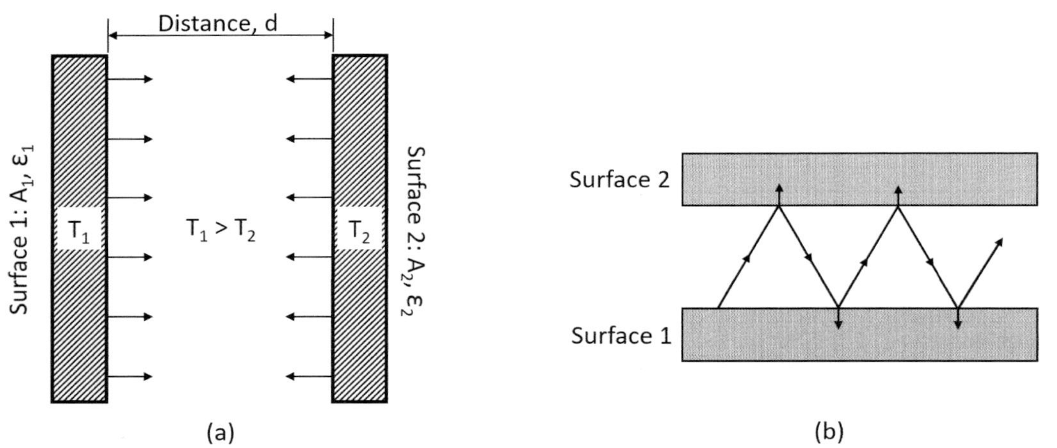

FIGURE 15.1 Radiation heat transfer between two parallel plates at different temperatures.

In this case, the radiative heat transfer is dominated by the temperature of the hotter surface and the emissivity of the colder surface. If both plates are thermally isolated from their surrounding environment, radiative exchange alone will eventually bring them into thermal equilibrium at the same temperature.

Equations (15.1) and (15.2) are widely used in simplified models to estimate radiative heat transfer between two surfaces at different temperatures. A common application is the thermal analysis of cylindrical helium dewars. For a detailed derivation of Equation (15.1), readers may refer to Figure 15.1 and Reference [1].

15.1.3 View Factor

When estimating radiative heat transfer between surfaces with irregular geometries, *view factors* must be included in the analysis [2, 3]. The view factor is a dimensionless parameter that represents the fraction of radiation leaving one surface that directly reaches another surface.

Mathematically, the differential view factor from an element area dA_1 on Surface 1 to an element area dA_2 on Surface 2 is given by

$$dF_{dA1\to dA2} = \frac{\cos\theta_1 \cos\theta_2}{\pi S^2} dA_2 = \frac{1}{\pi}\cos\theta_1 \, d\Omega_1 \tag{15.3}$$

where:

- dA_1 is the emitting (source) surface element,
- dA_2 is the receiving (target) surface element,
- $dF_{dA1\to dA2}$ is the fraction of radiation leaving d_{A1} that directly reaches d_{A2},
- θ_1 is the angle between the normal of d_{A1} and the line connecting the two surface elements,
- θ_2 is the angle between the normal of d_{A2} and the same connecting line (but at d_{A2}),
- S: straight-line separation between d_{A1} and d_{A2},
- $d\Omega_1$ is the solid angle subtended by d_{A2} as seen from d_{A1}.

The view factor is a fundamental concept in radiative heat transfer, particularly for systems involving complex or non-ideal geometries. Analytical determination view factors can be extremely challenging, and accurate evaluation typically relies on numerical techniques or specialized computational software.

15.2 MULTILAYER SUPERINSULATION

MLI consists of multiple layers of thin thermal insulator designed to reduce radiative heat transfer. Each layer includes a substrate coated with a highly reflective (i.e., low-emissivity) metal material, typically applied by vapor deposition to a thickness of 10^{-8} m or less, as shown in Figure 15.2. The typical emissivity values for MLI are approximately 0.03 at 78 K and 0.01 at 4.2 K [4]. When the reflective coating is applied to only one side of the substrate, it should face at the higher temperature surface in order to minimize radiative heat transfer.

The common substrate materials in MLI include Mylar, Kapton, and Teflon, with thicknesses typically ranging from 5 to 70 μm. The reflective coatings are most often aluminum, although silver or gold may be used for specialized applications. In standard MLI fabrications, the reflective material is applied to only one side of the substrate to avoid direct metal-to-metal contact between adjacent layers, which would significantly increase conductive heat transfer. In some specialized designs, coatings may be applied to both sides, but this requires careful control of interlayer contact.

FIGURE 15.2 Super reflective metallized polyester film. (Reprint with permission from ROL-VAC, LP. www.rolvac.com/) [5].

(a) Bridal veil (b) Remy

FIGURE 15.3 MLI with low-conductivity spacer materials. (Projects accomplished by authors at Janis Research Company. Image courtesy of FormFactor, Inc.)

MLI is installed exclusively between surfaces at different temperatures. It can be continuously wrapped around a cylindrical surface or assembled into stitched or layered "blanket" for covering flat or complex surfaces. MLI is typically not applied to surfaces intended to shield radiation from 4 K regions, as radiative heat transfer from 4 K or lower is usually extremely small.

For optimal performance, adjacent MLI layers should not be in direct contact, a condition commonly referred to as "floating." Because the metal coatings have high thermal conductivity, thin spacers with low thermal conductivity can be inserted between layers to maintain separation. Common spacer materials include Remy and polyester "bridal veil," as illustrated in Figure 15.3.

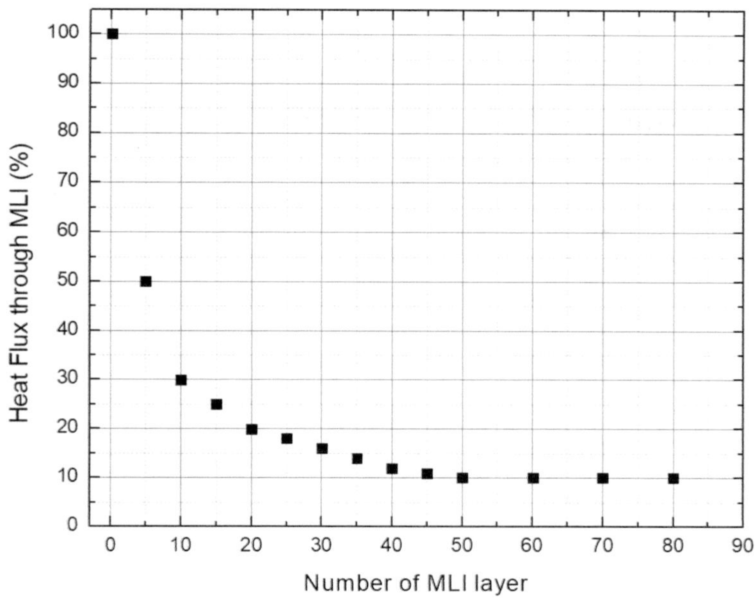

FIGURE 15.4 Radiation heat flux as a function of MLI layer number obtained from FEA simulation.

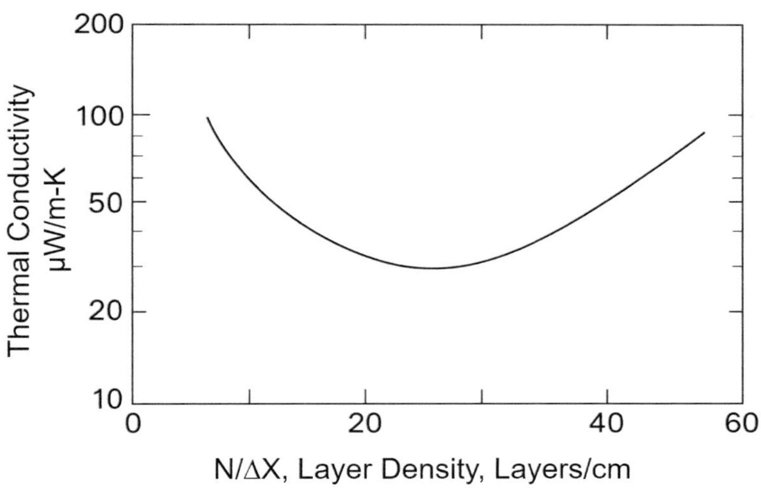

FIGURE 15.5 Variation of thermal conductivity with layers density for a typical MLI installation with boundary temperatures of 294 and 78K. *Source*: Flynn, T., *Cryogenic Engineering*, Second Edition, (2005), Marcel Dekker, Inc. [7]. (Reprint with permission from Taylor & Francis Group.)

15.3 OPTIMAL PERFORMANCE

Two parameters play critical roles in the MLI thermal performance: the total number of layers and the packing layer density that is defined as the number of layers per unit thickness. Both parameters must be optimized to minimize the total heat flux through the insulation.

To determine the appropriate number of layers, we performed a finite element analysis (FEA) of the effective radiation heat flux through MLI as a function of layer count, shown in Figure 15.4. The analysis demonstrates reasonable agreement with the performance observed in practical cryogenic systems. The results show that adding more than about 45 MLI layers does not significantly improve the performance, even when sufficient space is available. Figure 15.4 also suggests that, under normal operating conditions, the use of more than about 80 MLI layers is generally unnecessary.

In practical applications, MLI is often constrained by limited available space. As the packing density increases, adjacent layers are more likely to come into contact, introducing conductive heat transmission in addition to radiative transfer. As a result, the total heat flux through MLI is governed by a tradeoff between radiation and conduction.

The packing density must therefore be optimized to minimize the combined heat transfer. Figure 15.5 shows a typical relationship between total effective thermal transfer and MLI layer density. The curve exhibits a minimum at approximately 25 layers per cm, corresponding to the point of minimum total heat flux. At lower densities, heat transfer is dominated by radiation, whereas at higher densities, conduction through interlayer contacts becomes dominant.

Extensive studies have been conducted to determine the optimal packing density for different MLI products and system configurations [6]. The results suggest that a practical starting range for MLI design is 50–70 layers per inch, balancing thermal performance with mechanical robustness.

15.4 MLI INSTALLATION

15.4.1 General Principles

Proper installation is essential for achieving the expected thermal performance of MLI. The following general principles should be followed:

- Install MLI in a crinkled and loose manner, and do not smooth or compress the layers.
- Position MLI closer to cold surfaces rather than hot surfaces.
- Orient low-emissivity coatings toward hot surfaces.
- Use templates to tailor MLI sheets for irregular geometries.
- Provide venting holes or slits to allow trapped gases to escape during pump-down.

Examples of proper MLI installation are shown in Figure 15.6.

15.4.2 Common Mistakes and Corrective Actions

a) **Excessive Layer Density**

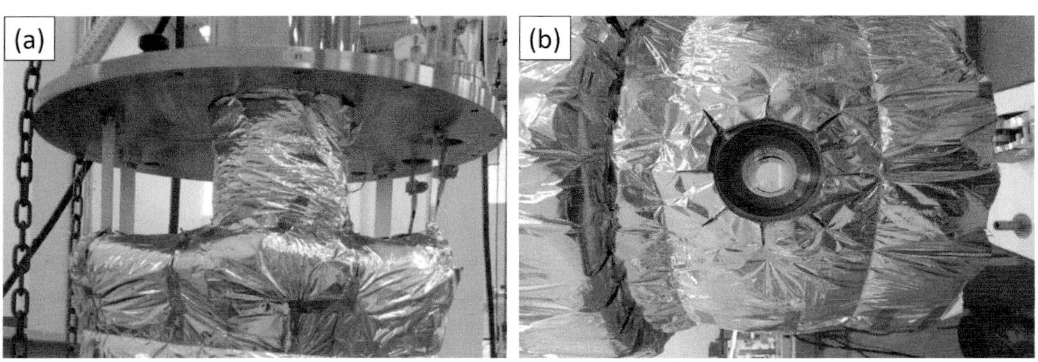

FIGURE 15.6 (a) Cover the penetration tube and (b) Wrap around opening (e.g. window port), (Projects accomplished by authors at Janis Research Company. Image courtesy of FormFactor, Inc.)

Wrapping the MLI too tightly increases the packing density well beyond 70 layers per inch, significantly increasing conductive heat transfer. This is the most common installation mistake. The corrective action is to install the MLI loosely and maintain the recommended density range.

b) **Exposed Flange Edges**
When wrapping cylindrical radiation shields, the flanges edges are often left uncovered, exposing them to higher-temperature radiation. MLI should be extended to fully cover flange edges.

c) **Melting MLI Edges Together**
When fabricating MLI blankets, the layer edges are often melted together with a soldering iron to hold MLI in place. This practice, however, creates direct thermal contact between the outermost and innermost layers, resulting in a thermal short. MLI layers should instead be stitched together loosely.

d) **Neglecting Penetration Tubes**
Penetration tubes, such as cryogen fill lines, vent lines, structural supports, pulse tubes, and regenerator tubes, often connect regions at different temperatures and are frequently left exposed to room-temperature radiation. Proper installation around penetration tubes requires:
   - Extending MLI layers to locations where the penetration tube temperature matches that of the extended layer, as shown in Figure 15.7(a), avoiding lateral temperature gradients within the superinsulation.
   - Wrapping MLI loosely around the penetration tube to prevent direct contact, as shown in Figure 15.7(b).
   - Installing a porous, low-thermal-conductivity buffer zone where appropriate, as illustrated in Figure 15.7(b).

e) **Lack of Perforation**
Residue gas trapped between layers can conduct heat. The MLI blankets must be perforated, either by small holes or slits, to allow effective evacuation.

f) **Absence of Charcoal Getters**
It is highly recommended to install a charcoal getter or molecular sieve on cold surfaces to maintain the desired vacuum in cryogenic systems with MLI.

g) **Uncrinkled Layers**
Smooth, flat layers increase interlayer contact and conduction. MLI layers should always be crinkled to reduce conductive heat transfer.

h) **Improper Attaching Methods**
MLI can be secured using tape, sewing, quilting, pins, or similar methods. However, these attaching methods can compress MLI and create thermal shorts. Care must be taken to minimize both the applied pressure and any thermal shorts.

15.4.3 Frequently Asked Questions

15.4.3.1 Baking Compatibility of MLI

The suitability of MLI for baking depends primarily on the substrate material. MLI with Mylar substrate should be kept below 200 °C to avoid degradation of the polymer film and its reflective coating. For example, heating Mylar-based films at 220 °C for 30 minutes results in approximately a 10% loss in tensile strength, exposure to 235°C for less than one minute causes brittleness and shattering.

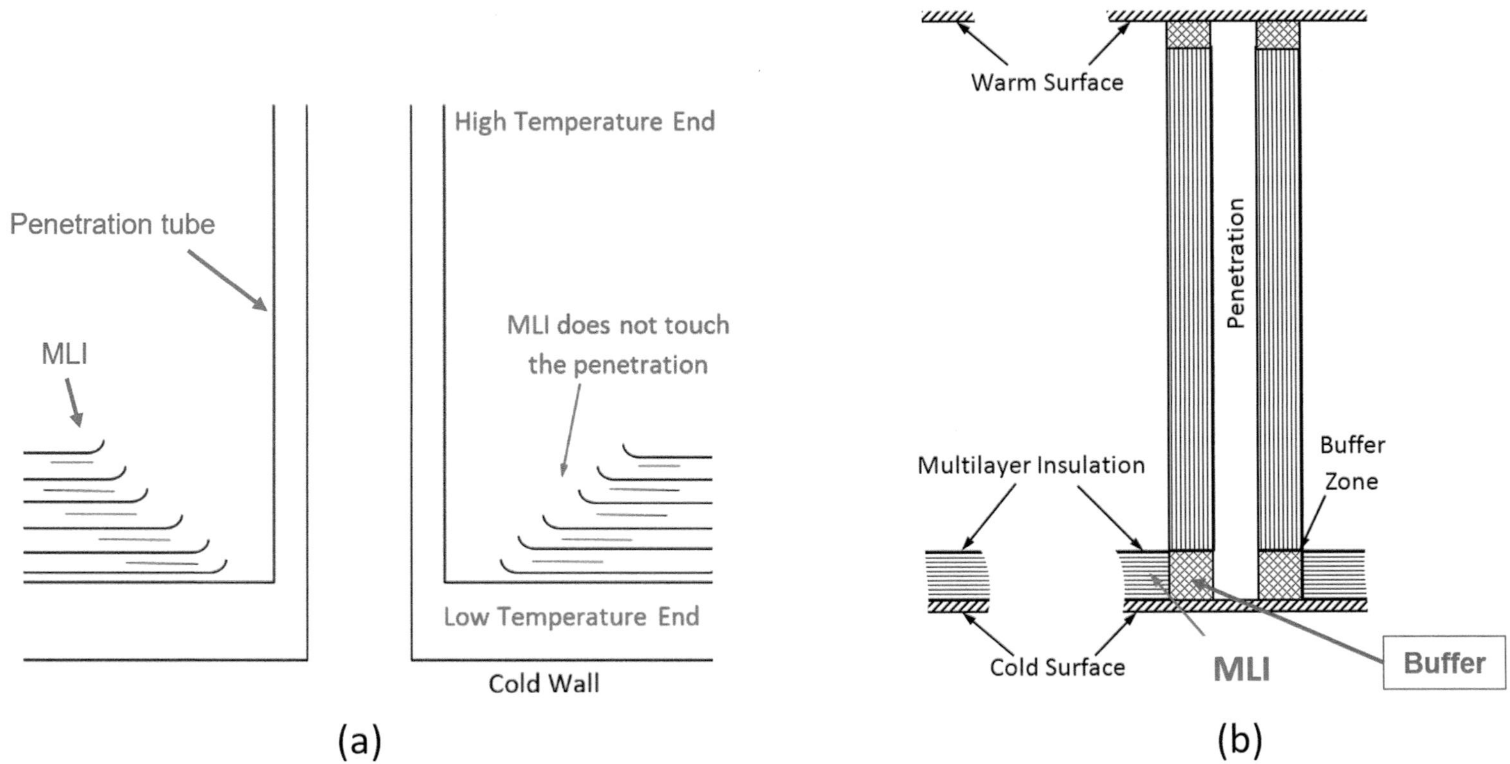

FIGURE 15.7 MLI installed on penetrations (Reprint with permission from Taylor & Francis Group). *Source*: Flynn, T., *Cryogenic Engineering*, Second Edition, (2005), Marcel Dekker, Inc. [7].

Experiments were conducted on the model CRYNSULATE single-side MLI sheet from ROL-VAC, LP [5]. No damage was observed after baking at 150 °C for 8 hours under vacuum. Similar result was observed after extended baking for up to and 16 days at the same temperature. In contrast, exposure to 350 °C resulted in complete destruction of the MLI material,

15.4.3.2 Use of MLI in UHV Systems

From a materials perspective, certain types of MLI are compatible with UHV environments. In particular, aluminum- or gold-coated Kapton-based MLI (Figure 15.8) exhibits very low outgassing rates and is therefore suitable for UHV applications [8]. In an experimental study, a commercial Mylar-based MLI sample was baked at 150 °C for 5 days in a vacuum chamber. Following bake-out, the chamber pressure reached 6.8×10^{-11} mbar at room temperature, without detectable degassing from the MLI sample after baking.

Despite this favorable material property, practical implementation of MLI in UHV systems remains challenging. When installed in cryogenic systems, MLI blankets typically have very large surface areas and readily adsorb water vapor. As results, achieving and maintaining UHV conditions

FIGURE 15.8 Gold coated Kapton MLI used in space science. (Reprint with permission from Taylor & Francis Group.) *Source*: Cryogenic Engineering and Technologies: Principles and Applications of Cryogen-Free Systems, Taylor & Francis Group, LLC, (2020), edited by Zuyu Zhao and Chao Wang [8].

can be difficult even after thorough bake-out procedures. As a practical compromise, UHV systems commonly employ only a small number of MLI layers, typically 3–5. The limited use of MLI can reduce radiative heat load by 30–50%, as shown in Figure 15.4, while keeping the total surface area low enough to allow efficient pumping on the system after baking.

REFERENCES

1. D. Quattrochi (Ed.), *Radiation Heat Transfer (Heat transfer by thermal radiation)*, MIT Course Notes (2006).
2. B.T. Chao, et al., *Advanced Heat Transfer*, Short Course held at the University of Illinois, Urbana, in cooperation with the Program of the State Technical Service Act and the American Society of Mechanical Engineers (August 24–28, 1967).
3. John R. Howell, and M. Pinar Mengüç, Radiative Transfer Configuration Factor Catalog: A Listing of Relations for Common Geometries, *Journal of Quantitative Spectroscopy and Radiative Transfer*, 112(5), 910–912 (2011).
4. W. Steven and Van Sciver, *Helium Cryogenics*, International cryogenics monograph series, General Editors K.D. Timmerhaus and A.F. Clark, Springer Science+Business Media, New York (1986).
5. ROL-VAC, LP 207 Tracy Road, Dayville, CT, USA Website: www.rolvac.com CRYNSULATE is a lightweight metalized polyester film used for MLI, with crinkled structure for spacing, single-side aluminium coating (~400 Å), emissivity ~0.03, and thickness ~6 μm.
6. Lebrun, P., Mazzone, L., Sergo, V., and Vullierme, B. Investigation and Qualifications of Thermal Insulation Systems between 80K and 4.2K, *Cryogenics*, 32, ICMC Supplement, 44–4 (1992).
7. T. Flynn, *Cryogenic Engineering*, Second Edition, Marcel Dekker, Inc. New York, US (2005).
8. Zuyu Zhao and Chao Wang (Eds.), *Cryogenic Engineering and Technologies: Principles and Applications of Cryogen-Free Systems*, CRC Press (Taylor & Francis Group) Boca Raton, FL, USA (2020).

16 Vibration Reduction

In cryogenic systems, vibration reduction is critical for most applications, especially for those requiring high geometric or electronic resolution, such as scanning probe microscopy. Vibrations can arise from both internal and external sources. Internal vibrations typically arise from the circulation of functional cryogens—such as the boiling or evaporation of refrigerants, or the flow of refrigerants in continuous-operating system. However, previous studies have shown that internal vibration sources in both dilution refrigerators and He-3 systems are minimal and generally do not pose a significant impact on system performance [1].

For external vibration, in addition to common vibration sources such as building structural vibrations or vacuum pump operation, the dominant contribution originate from the cryocooler and its associated compressor in cryogen-free cryostats. Detailed discussions regarding the vibration from both Gifford–McMahon (GM) and PTR cryocoolers, including their operating principles, vibration mechanisms, measurement techniques, and mitigation strategies, are presented in Reference [2].

This chapter addresses a broader range of approaches for mitigating external vibrations in cryogenic environments, applicable to both "wet" and cryogen-free systems. Most of the methods described have been successfully implemented in our practical projects.

16.1 CONVENTIONAL VIBRATION REDUCTION

A variety of conventional approaches have been developed to mitigate external vibrations in cryogenic systems and are widely used in laboratory and industrial environments.

16.1.1 PTR Vibration Isolation

- A spring-loaded welded bellow is installed between the PTR cold head and the cryostat (Figure 16.1(a)) to absorb vibration energy transmitted from the cryocooler.
- The PTR motor, the only moving component in the PTR cold head, is mounted remotely rather than directly on the cryostat (Figure 16.1(b)). Positioning the motor between the cold head and the compressor (helium high-pressure lines) reduces both mechanical vibration and electrical noise.
- Flexible thermal links, typically made of braided copper wires, connect the PTR cold stages to the cryostat, providing efficient thermalization while minimizing vibration transmission (Figure 16.1(c)). When it is used alone, this approach can reduce vibration amplitudes to approximately 1.0 μm in a PTR-equipped cryostat.

DOI: 10.1201/9781042000050-16

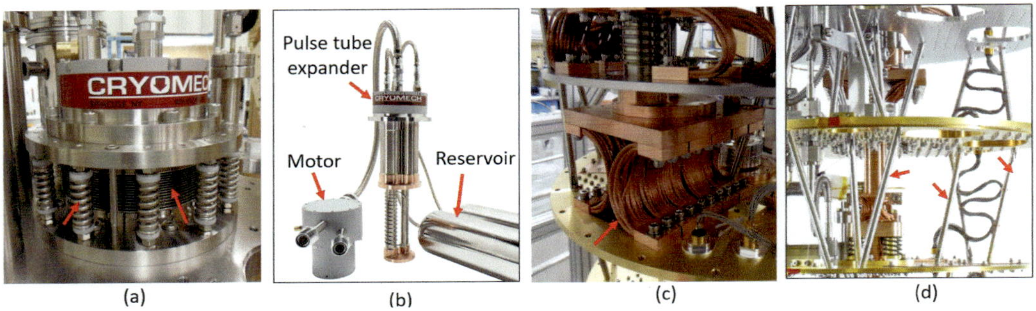

FIGURE 16.1 Four commonly used methods for PTR vibration isolation. (a), (c), and (d) Projects accomplished by authors at Janis Research Company. Image courtesy of FormFactor, Inc.; (b) Reprint with permission from Bluefors Cryocooler Technologies Inc. https://bluefors.com/contact/bluefors-syracuse

- Rigid mechanical supports help limiting relative movement of cryostat components (Figure 16.1(d)). The triangular truss shown is a structurally efficient design widely used in both "wet" and cryogen-free cryostats.

16.1.2 Vibration Damping

- Figure 16.2(a): Gimbals provide a highly effective method for isolating cryostats from environmental vibrations [3] and are often employed for vibration-sensitive measurements in the millikelvin temperature range.
- Figure 16.2(b): Mounting the experimental setup on a vibration isolation system equipped with air cushions or active damping mechanisms is also a common practice. These systems further reduce vibrations transmitted to the cryostat, thereby improving overall stability.
- Figure 16.2(c): Constraining the PTR high-pressure lines between sandbags provides a simple and effective method for reducing high-frequency vibrations originating from the PTR compressor. In addition, vendors can also offer super-flexible hoses specifically designed to further suppress vibration transmission.

16.1.3 Wall-Mounted Pipes and Concrete Blocks

To minimize vibrations originating from the pumping station, pumps are installed in a separate room, with the pumping lines rigidly secured at wall penetrations. Concrete blocks are often added along the piping to further suppress vibration transmission (Figure 16.3).

16.1.4 Heavy-Weight Pipes

High-frequency vibrations can propagate to the cryostat through rigid piping. Incorporating sections of heavier piping increases the effective mass of the system (Figure 16.4), thereby lowering its natural frequency and shifting it away from the dominant excitation frequencies. This approach also provides partial attenuation of vibration transmission.

16.1.5 Suspension

Suspending components such as pumps or remote motors from overhead structures provides an effective means of decoupling these vibration sources from the cryostat (Figure 16.5).

(a) (b) (c)

FIGURE 16.2 Three commonly used vibration-damping methods that are widely applied in many experimental systems and are not limited to cryogenic applications.

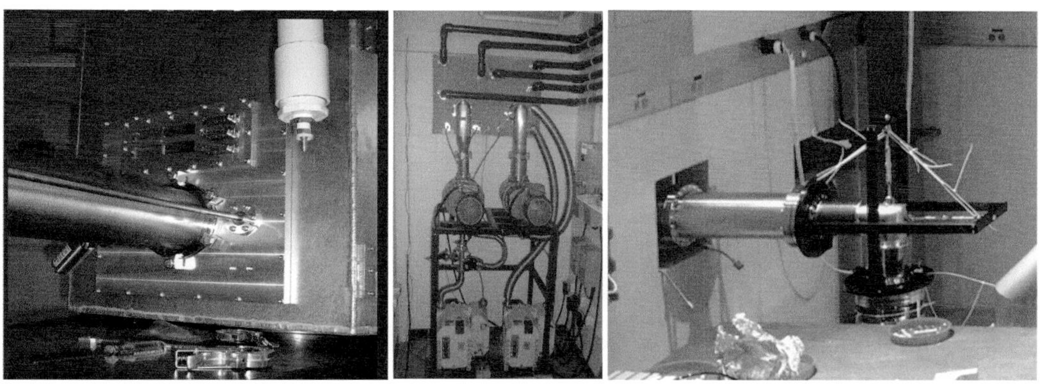

FIGURE 16.3 Vibration isolation with wall-mounted pipes and concrete blocks. (Projects accomplished by authors at Janis Research Company. Image courtesy of FormFactor, Inc.)

16.1.6 Eddy Current Damping

When a conductive component vibrates relative to the magnetic field of an eddy-current damper, eddy currents are induced within the conductor. The interaction between these currents and the magnetic field generates a non-contact damping force that opposes the motion, thereby dissipating vibrational energy.

16.1.7 Multilevel Vibration Reduction

Figure 16.6 illustrates a multilevel vibration reduction setup designed for STM applications operating at low millikelvin temperatures. The system employs a three-stage vibration isolation scheme, as detailed in Reference [1].

- Stage 1: A 110-ton concrete mass with a double-keel design, supported by six pneumatic isolators.
- Stage 2: A 6-ton granite table mounted on four pneumatic isolators.
- Stage 3: An aluminum ring suspended by four stainless steel rods and equipped with four pneumatic isolators.

FIGURE 16.4 Installation of heavy-weight pipes. (Projects accomplished by authors at Janis Research Company. Image courtesy of FormFactor, Inc.)

FIGURE 16.5 Suspending items from above. (Projects accomplished by authors at Janis Research Company. Photo courtesy of FormFactor, Inc.)

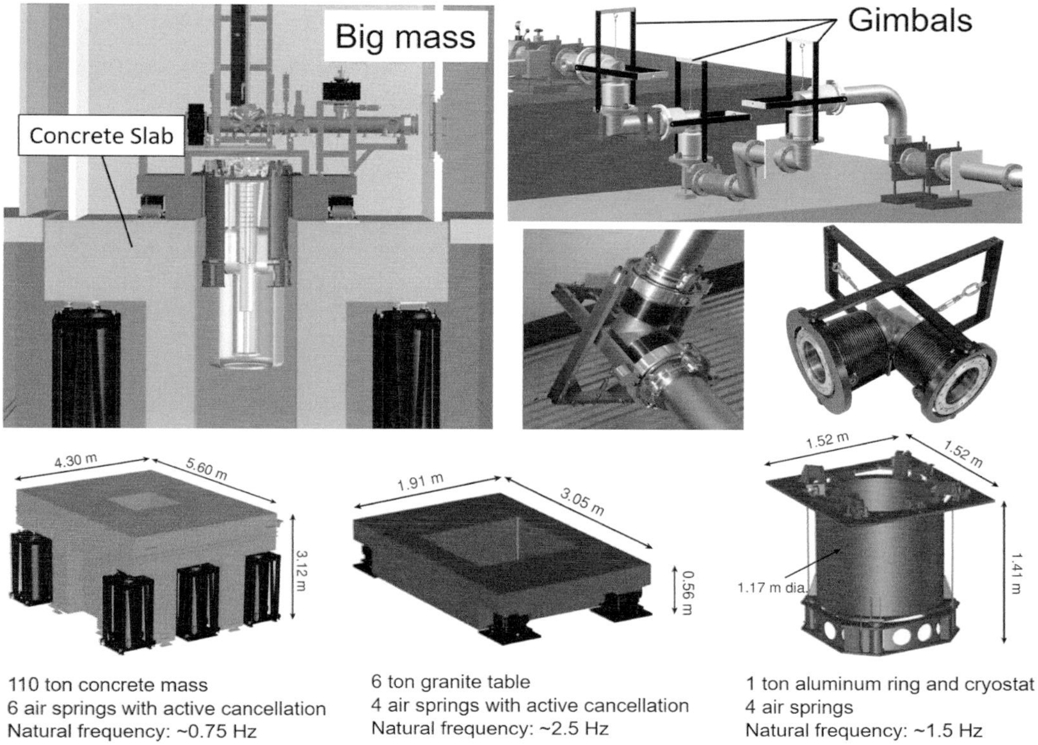

FIGURE 16.6 Multilevel vibration reduction set up for STM application at millikelvin [1]. (Reprint with permission from the author, REVIEW OF SCIENTIFIC INSTRUMENTS, 81, 121101 (2010), *Invited Review Article: A 10 mK scanning probe microscopy facility,* Young Jae Song, Alexander F. Otte, Vladimir Shvarts, Zuyu Zhao, Young Kuk, Steven R. Blankenship, Alan Band, Frank M. Hess, and Joseph A. Stroscio).

Each stage features a central opening to accommodate the cryostat. Transfer function measurements indicate that the combined attenuation of the three stages exceeds 100 dB at 25 Hz, achieving the system's design goal.

16.1.8 Separate Foundations

Vibrations from nearby structures, such as adjacent buildings, can compromise highly sensitive measurements. Constructing an independent foundation for the laboratory effectively isolates it from these external vibration sources.

16.1.9 Crushed Stone and Gravel Layer

A layer of crushed stone or gravel, typically a few inches thick, is placed between the compacted soil and the concrete foundation pit. This layer serves multiple purposes: it dampens vibrations from surrounding structures by absorbing mechanical energy and helps prevent water accumulation beneath the foundation. This simple and effective technique is particularly beneficial for vibration-sensitive setups.

16.2 EFFECTIVE VIBRATION REDUCTION IN CRYOGEN-FREE SYSTEMS

This section presents three examples of effective vibration mitigation strategies in cryogen-free systems.

16.2.1 Cryogen-Free Cryogenic Mössbauer Systems with GM Cryocoolers

Mössbauer spectroscopy, which employs gamma rays to probe the electronic and magnetic structure of materials in transmission geometry, is highly sensitive to environmental vibrations. Such vibrations can degrade spectral resolution and reduce the signal-to-noise ratio, particularly at low temperatures. A notable implementation, shown in Figure 16.7, features a cryogen-free, low-temperature Mössbauer cryostat integrated with a GM cryocooler [7].

- The cryostat employs a two-stage GM cold head in an inverted orientation, with the motor at the bottom and the second stage at the top. Although a PTR generally operates more quietly, it is unsuitable for this application due to the required mounting orientation (motor above and the second stage below). Additional details on the orientation dependent performance of PTR are discussed in Section 16.4.

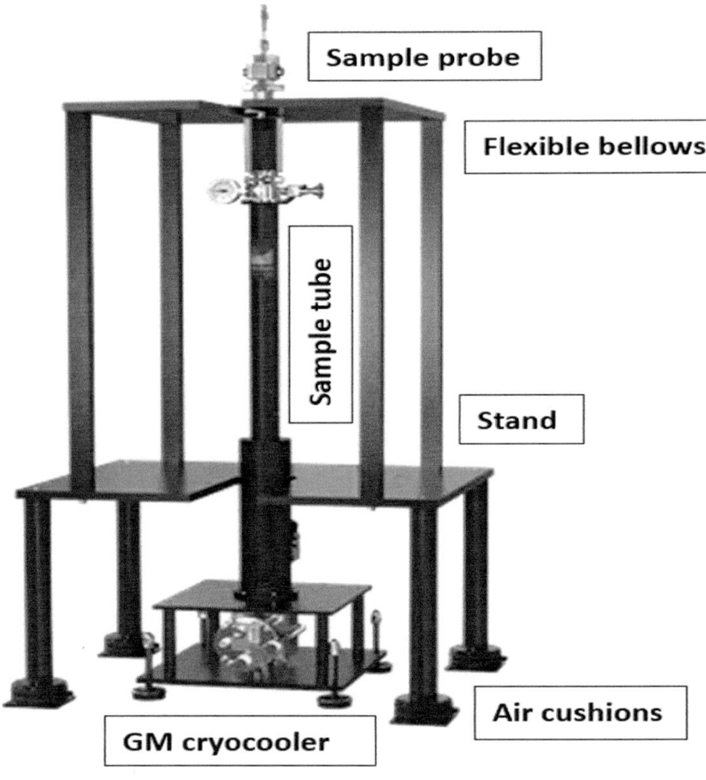

FIGURE 16.7 Cryogenic cryogen-free Mossbauer cryostat. (Reprint with permission from Lake Shore Cryotronics, Inc.) [4].

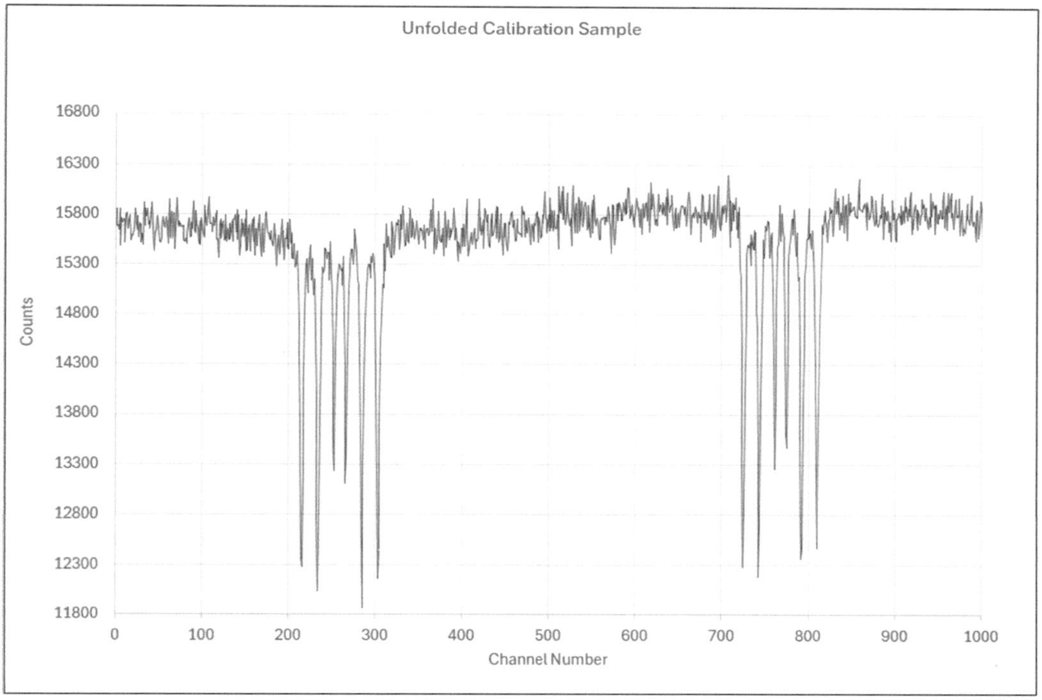

FIGURE 16.8 Mössbauer spectra with the CCR is running. (Sample: α-Fe; source: ^{57}Co in an Rh matrix) (Reprint with permission from Lake Shore Cryotronics, Inc.) [4].

- The system incorporates an exchange-gas sample tube and a sample probe, with the probe suspended inside the tube, preventing direct contact between the sample holder and the tube. To reduce vibration transmission from the cryocooler, bellows are installed between the sample probe and the top flange of the sample space.
- A heavy stand with integrated air cushions provides additional vibration isolation.

Figure 16.8 shows the Mössbauer spectra of α-Fe while the CCR was operating. No spectral broadening was observed, indicating that vibrations from the cryocooler were effectively suppressed.

Analysis indicated that further reductions can be achieved by:

- Using taller cushions, provided the cryostat remains mechanically stable.
- Employing bellows with a lower spring constant and a smaller cross-sectional area.
- Increasing the cryostat mass by adding additional weight.

16.2.2 Vibration Minimization for a 100 microkelvin NADR System

Vibration introduces heat and noise, disrupting the delicate conditions required for microkelvin research. A cryogen-free nuclear adiabatic demagnetization refrigeration (NADR) system with a copper nuclear stage has been successfully developed through the implementation of various vibration-reduction strategies described in Section 16.1 [5, 6] (Figure 16.9).

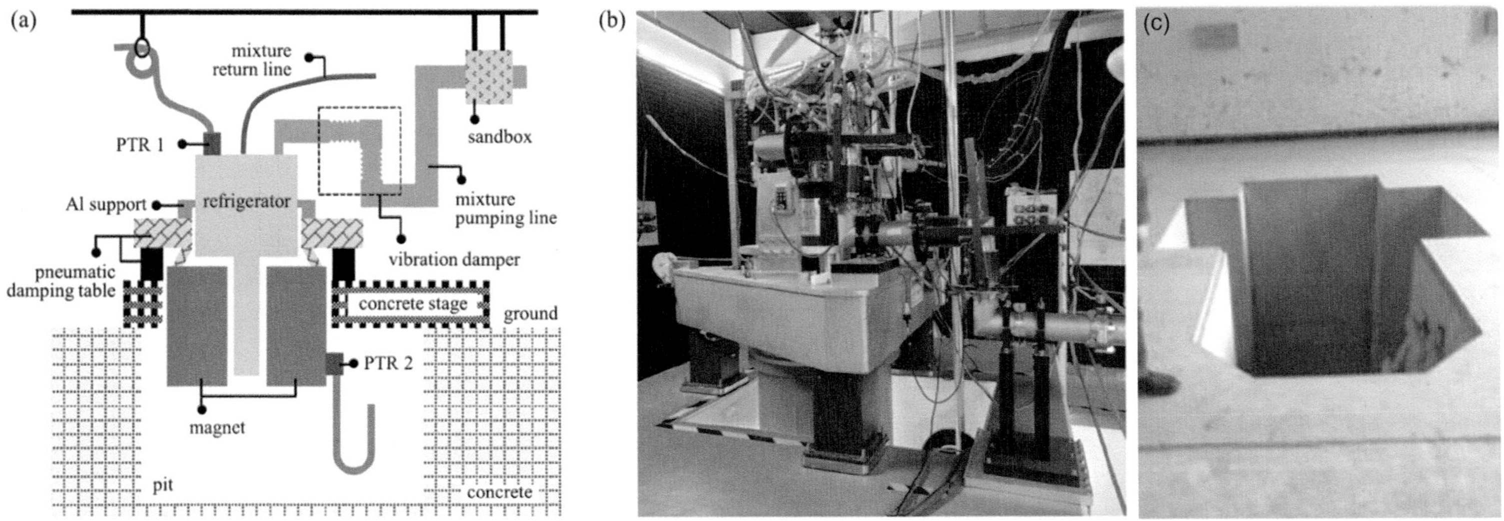

FIGURE 16.9 Cryogen-free 100 micro-Kelvin NADR system (a) shows a schematic diagram of the overall system structure along with the external vibration reduction framework (b) presents a photograph of the actual system. (Projects accomplished by authors at Janis Research Company. Photo courtesy of FormFactor, Inc.) (Reprint with permission from Jiaojie Yan, Ph.D. Thesis, *Construction of ultra-low temperature environment and transport measurements,* Directed by Professor Xi Lin, College of Physics, Beijing University, Chain, May, 2022) [5, 6]. (c) Cement platform supporting the system. (Projects accomplished by authors at Janis Research Company. Photo courtesy of FormFactor, Inc.)

- A cement platform (~0.5 m thick) surrounds the pit and extends to its top.
- An optical platform is installed above the through-hole in the cement platform, supported by three air-floating legs fixed to the cement platform.
- The He-3/He-4 mixture pumping line is suspended from the ceiling and routed through a sandbox.
- A universal vibration isolator is placed between the mixture recovery pipeline and the dilution refrigerator.
- Mixture return lines and high-pressure helium hoses are designed with a U-shaped geometry to reduce stress and coupling with external vibrations.
- Double cryogen-free superconducting magnets—one for the NADR and the other for the experiment—are mounted on a non-magnetic elevator, functioning as a completely independent unit.

This system achieves temperatures below 100 μK for the first time in a dry refrigerator with stable operation maintained for over 10 hours [5, 6].

16.2.3 Low Vibration Cryostat with Novel Design for Versatile Applications

16.2.3.1 Cold Head Suspension Vibration Reduction Technology

The conventional method for mounting a cold head onto a cryostat employs a welded bellows connection between the cryocooler and the cryostat, as illustrated in Figure 16.1(a). Experimental tests conducted by Physike Technology Co., Ltd. [7] show that this configuration reduces vibration levels by only about 50%. This limited effectiveness arises because the cryocooler mounting flange is subjected to a total downward force exceeding 100 kg due to atmospheric pressure, which creates an efficient mechanical path for vibration transmission.

To address this limitation, a patented cold-head suspension vibration-reduction technology incorporates two welded bellows—one above and one below the cold head (Figure 16.10). The lower bellows remains nearly stress-free during operation, significantly suppressing vibration transmission from the refrigerator, through the mounting flange, to the cryostat.

This technology has been successfully employed in various types of vibration-sensitive cryogenic systems, such as STM cryostats, He-3 cryostats, and optical microscopy cryostats, and is expected to be applied to dilution refrigerator systems in the near future.

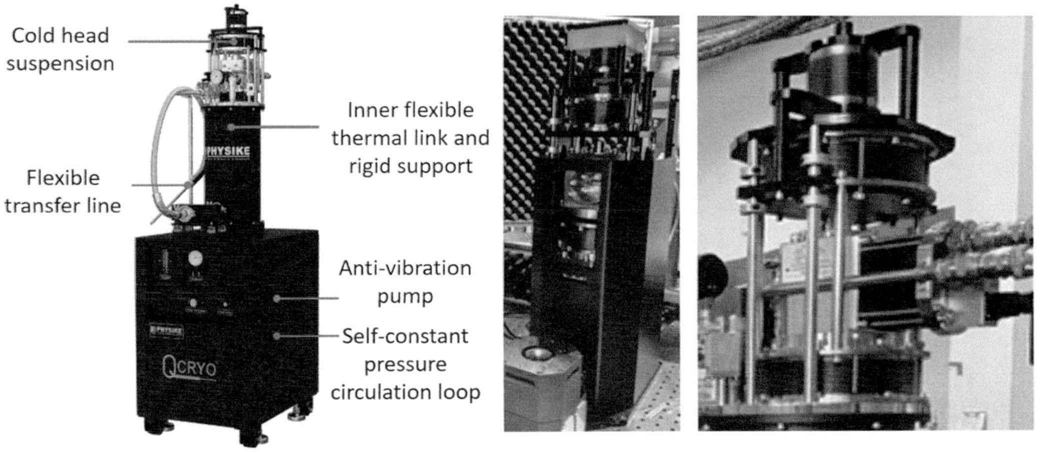

FIGURE 16.10 Cryostat with cold-head suspension vibration-reduction. (Reprint with permission from Physike Technology Co., Ltd.)

16.2.3.2 Split Design

The "split design" of a cryostat (Figure 16.11)—where the cryocooler, including its cold head and compressor, is physically separated from the cryostat containing the sample—provides a fundamental means of minimizing vibration transmission. In this configuration, helium gas at ambient temperature is first liquefied by the cryocooler and then delivered to the cryostat through a vacuum-jacketed, flexible transfer line. The liquid helium can be further cooled by reducing its vapor pressure.

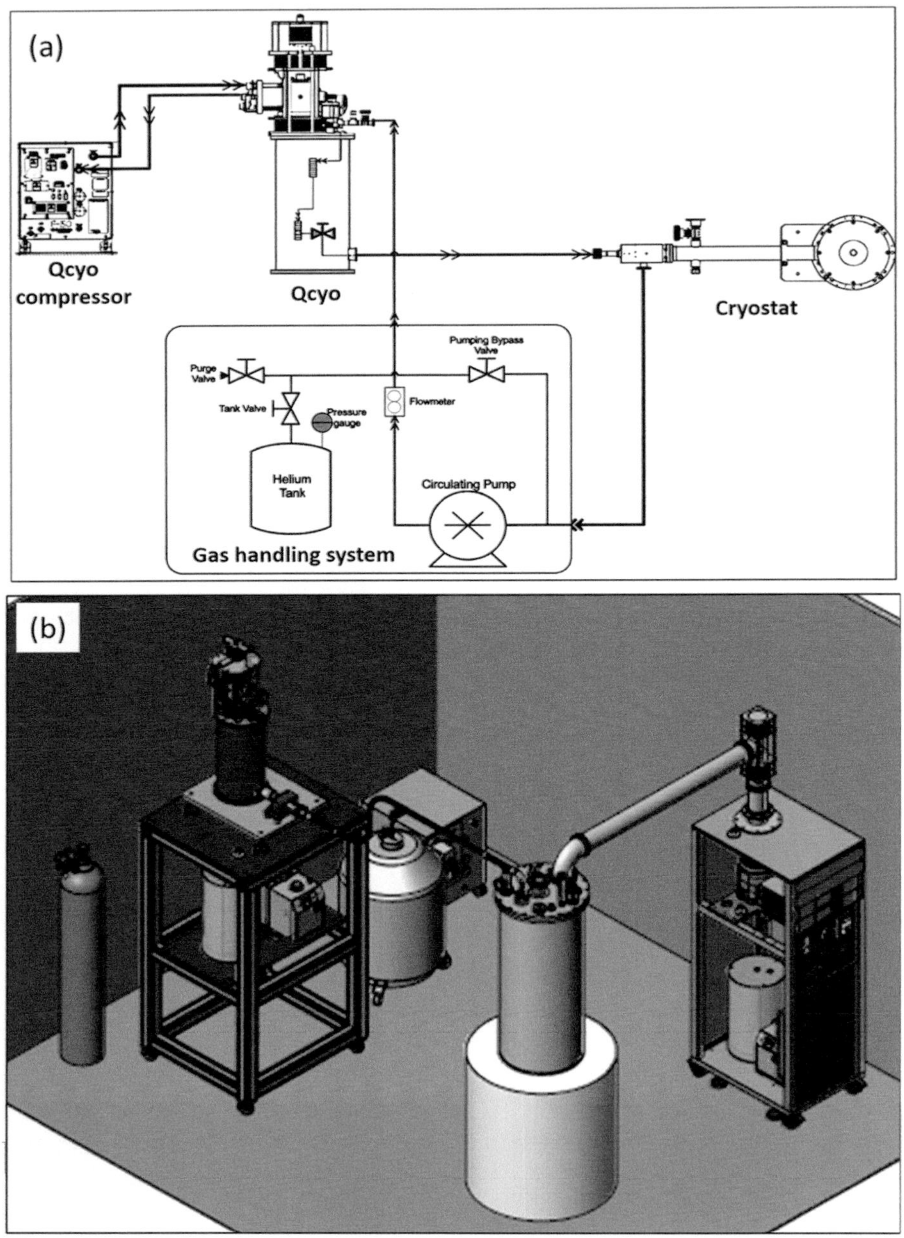

FIGURE 16.11 (a) Flow diagram of model Qcryo cryogen free microscopy cryostat with split design (b) Continuous flow type cryogen-free He-3 cryostat with split design. (Reprint of both figures with permission from Physike Technology Co., Ltd.)

By spatially decoupling the cryostat from the 4 K cold head and helium compressor, the split design effectively suppresses the transmission of mechanical vibrations, making it particularly suitable for vibration-sensitive cryogenic experiments. For example, in the Model Qcryo-S-500 system, vibration measurements at the sample position using a Polytec VibroGo laser vibrometer [8] at 2.8 K show peak-to-peak amplitudes below 5 nm in both the parallel and vertical directions.

These technologies have also been adopted in a continuous cryogen-free STM systems operating below helium temperatures. Accumulating experience within the community (e.g. CAS and ACME [9, 10]) indicates that, when carefully implemented, such approaches can achieve tunneling noise levels and spectroscopic performance comparable to those of traditional liquid-helium-based STM systems [11, 12].

16.3 VIBRATION MEASUREMENTS

16.3.1 Eddy Current Noncontact Displacement Sensor

An eddy current displacement sensor measures the position and vibration of a conductive target without physical contact (Figure 16.12). An alternating current drives the sensing coil, generating a magnetic field that induces eddy currents in the nearby target. As the target moves or vibrates, changes in the eddy currents are converted into a voltage output proportional to the target's displacement or vibration. Note that:

- The target must be electrically conductive, typically metallic, to support the generation of eddy currents.
- This is a non-contacting device, requiring a small but sufficient gap—typically a few millimeters—between the sensor and the target.
- The sensor can measure both vibrations and low-frequency or static displacements (DC motion).

16.3.2 Laser Doppler Vibrometer (LDV)

A laser Doppler vibrometer is also a noncontact instrument for surface vibration measurements. It operates by directing a laser beam onto the target from a convenient distance and determining the vibration amplitude and frequency from the Doppler shift of the reflected beam. A unique advantage

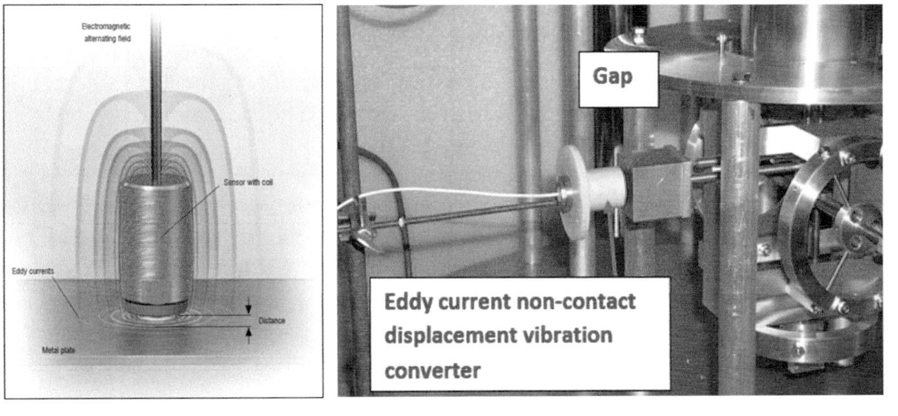

FIGURE 16.12 Non-Contact Eddy Current Vibration and Displacement Measurement Converter. (Projects accomplished by authors at Janis Research Company. Photo courtesy of FormFactor, Inc.) *Source*: Information from www.bestech.com.au/blogs/eddy-current-sensor-principle/

of this instrument is that measurements can be performed through optical windows, allowing direct assessment of cryogenic objects without disturbing their environment. Typical example of Commercial LDVs include features as:

- Laser type: helium-neon
- Wavelength: 633 nm
- Variable working distance: typically from 0.2 m to 30 m
- Resolution: typically a few nanometers (Picometer resolution equipment is available)
- Vibration frequency range: from DC up to 20 MHz

16.3.3 Accelerometer

An accelerometer is a vibration sensor that measures acceleration of the surface to which it is attached. A common type is the piezoelectric accelerometer, which consists of a piezoceramic crystal sandwiched between a seismic mass and a base. When the base experiences acceleration, the inertia of the mass stresses the crystal, generating a proportional electrical charge. Note that:

- It is inherently AC-coupled and therefore cannot measure constant (DC) acceleration.
- The low-frequency response is typically limited to a few hertz.
- The upper frequency limit is usually kept well below the mechanical resonance of the device.

16.3.4 Geophone

A geophone is typically a moving-coil velocity transducer in which a spring-suspended coil moves relative to a fixed permanent magnet attached to the case. When the base of the device moves, the relative motion between the coil and the magnet induces a voltage proportional to velocity. The key features include:

- Usually single-axis devices—either vertical or horizontal—so correct orientation is essential for accurate measurements.
- Designed with low resonant frequencies, typically around 3–4 Hz, making them well-suited for detecting low-frequency vibrations [1].

16.4 QUESTIONS AND ANSWERS ON PRACTICAL APPLICATIONS OF PTR

This section summarizes a few technical topics of cryocoolers with common interest [13]. The goal is to share insights, clarify performance characteristics, and address practical considerations related to cryocooler operation and integration, based on real-world experience and expert input.

Q1: Both SHI model RDK408 GM cryocooler and Bluefors Cryocooler Technology model PT 410 pulse tube cryocooler (PTR) have 1.0 W cooling power at 4 K. The GM cryocooler has a typical ±13 μm displacement, while the PTR has ±10 μm. Why does the PTR generate smaller vibration?

A: Although both GM and pulse tube cryocoolers have similar displacement amplitudes at the cold head, the vibration experienced by the cryostat is primarily determined by the acceleration of moving parts, which translates into dynamic force transmitted to the cryostat.

- In a GM cryocooler, mechanical displacers are actively driven by a motor, leading to significant moving mass undergoing rapid acceleration. This results in large reaction forces—and therefore higher vibrations—being transferred to the cryostat.
- In contrast, a PTR has no moving parts in the cold head so it experiences minimal acceleration. Despite similar displacement magnitudes, the acceleration (and thus dynamic force) in a PTR is much lower than that in a GM cryocooler.

Q2: Would you comment on the function of the remote motor?

A: The remote motor serves two key functions.

- It reduces mechanical vibration transmitted to the cold head by physically separating the motor from the cold head assembly, which is especially beneficial for vibration-sensitive experiments.
- It helps to minimize electromagnetic interference (EMI) near the cryostat. This reduction in EMI is particularly important for dilution refrigerator (DR) systems operating at millikelvin temperatures, where even small amounts of electrical noise can negatively impact performance.

Q3: You mentioned that we should use flexible thermal links from the first and second stages to our dilution refrigerator. With your flexible connections, do we still need additional vibration isolators from the vendor? What is the vibration level with this setup?

A: Using flexible thermal links reduces the vibration level transmitted to the cryostat (e.g. He-3 system or dilution refrigerator) to approximately 1 μm. Additional vibration isolation measures need to be implemented to further suppress any residual mechanical vibration for highly vibration-sensitive applications.

Q4: What is the acoustic noise level of water cooled PT405 or PT410?

A: The acoustic noise level from a water-cooled PT405 or PT410 cold head is approximately 56 dB, measured at a distance of 1 m. The compressor produces a higher noise level of around 70 dB at the same distance. In comparison, air-cooled PTRs tend to be slightly noisier due to the operation of cooling fans, which can add several decibels to the overall acoustic profile.

Q5: What is the "magnetic distortion from the 2nd stage"? How big is the "magnetic distortion"?

A: The rare earth material in the second stage regenerator can generate a small oscillating magnetic field. The "magnetic distortion" from the second stage is approximately 200 nano-tesla.

Q6: What is the impact on PTR from the magnetic field?

A: To ensure normal PTR performance, the magnetic field around the rotary motor in the cold head should be less than 500 Gauss. The magnetic field of < 1T around the second stage regenerator should not affect the cooling performance. If the magnetic field is > 1T, it will slightly reduce the 4 K performance by a few percent, but less than 25%. When the remote valve is not exposed to a magnetic field, the impact of a "weak" magnetic field (e.g., an 8 T magnet located a few inches away) on the PTR first stage is negligible. More details are provided in [2] and [13].

Q7: While GM cryocoolers can operate in any orientation, how does the performance of a pulse tube refrigerator (PTR) vary with its operating orientation?

A: PTRs generally perform best when operated with the cold head oriented vertically upward. The cooling power tends to decrease as the cold head deviates from this vertical orientation due to the gas gravitational effect as illustrated in Figures 16.13 and 16.14. Readers may refer to reference [14, 15] for further details.

Q8: Summary of main characteristics between PTR and GM cryocooler

A: Please review the following Table 16.1 as a reference

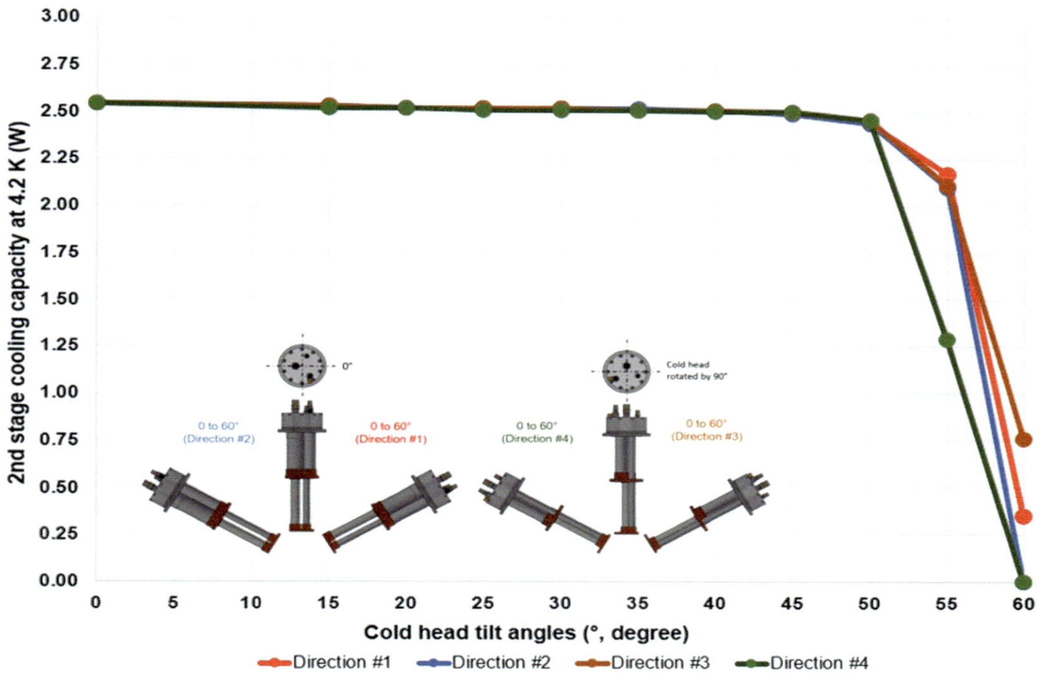

FIGURE 16.13 PT425-RM cooling capacities of the second stage at 4.2 K changing over tilt angles. (Reprint with permission from Bluefors Cryocooler Technologies Inc. 6682 Moore Rd., Syracuse, NY 13211, USA.)

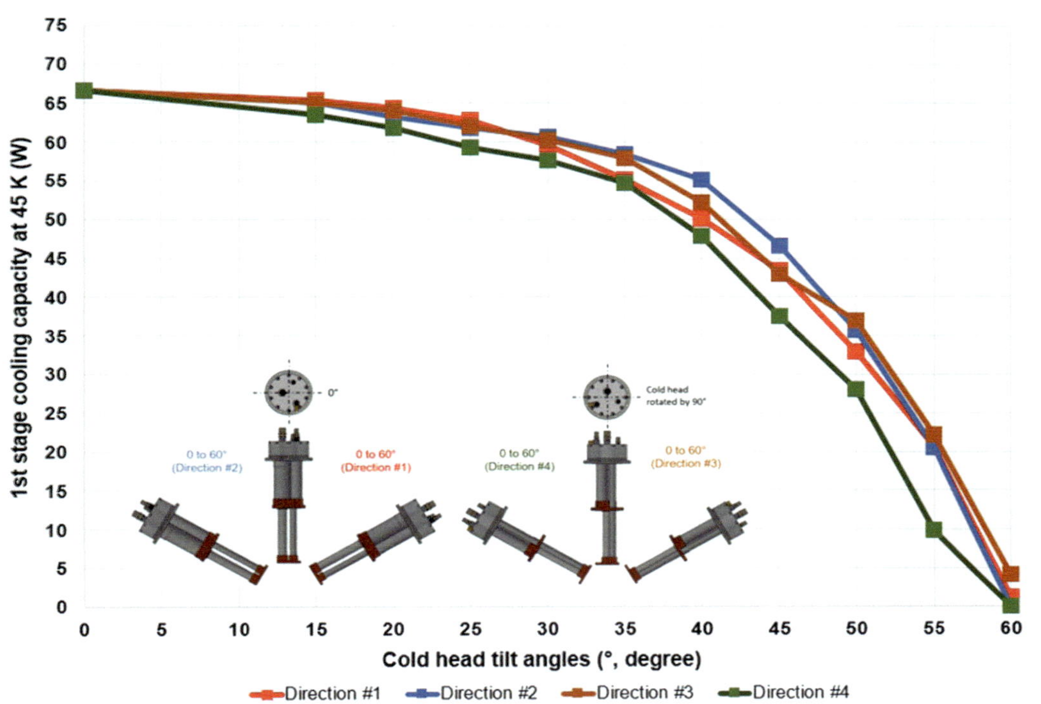

FIGURE 16.14 PT425-RM cooling capacities of the 1st stage at 45 K changing over tilt angles. (Reprint with permission from Bluefors Cryocooler Technologies Inc. 6682 Moore Rd., Syracuse, NY 13211, USA.)

TABLE 16.1
Summary of Main Characteristics Between PTR and GM Cryocooler

	Model RDK 408 GM Cryocooler	Model PT410 PTR
Vibration level		
Displacement	±13 μm	± 10 μm
Mounting force	40 lb.	1.0 lb.
Acceleration	9 m/s^2	0.09 m/s^2
Magnetic distortion from the second stage regenerator	± 200 nT	± 20 nT
Cold head operating orientation	Free	Vertical position with cold end down

REFERENCES

1. Young Jae Song, Alexander F. Otte, Vladimir Shvarts, Zuyu Zhao, Young Kuk, Steven R. Blankenship, Alan Band, Frank M. Hess, and Joseph A. Stroscio, Invited Review Article: A 10 mK Scanning Probe Microscopy Facility, *Review of Scientific Instruments*, 81, 121101, (2010).
2. Zuyu Zhao and Chao Wang (Eds.), *Cryogenic Engineering and Technologies: Principles and Applications of Cryogen-Free Systems*, CRC Press (Taylor & Francis Group), Boca Raton, FL, USA, (2020).
3. W.P. Kirk and M. Twerdochlib, Improved Method for Minimizing Vibrational Motion Transmitted by Pumping Lines, *Review of Scientific Instruments*, 49(6), 765–769, (1978).
4. Lake Shore Cryotronics, Inc. 480 Olde Worthington Rd, Westerville, OH, USA Website: www.lakeshore.com
5. Jiaojie Yan, Jianing Yao, Vladimir Shvarts, Rui-Rui Du, and Xi Lin, Cryogen-Free Hundred Microkelvin Refrigerator, *Review of Scientific Instruments*, 92(2), 024001, (2021).
6. Jiaojie Yan, Construction of Ultra-Low Temperature Environment and Transport Measurements, Ph.D. Thesis, Directed by Professor Xi Lin, College of Physics, Beijing University, Chain, (2022).
7. Physike Technology Co., Ltd. Guang Hua Innovation Park, Building 19, 18 East Anningzhuang Rd., Beijing, China Email: sales@physike.com Website: www.physike.com
8. Polytec GmbH Headquarters: Polytec-Platz 1–7, 76337 Waldbronn, Germany Email: info@polytec.de US Office: 16400 Bake Parkway, Irvine, CA, USA Email: info@polytec.com Website: www.polytec.com
9. Beijing National Center for Condensed Matter Physics and Institute of Physics, Chinese Academy of Sciences No. 8, South Third Street, Zhongguancun, Haidian District, Beijing 100190, People's Republic of China Website: www.iop.cas.cn
10. ACME (Beijing) Technology Co., Ltd. Building 7, No. 11 Xingke East Street, Yanqi Economic Development Area, Huairou District, Beijing 101407, China Website: www.acme-bj.com
11. Ruisong Ma, Hao Li, Chenshuai Shi, Fan Wang, Le Lei, Yuanzhi Huang, Yani Liu, Huan Shan, Li Liu, Shesong Huang, Zhi-Chuan Niu, Qing Huan, Hong-Jun Gao, Development of a Cryogen-Free Sub-3 K Low-Temperature Scanning Probe Microscope by Remote Liquefaction Scheme, *Review of Scientific Instruments*, 94, 093701, (2023).
12. Ruisong Ma, Zhiwei Liu, Dairong Liu, Hao Li, Chenshuai Shi, Jianfeng Tao, Yuanzhi Huang, Li Liu, Zhenchao Dong, Bing Wang, Shijing Tan, Qing Huan, A Cryogen-Free Low Temperature Scanning Probe Microscope System for Ångström-Resolved Spectroscopic Imaging, *Advanced Scientific Instruments* https://doi.org/10.1016/j.asi.2026.100011i (in press; available online 28 February, 2026, 100011)
13. Private communication with Bluefors Cryocooler Technologies Inc.
14. Bluefors Cryocooler Technologies Inc. 6682 Moore Rd, Syracuse, NY, USA Website: www.bluefors.com
15. Xihuan Hao, Brent Zerkle, Joe Cosco, Rich Dausman, *Experimental Investigation on the Tilt Angle Performance of 4 K Two-Stage Pulse Tube Cryocoolers for Cosmic Microwave Background (CMB) Telescopes*, Bluefors Cryocooler Technologies, Inc., Syracuse, NY, USA, 13211, (2024).

17 Cryogenic Electrical Wiring and Cabling and Vacuum Feedthroughs

17.1 CRYOGENIC WIRING AND RF CABLES

Wiring and high-frequency transmission lines—such as radio-frequency (RF) cables—are essential components for conducting reliable experiments at cryogenic temperatures. The types of wires and coaxial cables discussed in this chapter have been successfully implemented in a wide range of cryogenic systems, including He-3 cryostats and dilution refrigerators (DRs) operating at millikelvin temperatures. For additional technical details, readers are referred to Reference [1].

17.1.1 DC Wires

DC wires are used to transmit direct current (DC) signals in cryogenic systems. The selection of appropriate cryogenic wiring for a given application depends on multiple factors (Table 17.1), including the conductor material, electrical resistivity, thermal conductivity, insulation type, breakdown voltage, operating current, maximum operating temperature, and the intended function within the system. To minimize electromagnetic interference, thin stainless-steel wires are often braided around the DC wire to serve as an electromagnetic shield. This braided shielding typically covers approximately 90% of the wire surface and is usually grounded to the cryostat to reduce RF noise.

17.1.2 Coaxial Cables (RF Cables)

Coaxial cables, also referred to as RF cables, are composed of a central inner conductor, a dielectric insulating layer, and a concentric outer conductor, as illustrated in Figure 17.1. These cables are widely used for high-frequency signal transmission, typically ranging from the megahertz (MHz) to gigahertz (GHz) frequencies. Importantly, both the inner and outer conductors participate in the signal transmission path: the outer conductor carries return currents and helps maintaining signal integrity [2]. In addition to high-frequency signals, RF cables are also capable of transmitting DC signals when required.

The selection of suitable RF cables for a given cryogenic system or application depends on several key parameters, including characteristic impedance, conductor and shielding materials, thermal conductivity, breakdown voltage, operating frequency, insulation type, and maximum operating temperature.

One of the most critical properties of an RF cable is its *characteristic impedance*, denoted by Z_0. This parameter is defined as the ratio of voltage amplitude to current of a single electromagnetic wave propagating along an infinitely long coaxial cable in the absence of reflections. The characteristic

DOI: 10.1201/9781042000050-17

TABLE 17.1
Commonly Used Cryogenic Wires

Model	Description	Specifications	Vender
N13-38M-100*	38 AWG Shielded Twisted NbTi Pair	Wires: 38 AWG NbTi with 70 Copper-30 Nickel Clad .003" Translucent FEP to .010" +/- .001" OD Jacket: 0.005" Translucent FEP Shielding: 44 AWG Soft Type 316L Stainless Steel Braid to .039" +/- .003" OD	New England Wire Technologies 130 N Main St, Lisbon, NH 03585 info@newenglandwire.com
N13-38E-103*	38 AWG Shielded Twisted Manganin Pair	Wires: heavy polyurethane insulated manganin wires Jacket: 0.005" TEFZEL to 0.020" +/- 0.001" o.d. Shielding: 44 AWG Cu-Ni braided shield, 90% minimum coverage; 0.029" +/- 0.002" o.d.	New England Wire Technologies 130 N Main St, Lisbon, NH 03585 info@newenglandwire.com
N12-36M-131*	36 AWG shielded NbTi sin gle wire	Wires: 36 AWG NbTi with 70 Copper-30 Nickel cladding Jacket: 0.006" FEP to 0.017" ±0.002" OD translucent Shielding: Braid 44 AWG type 316L SS 90% min coverage	New England Wire Technologies 130 N Main St, Lisbon, NH 03585 info@newenglandwire.com
N12-38R+ 00002-0*	38 AWG shielded single wire	Wires: 38 AWG hard bar NEWALOY 75 Jacket: 0.0045" PFA TO 0.013" +/- 0.001" OD – translucent Shielding: Braid 44 AWG type 316L SS 90% min coverage to 0.022" +/- 0.001" OD	New England Wire Technologies 130 N Main St, Lisbon, NH 03585 info@newenglandwire.com
N23-44M-100-0* (low noise wires)	44 AWG Shielded Twisted NbTi Pair	Wires: 44 AWG NbTi with 70 Copper-30 Nickel Clad .010" FEP to .022" +/- .002" OD – Translucent, Low-noise carbon powder coating, Jacket: .005" FEP Jacket to .064" +/- .004" OD – translucent Shielding: 44 AWG Soft Type 316L Stainless Steel Braided shield 90% min coverage	New England Wire Technologies 130 N Main St, Lisbon, NH 03585 info@newenglandwire.com
N13-30E-100	38 AWG Shielded Twisted Copper Pair	Wires: 38 AWG heavy polyurethane-nylon insulated copper wires Jacket: 0.006" translucent ETFE Shielding: 44 AWG type 304 stainless steel braided shield	New England Wire Technologies 130 N Main St, Lisbon, NH 03585 info@newenglandwire.com
N12-38R-100-0*	38 AWG shielded twisted pair	Wires: 38 AWG T75H, 4.5PFA NEWALOY 75 Jacket: 0.0045" PFA TO 0.013" +/- 0.001" OD – translucent Shielding: Braid 44 AWG type 316L SS 90% min coverage to 0.022" +/- 0.001" OD	New England Wire Technologies 130 N Main St, Lisbon, NH 03585 info@newenglandwire.com
CFW2002932*	36 AWG manganin twisted pair	Manganin, twisted, 2 x 0.005"; Manganin 290; HML NATRL insulation coating (no shield)	California Fine Wire Co. 338 S 4th St, Grover Beach, CA 93433 https://calfinewire.com/

Source: Author's note.

Items marked with * have been proven UHV compatible based on prior use.

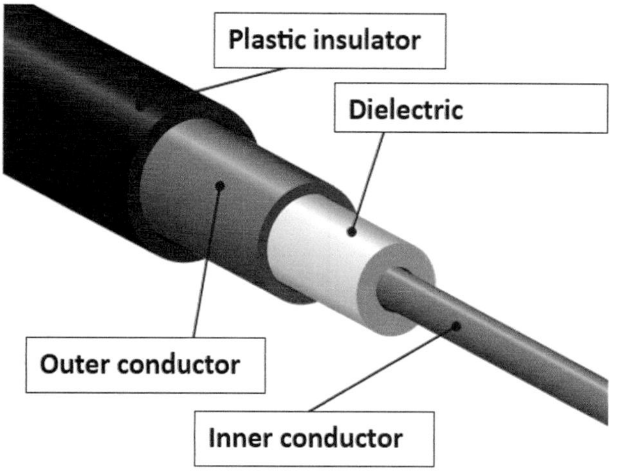

FIGURE 17.1 RF cable. (Reprint from www.shutterstock.com/search/semi-rigid-coaxial-cable; public domain.)

impedance is determined by the geometric dimensions of the inner and outer conductors and by the dielectric properties of the insulating material. It is expressed as

$$Zo = \frac{138 log_{10}(\frac{D}{d})}{\sqrt{\varepsilon}} \quad (17.1\ ) \qquad (17.1)$$

where

Z_0: characteristic impedance (Ω)
D: inner diameter of the outer conductor (or the outer diameter of the dielectric)
d: outer diameter of the inner conductor
ε: dielectric constant of the insulating material between the inner and outer conductors

As an example, the UT-85 semi-rigid coaxial cable is a widely used in cryogenic applications with the following parameters:

- Inner diameter of the outer conductor, D: 0.066 in
- Outer diameter of the inner conductor, d: 0.0201 in
- Dielectric material: Polytetrafluoroethylene (PTFE, known as Teflon)
- Dielectric constant of PTFE, ε: 2.1
- Characteristic impedance, Z_0: 50 ± 2.0 Ω.

Cable attenuation, typically expressed in decibels per unit length (dB/m), quantifies the power loss that occurs as signals propagate through an RF cable. This power loss depends on factors such as the cable material properties, length, signal frequency, and geometric configuration. In addition to attenuation, other key electrical properties—such as inductance, capacitance, and cutoff frequency—are also critical for determining cable performance in specific high-frequency and cryogenic applications.

A voltage difference between the inner and outer conductors of an RF cable can result in a small leakage current flowing through the dielectric material. Even extremely small leakage currents can lead to significant measurement errors, particularly in sensitive cryogenic experiments. Further discussion on this issue is provided in Section 17.5.

Commonly used cryogenic RF cables are summarized in Table 17.2.

TABLE 17.2
Often used cryogenic RF cables

Model	Description	Specifications	Vender
N12-50F-257-0	38 AWG flexible coaxial cable	Outer conductor: 44 AWG type 316L stainless steel braided shield, 90% minimum coverage Dielectric: 0.004" clear FEP to 0.013" +/- 0.001" o.d. Inner conductor: 38 AWG 17/50 type of 316L stainless steel Impedance: 51.6 +/- 5 ohm Insulation: 0.005" clear FEP jacket to 0.032" +/ - 0.002" o.d.	New England Wire Technologies 130 N Main St, Lisbon, NH 03585 info@newenglandwire.com
N12-30B+ 00001-0	30 AWG flexible coaxial cable	Outer conductor: 44 AWG type 316L stainless steel braided shield, 90% minimum coverage Dielectric: 0.009" PFA to 0.028" +/- 0.002" o.d. Inner conductor: 30 AWG solid bare copper Impedance: 50.0 +/- 5 ohm Insulation: 0.010" PFA jacket to 0.057" +/- 0.003" o.d	New England Wire Technologies 130 N Main St, Lisbon, NH 03585 info@newenglandwire.com
N12-36B-110-0	36 AWG flexible coaxial cable	Outer conductor: 44 AWG type 316L stainless steel braided shield, 90% minimum coverage Dielectric: 0.0045" FEP to 0.014" +/- 0.001" o.d. Inner conductor: 36 AWG bare copper Impedance: 50.0 +/- 5 ohm Insulation: 0.005" clear FEP jacket to 0.033" +/ - 0.002" o.d.	New England Wire Technologies 130 N Main St, Lisbon, NH 03585 info@newenglandwire.com
UT-85-BSS	Semi-rigid coaxial cable	Outer conductor: 0.0865"+/-0.001" 304 stainless steel Dielectric: 0.066 " PTFE Inner conductor: 0.0201"+/-0.0005" SP BeCu Impedance: 50.0 +/- 2.0 ohm	Microstock, Inc West Point, PA 19486 www.microstock-inc.com/
UT-85-SS-SS*	Semi-rigid coaxial cable	Outer conductor: 0.0865"+/-0.001" 304 stainless steel Dielectric: 0.066 " PTFE Inner conductor: 0.0201"+/-0.0005" 304 stainless steel Impedance: 50.0 +/- 2.0 ohm	Microstock, Inc West Point, PA 19486 www.microstock-inc.com/

Source: Author's note.

Items marked with * have been proven UHV compatible based on prior use.

17.2 COMMERCIAL DC FEEDTHROUGH AND CONNECTORS

Figure 17.2 illustrates two most commonly used commercial electrical feedthroughs, both of which are non-UHV room temperature components. Detoronics feedthroughs are widely used due to their low cost and ease of use. In contrast, Fischer feedthroughs are used when a more compact form factor is required or when nonmagnetic construction is critical for magnetically sensitive experiments.

Figure 17.3 illustrates commonly used commercial UHV compatible, room-temperature electrical feedthroughs. These feedthroughs typically employ multipin Sub-D or ball-style connectors and are designed to provide reliable electrical connections while maintaining UHV integrity. They are widely used in cryogenic and surface-science applications where vacuum compatibility, low outgassing, and long-term stability are essential.

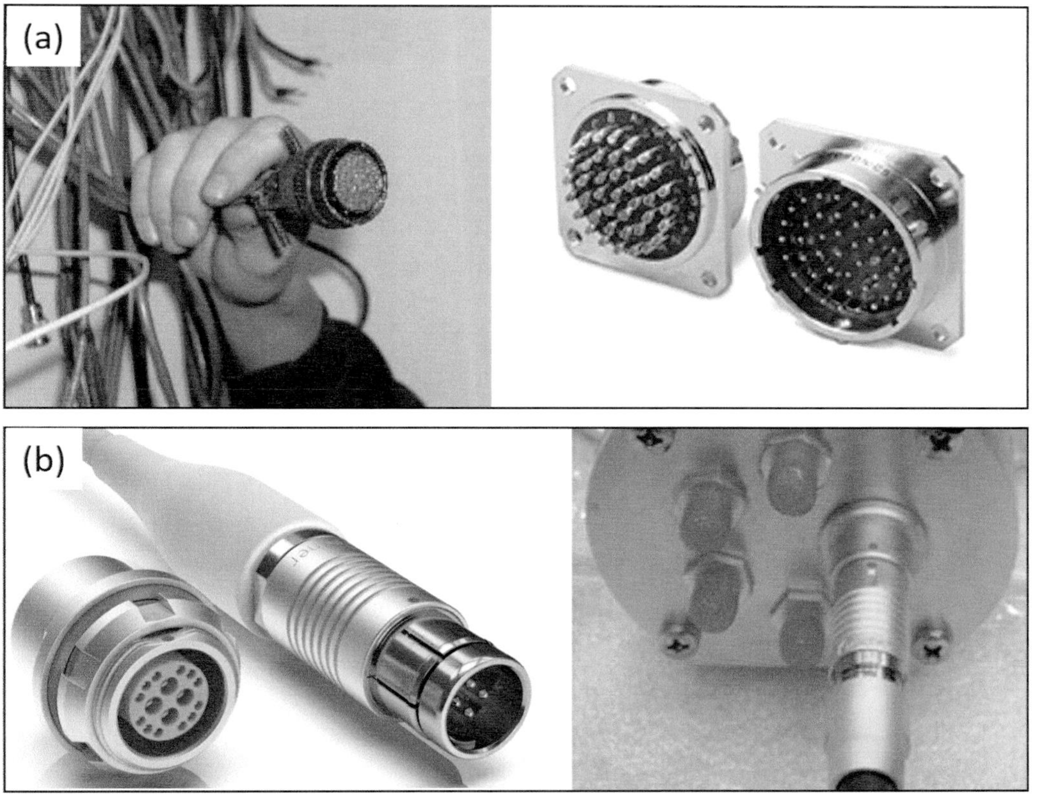

FIGURE 17.2 (a) Detoronics multipin connector and receptacle. (Projects accomplished by authors at Janis Research Company. Image courtesy of FormFactor, Inc.) Information source: Detoronics Corporation, www.Detoronics.com; (b) Fischer multi-pin connect. (Projects accomplished by authors at Janis Research Company. Image courtesy of FormFactor, Inc.) Information source: Fischer Connectors SA, https://fischerconnectors.com/en.

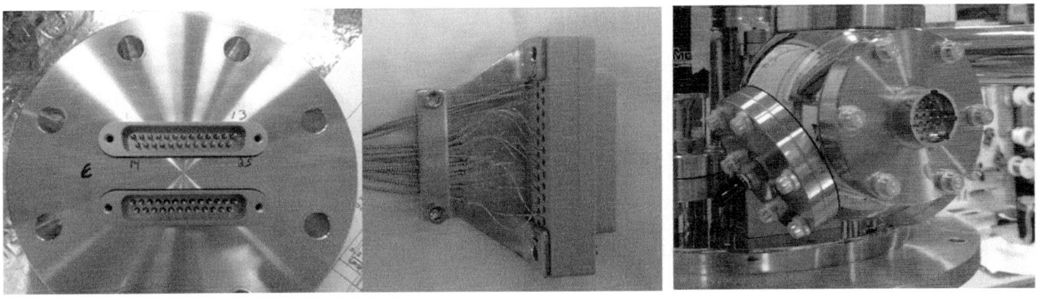

(a) (b)

FIGURE 17.3 (a) UHV compatible multipin sub-D feedthrough and connector. (Projects accomplished by authors at Janis Research Company. Image courtesy of FormFactor, Inc.) Information source: MDC Precision, www.mdcprecision.com/; (b) UHV compatible Ball feedthrough (Projects accomplished by authors at Janis Research Company. Image courtesy of FormFactor, Inc.) Information source: Accu-Glass Products, Inc., www.accuglassproducts.com.

17.3 COMMERCIAL RF FEEDTHROUGH AND CONNECTORS

17.3.1 BNC Feedthrough and Connector

BNC (Bayonet Neill–Concelman) connector (Figure 17.4) is a compact, quick-connect RF interface designed for use with coaxial cables. It is widely employed in video and radio-frequency applications and typically supports operating frequencies up to approximately 2 GHz and voltages up to 500 V. The connector utilizes a bayonet-style locking mechanism, where two lugs on the female connector engage corresponding slots in the male connector shell, providing a secure and reliable connection.

17.3.2 SMA and SMK Bulkhead and Connector

The SMA (SubMiniature version A) connector (Figure 17.5) is a widely used miniature RF interface that supports frequencies from DC up to 18 GHz. Due to its threaded coupling mechanism, SMA connectors provide excellent mechanical stability and repeatable electrical performance, making them well suited for cryogenic and high-vibration environments.

The SMK (SubMiniature version K) connector features a similar mechanical design to the SMA but incorporates an air dielectric and a Noryl support bead. These design enhancements allow the SMK connector to operate at significantly higher frequencies, extending up to 40 GHz.

17.3.3 SSMC Bulkhead and Connector

The SSMC (SubMiniature Coax) connector (Figure 17.6) is an ultra-compact RF interface designed for high-density interconnect applications. Its bulkhead and mating connectors create a subminiature interface capable of supporting both DC and RF signal transmission, with an operating frequency up to 12.4 GHz. Due to their small footprint, SSMC connectors are particularly suitable in space-constrained cryogenic systems.

17.3.4 SMP Interface

17.3.4.1 SMP Connector

The SMP (SubMiniature Push-on) connector is a compact RF interface that supports frequencies from DC up to at least 40 GHz. Its compact size, broadband performance, and ease of

FIGURE 17.4 BNC feedthrough and connector. (Projects accomplished by authors at Janis Research Company. Image courtesy of FormFactor, Inc.) Information source: Accu-Glass Products, Inc., www.accuglassproducts.com.

installation have made it increasingly popular in modern microwave and cryogenic applications. The connector employs a push-on coupling mechanism and often uses a female-to-female adapter—commonly called a *bullet*—as shown in Figure 17.7(a).

Because the coupling is not threaded, SMP connectors can tolerate slight radial misalignment. However, excessive misalignment may result in insertion-loss and return-loss degradations on the order of a few tenths of a decibel.

17.3.4.2 SMP-Terminated RF Cable

Commercial RF cables are often supplied with SMP connectors pre-installed on both ends. For NbTi superconducting RF cables, SMP connectors are commonly attached using the crimping method. It is advisable to verify electrical continuity between the connectors and the cable, especially after repeated thermal cycling between room temperature and cryogenic temperatures.

17.3.4.3 Catcher's Mitts and RF Cables Installation

A Catcher's Mitt (Figure 17.7(b)) is a male-to-male adapter designed for SMP connectors. It consists of a hexagonal nut with a smooth bore on one side and male mounting threads on the other. In practice, the SMP connector is typically easier to push onto the smooth-bore side.

When mounted on a cold plate as a bulkhead, the Catcher's Mitt provides RF cable access from both sides of the plate. In addition, it serves as an effective thermal anchoring stage, helping to intercept and manage heat flow along RF cables in cryogenic systems.

Catcher's Mitts and RF cables are installed by threading the Catcher's Mitt into the cold plate, with the hexagonal side typically seated against the underside of the plate, followed by connecting RF cables to both sides of the Catcher's Mitt, as illustrated in Figure 17.7(c) and 17.7(d).

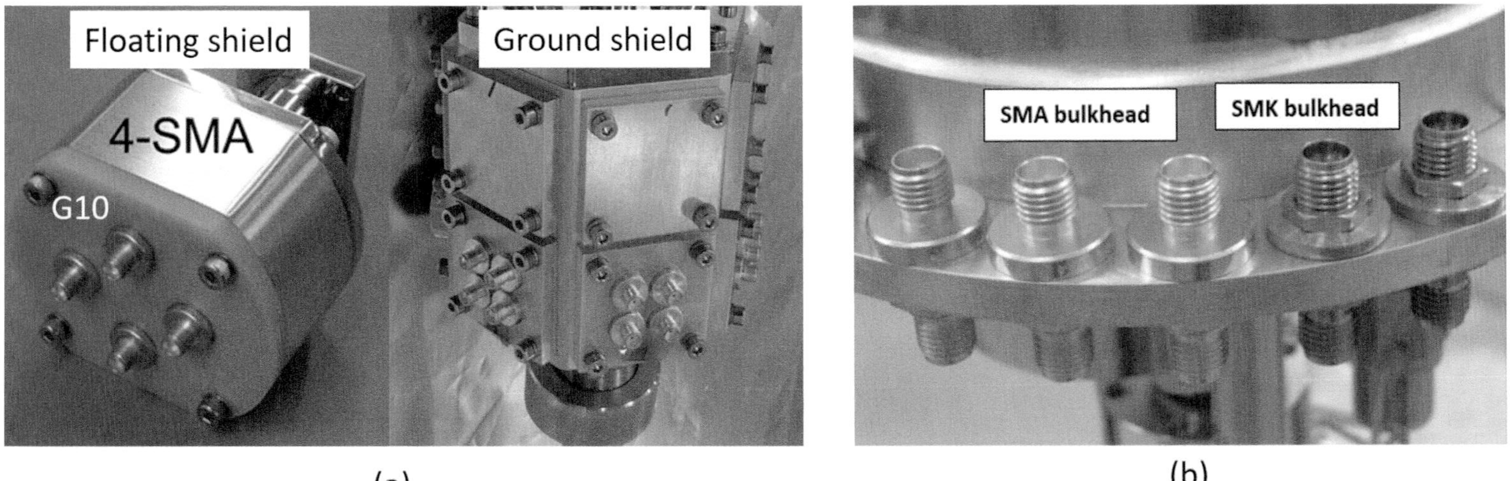

FIGURE 17.5 (a) SMA feedthrough. (Projects accomplished by authors at Janis Research Company. Image courtesy of FormFactor, Inc.) Information source: Pasternack Enterprises Inc., www.pasternack.com; (b) SMA and SMK bulkhead. (Projects accomplished by authors at Janis Research Company. Image courtesy of FormFactor, Inc.) Information source: Fairview Microwave, Inc.; www.fairviewmicrowave.com/.

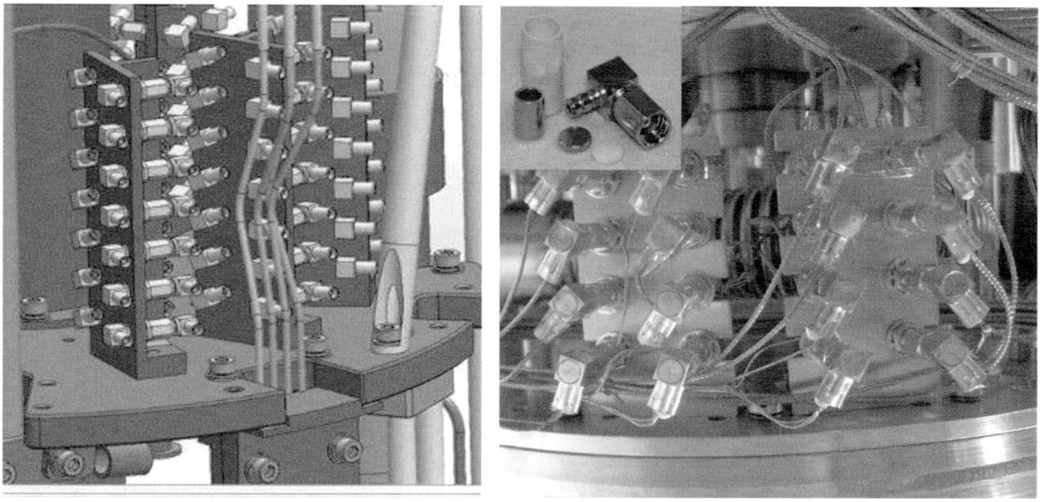

(a) (b)

FIGURE 17.6 (a) SSMC interface assembly. (Projects accomplished by authors at Janis Research Company. Image courtesy of FormFactor, Inc.) Information source: Pasternack Enterprises Inc., www.pasternack.com; (b) SSMC right angle connector. (Projects accomplished by authors at Janis Research Company. Image courtesy of FormFactor, Inc.) Information source: Fairview Microwave, Inc.; www.fairviewmicrowave.com/.

(a) (b) (c) (d)

FIGURE 17.7 (a) RF cable with female-to-female SMP connector installed. (Information from Fairview Microwave, Inc.; www.fairviewmicrowave.com/), (b) Catcher's Mitts, (c) Catcher's Mitts installed on the cold plate, with the hexagonal nut positioned on the underside of the mounting plate. (Projects accomplished by authors at Janis Research Company. Image courtesy of FormFactor, Inc.), (d) RF cable assembly connected to both sides of the Catcher's Mitt. (Projects accomplished by authors at Janis Research Company. Image courtesy of FormFactor, Inc.)

17.4 WIRE AND CABLE THERMAL ANCHOR

Effective thermal anchoring of wires and RF cables is essential for both the performance and long-term stability of cryogenic systems. By securing these components at intermediate temperature stages—such as 77 K, 50 K, and 4 K—thermal conduction from warmer to colder regions can be significantly reduced. This staged anchoring approach minimizes the heat load on the coldest components, which is particularly critical in ultra-low-temperature systems operating in the millikelvin regime.

In addition to thermal management, proper anchoring also helps damp vibrations transmitted along cables and wires, thereby reducing mechanical noise that can compromise sensitive measurements.

The following section describes commonly employed techniques for thermally anchoring wires and RF cables at various temperature stages.

17.4.1 Wires and Flexible Coaxial Cables Thermal Anchor

17.4.1.1 Thermal Anchor Bobbin

A common technique for thermally anchoring wires and flexible coaxial cables is the use of a copper thermal anchor bobbin. In this approach, the wires or cables are wounded around a copper bobbin, which serves as a thermal sink to efficiently thermalize the windings to the anchor stage. The windings are secured with a thermally conductive epoxy, such as STYCAST 2850FT, ensuring good thermal contact between windings and the copper surface while also providing mechanical stability. Figure 17.8(a) illustrates a typical thermal anchor bobbin, showing the flexible coaxial cable routing and bonding configuration used to maximize heat exchange.

17.4.1.2 Thermal Anchor Stage with Multicontact

Another effective thermal anchoring is to secure a bundle of wires and flexible coaxial cables within a copper multicontact clamp. The wires and cables are inserted into the clamp and fixed in place with a thermally conductive epoxy, such as STYCAST 2850FT, providing thermal contact between each conductor and the copper interface. After the epoxy cures, the multicontact is mechanically fastened to a cold plate using nuts on both sides of the plate. This anchoring offers a robust mechanical connection while ensuring efficient thermal coupling to the thermal stage (Figure 17.8(b) and (c)).

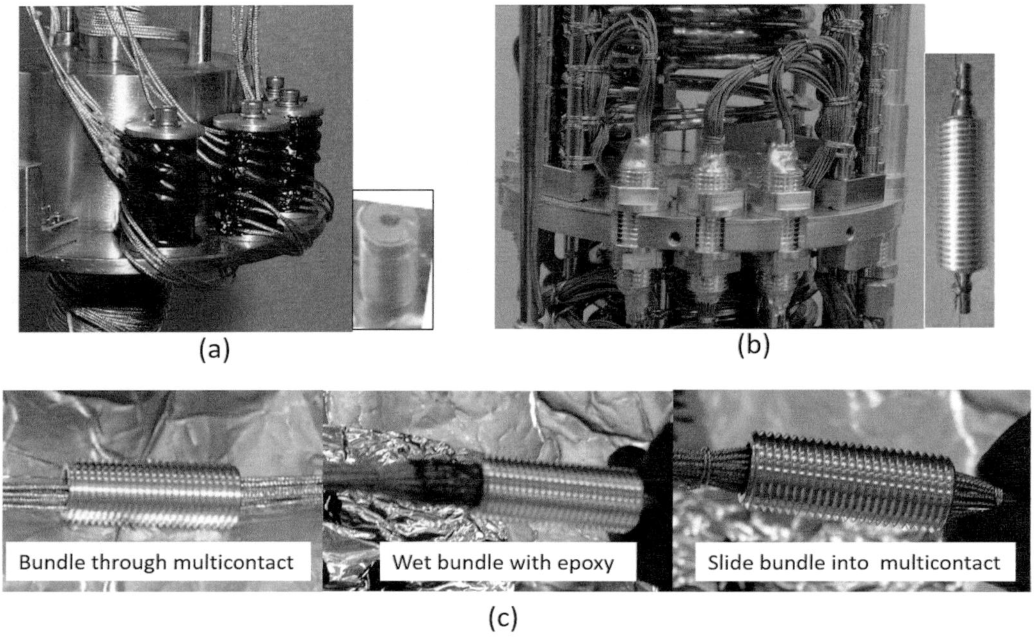

FIGURE 17.8 (a) Thermal anchor with bobbin, (b) Thermal anchor wires with multicontact, and (c) Thermal anchor stage with multicontact. (Projects accomplished by authors at Janis Research Company. Image courtesy of FormFactor, Inc.)

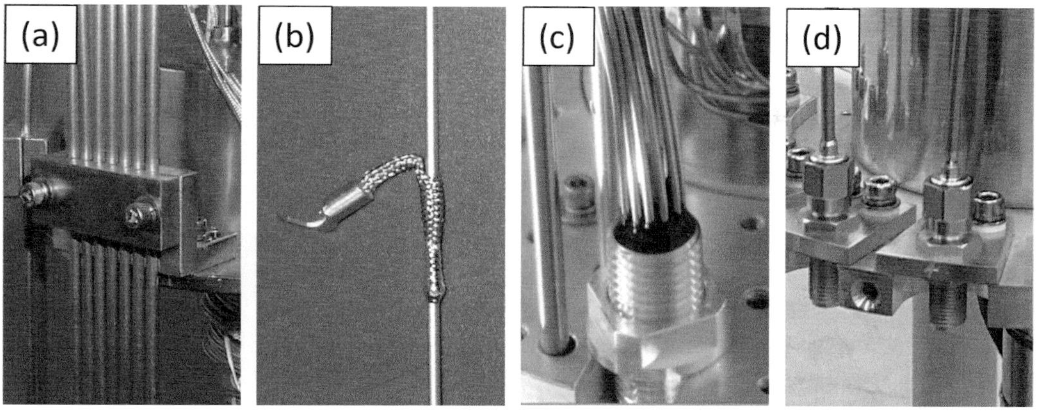

FIGURE 17.9 Thermal anchor of semi-rigid coaxial cables. (Projects accomplished by authors at Janis Research Company. Image courtesy of FormFactor, Inc.)

17.4.2 Semi-rigid Coaxial Cables Thermal Anchor

Semi-rigid coaxial cables require reliable thermal anchoring to reduce heat load to the cold stages while maintaining signal integrity. Several effective approaches are commonly used:

- Direct clamping: The coaxial cables are secured directly to a cold stage using copper clamps or brackets, providing strong thermal contact (Figure 17.9(a)).
- Flexible copper braid thermal links: A short length of flexible copper braid is soldered or clamped between coaxial cable and cold stage. This provides a low-resistance thermal path while maintaining mechanical decoupling (Figure 17.9(b)).
- Multicontact clamp (epoxy-anchored): A copper multicontact clamp can be used as thermal anchoring interface. Coaxial cables are embedded inside a multicontact and secured with thermally conductive epoxy. Finally, the connector is then fastened to the cold plate using nuts on both sides of the plate (Figure 17.9(c)).
- Bulkhead feedthroughs or "Catcher's Mitts": Semi-rigid cables are interconnected by or terminated at bulkheads or Catcher's Mitts, which are thermally anchored to the corresponding temperature stage. These components provide both thermal anchoring and mechanical stability (Figure 17.9(d)).

Each method offers a tradeoff among thermal performance, mechanical robustness, and ease of assembly, and the optimal choice depends on the specific requirements and constraints of the cryogenic system.

17.5 TRIAXIAL CABLE

A triaxial cable consists of three concentric conductors, a core, an intermediate guard, and an outer shield, as illustrated in Figure 17.10. Triaxial cables are commonly used for ultra-low-current measurements, typically in the picoampere to femtoampere range, such as during semiconductor wafer-level testing in probe systems.

To suppress leakage currents between conductors, the guard conductor is actively driven to the same voltage as the inner signal core using a buffer amplifier. By eliminating the voltage differential between the core and the guard, this technique significantly reduces parasitic leakage currents that could otherwise compromise measurement accuracy.

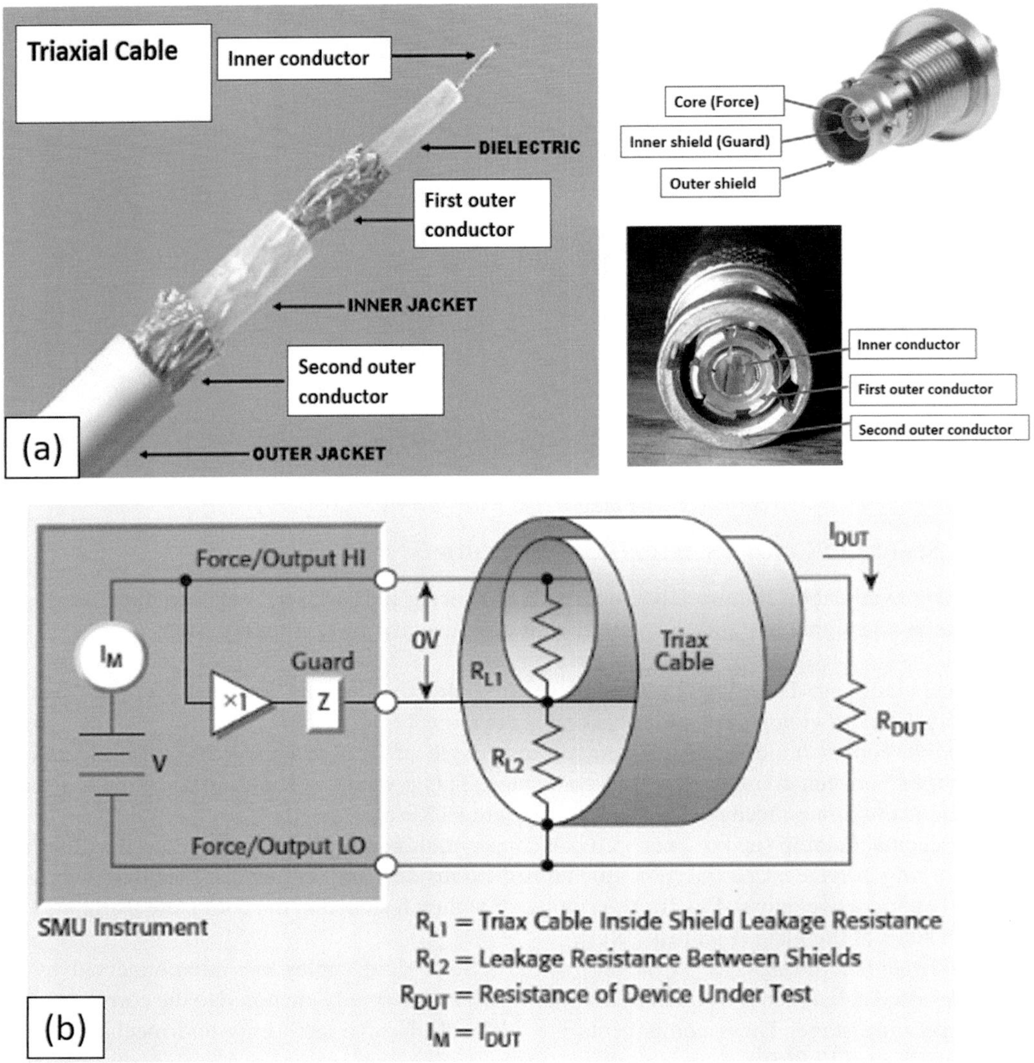

FIGURE 17.10 Triaxial cables and measurement set-up. Information source: https://images.google.com search for triaxial cable measurement set-up.

A high-impedance resistor (R12) is installed between the guard and the outer shield, as shown in Figure 17.10. Although a small leakage current may flow between the guard and the outer shield, it does not impact measurement accuracy because these current bypasses the device under test and therefore does not contribute to the measured signal.

17.6 CUSTOMIZED HERMETIC CRYOGENIC FEEDTHROUGH

Hermetic cryogenic feedthroughs are essential components in custom cryostats. They must maintain vacuum integrity and provide reliable electrical connections, even after repeated thermal cycling between ambient and cryogenic temperatures. Proper thermal anchoring of the wires or cables connected to the feedthrough is equally critical to prevent unwanted heat from entering the cryogenic system.

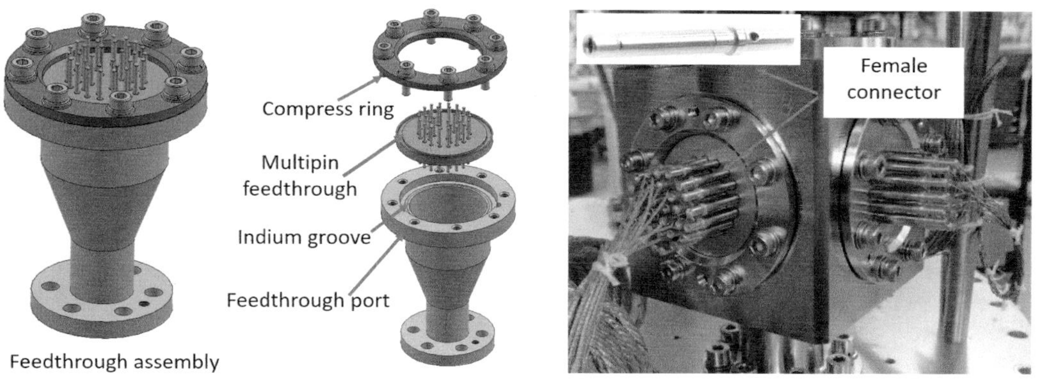

FIGURE 17.11 Demountable indium sealed multi-pin feedthrough. (Projects accomplished by authors at Janis Research Company. Image courtesy of FormFactor, Inc.)

Although several commercial vendors offer cryogenic feedthroughs and related components [3], many applications, such as the hermetic sample chamber described in Chapter 8, require customized solutions to meet specific electrical, thermal, and mechanical requirements.

This section describes the design and fabrication of three types of homemade hermetic cryogenic feedthroughs, indium-sealed multi-pin feedthrough, epoxy-based cold feedthrough, and custom SMA feedthrough assembled from commercial components, each tailored to different experimental requirements. These designs support both DC and high-frequency signal transmission, providing flexible solutions for a wide range of cryogenic applications.

17.6.1 Demountable Indium-Sealed Multipin Feedthrough

A demountable indium-sealed multipin feedthrough, based on weldable straight-pin feedthroughs, provides a reliable and reusable method for establishing electrical connections across a cryogenic vacuum boundary (See Figure 17.11).

Step 1: Feedthrough selection
Select a weldable multi-pin feedthrough suitable for the desired application. In our projects, CeramTec 32-Straight-Pin feedthroughs were used [3] (https://catalog.ceramtec.us/viewitems/feedthroughs/multipin-header-feedthroughs).

Step 2: Indium seal groove preparation
Machine a circular groove into the mating surface of the feedthrough. The groove should be sized to securely accommodate a 1.0 mm diameter indium wire, which is recommended for achieving a robust hermetic seal.

Step 3: Indium wire installation
Place a 1.0 mm diameter indium wire into the machined groove, ensuring it is seated evenly without gaps.

Step 4: Feedthrough positioning
Orient the feedthrough with the flat sealing surface facing the indium wire and carefully align it with the indium-filled groove on the mating surface to ensure proper contact.

Step 5: Compression-ring sealing
Install a stainless-steel compression ring over the feedthrough and tighten the mounting bolts evenly. This applies controlled pressure, deforming the indium wire to form a vacuum-tight seal.

Step 6: Internal soldering
Solder the internal signal wires to the detachable female connector pins.

Step 7: Electrical connection
Mate the detachable female connector pins to the corresponding male pins on both sides of the feedthrough, completing the signal path across the vacuum boundary.

This design uses commercially available components and provides a convenient, reliable, and serviceable solution for establishing electrical connections across a cryogenic vacuum boundary. It is important to note that the hermetic seals on the multipin feedthrough are fragile. Wires should not be welded or soldered directly to the feedthrough pins, as excessive heat can weaken the vacuum seal, potentially causing leaks or mechanical failure.

17.6.2 Hermetic Epoxy Cold Feedthrough for Wire Bundle

Epoxy-based feedthroughs offer a compact and versatile solution for routing wire bundles into cryogenic vacuum environments.

Step 1: Materials and Components
Prepare the following components:

- Stainless-steel tube
  - Outer diameter: 0.125 in
  - Length: 1.5–2.0 in
- Small indium-sealed flange
  - Material: Stainless steel or brass
  - Function: Mounting interface for the feedthrough assembly

(Design details of the indium-sealed flange are presented in Chapter 18. VCR or Conflat flanges may be substituted depending on application requirements.)

- Plastic tubing
  - Outer Diameter: ~0.5 in
  - Length: 4–6 in
  - Function: Mold or containment sleeve during epoxy curing
- Stycast 1266 epoxy
  - Type: Two-component epoxy resin
  - Properties: Low-temperature compatible, durable, cryogenically stable, excellent sealing performance
- Glass desiccator with pump-out port
  - Function: Vacuum degassing of mixed epoxy
- Roughing Pump
  - Function: Ultimate vacuum level (typically 10^{-3}–10^{-2} Torr), supporting epoxy degassing process and preparing for integration into high-vacuum systems

Step 2: Epoxy Cup Fabrication

- Epoxy Mixing
  Mix 100 g of Stycast 1266 resin and 28 g of catalyst, following the manufacturer's specified ratio. Stir thoroughly until the mixture is fully homogeneous.
- Vacuum degassing
  Place the mixed epoxy into a glass desiccator connected to a roughing pump.
  - Pump until the epoxy begins vigorous outgassing, then stop pumping
  - Resume pumping after bubbling subsides and the mixture appears quiescent.
  - Repeat until no visible bubbles remain.

- Molding and curing
  Pour the degassed epoxy into the prepared plastic tubing mold and allow it to cure undisturbed at room temperature for the manufacturer-recommended duration. Prior to filling, the tube must be clean and free of contaminants.
- Machining
  After curing, remove the plastic tubing and machine the epoxy cylinder on a lathe to the desired cup geometry (see Figure 17.12).

  All mixing, curing, and machining steps should be carried out in a clean, dust-free environment to minimize contamination. The manufacturer's technical data sheets must always be referred to ensure proper curing conditions and to follow recommended safety and handling practices.

  During machining, low spindle speeds should be used to limit heat buildup. A continuous flow of water or alcohol coolant is recommended to maintain a low epoxy temperature. Overheating must be avoided, as it can cause cracking, surface defects, or warping of the material.

Step 3: Epoxy Cold Feedthrough Fabrication

- Mechanical preparation (Figure 17.13(a))
  - Taper one end of the stainless-steel tube to reduce wall thickness to mitigate thermal stress between the epoxy cup and the tube during cooldown.
  - Attach the flat (non-tapered) end of the tube to the indium-sealed flange by welding or soldering depending on the requirements.
  - Thoroughly clean the tube with acetone to remove oils, grease, or particulates.
- Wire installation and epoxy filling (Figure 17.13(b) and 17.13 (c))
  - Feed the wire bundle carefully through the tube.
  - Oriented the epoxy cup upward; insert the stainless-steel tube into the epoxy cup, positioning the tapered end approximately 3 mm (0.125 in) above the bottom of the

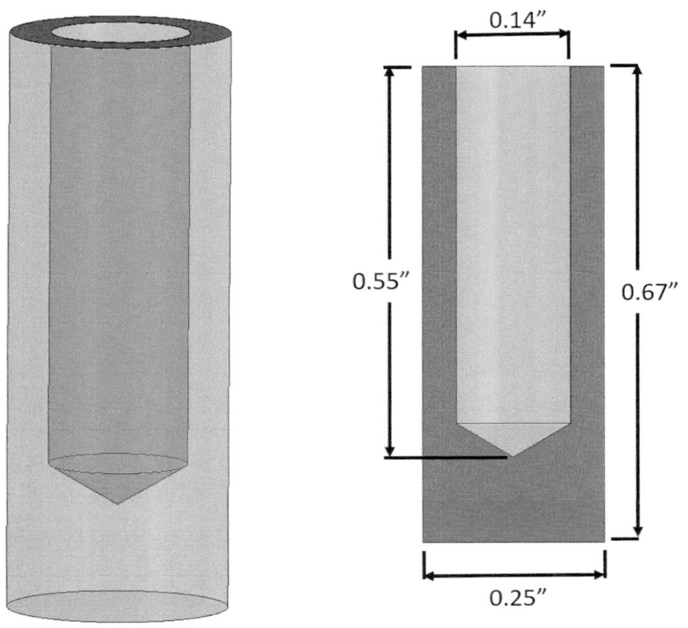

FIGURE 17.12 Epoxy cup.

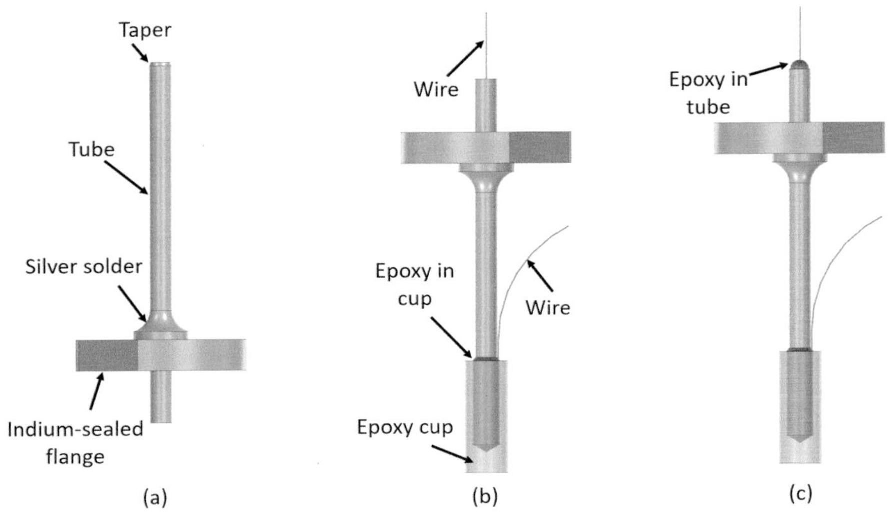

FIGURE 17.13 1266 Stycast epoxy cold feedthrough fabrication.

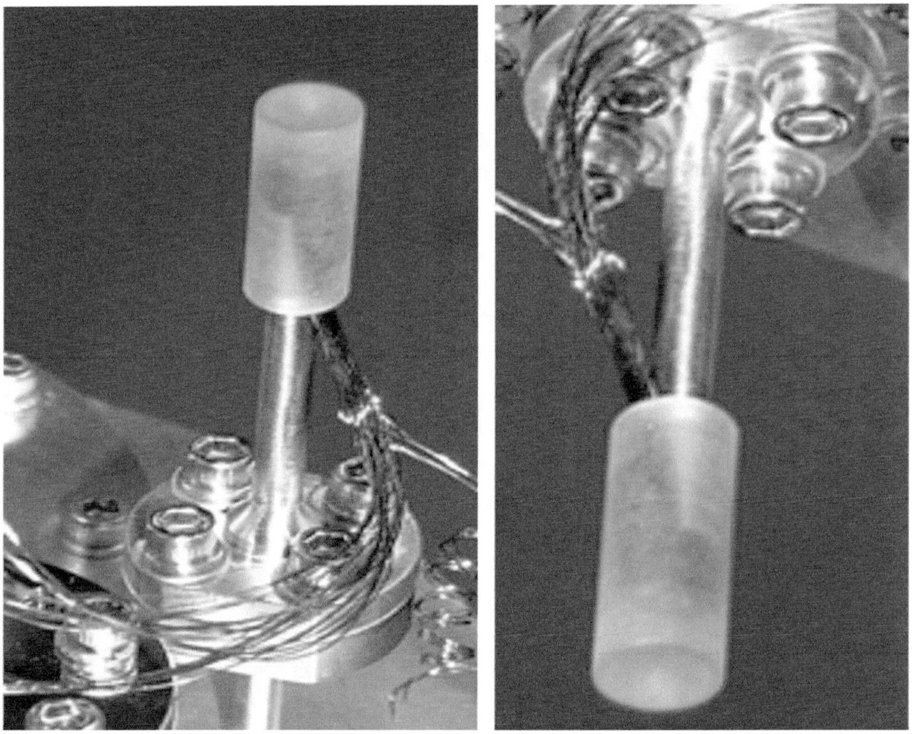

FIGURE 17.14 1266 Stycast epoxy cold feedthrough installed on cryostat. (Projects accomplished by authors at Janis Research Company. Image courtesy of FormFactor, Inc.)

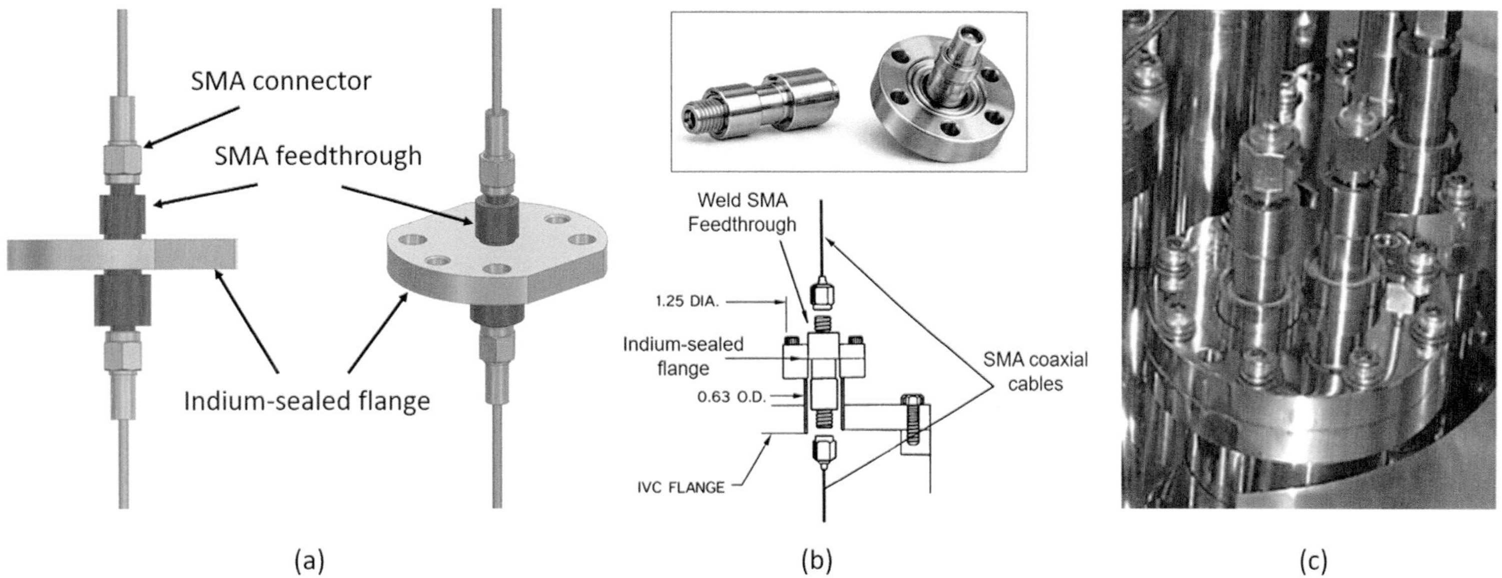

FIGURE 17.15 (a) Hermetic cryogenic SMA feedthrough design, (b) Double-ended hermetic SMA feedthrough, and (c) Hermetic cryogenic SMA feedthrough assembly. (Projects accomplished by authors at Janis Research Company. Image courtesy of FormFactor, Inc.)

cup. Make sure that the wires do not contact the tapered edge. (*The contact may damage the wire insulation and create a risk of electrical shorting*)
- Pour freshly mixed and degassed Stycast 1266 into the cup, fully encapsulating the wires and forming a void-free seal.
- After the epoxy in the cup has cured, fill the stainless-steel tube with epoxy to secure the wires.

After the epoxy has completely cured, the epoxy cold feedthrough is now complete and ready for use (Figure 17.14).

The feedthrough design enables efficient use of space by accommodating multiple wires within a compact footprint—for an example we have successfully installed 50 fine wires in a single feedthrough. This configuration is particularly advantageous in cryogenic and vacuum environments, where minimizing component size is essential while maintaining full functionality and reliability.

Stycast 1266 epoxy allows a small amount of helium permeation at room temperature, with typical leak rates on the order of 10^{-6} to 10^{-5} Torr·L/s. When the vacuum space is surrounded by liquid helium or helium-rich gas, this low-level permeation may be mistaken for a true leak. However, at temperatures below 78 K (liquid nitrogen temperature), the leak rate decreases significantly, to approximately 10^{-8} Torr·L/s or lower. This improvement makes the design highly effective for cryogenic applications where maintaining vacuum integrity is critical.

In the practical applications, sudden temperature changes can induce thermal shock and potentially cause cracking of the Stycast epoxy during system cooldown. Recommended measures to mitigate thermal shock include:

- Gradual cooling: Implement controlled, stepwise cooldown procedures to allow the epoxy to adjust to temperature changes with minimal stress. For example, introduce cryogen slowly onto the epoxy feedthrough during initial cooling.
- Vacuum isolation: Whenever possible, position the epoxy cup in a vacuum space to avoid direct contact with cryogens, reducing thermal gradients across the epoxy.

17.6.3 Hermetic Cryogenic SMA Feedthrough

To support high-frequency signal transmission in cryogenic environments, cryogenically rated hermetic SMA feedthroughs can be integrated by welding a commercial double-ended SMA feedthrough—such as those from MDC Precision—onto a Conflat or indium-sealed flange (see Figure 17.15). This approach provides a stable, reliable connection at low temperatures, ensuring the system remains both hermetic and electrically functional.

REFERENCES

1. Zuyu Zhao and Chao Wang (Eds.), *Cryogenic Engineering and Technologies: Principles and Applications of Cryogen-Free Systems*, CRC Press (Taylor & Francis Group) Boca Raton, FL, USA, (2020)..
2. Frank Gustrau, *RF and Microwave Engineering: Fundamentals of Wireless Communications*, Wiley, (2025).
3. Qualified vendor list: Accu-Glass Products, Inc. (www.accuglassproducts.com) CeramTec (www.ceramtec.com) Pfeiffer Vacuum (www.pfeiffer-vacuum.com) Key High Vacuum Products, Inc. (www.keyhigh.com) Lake Shore Cryotronics, Inc. (www.lakeshore.com) Insulator Seal Inc. (insulatorseal.com) MDC Precision (www.mdcprecision.com) Pasternack Enterprises Inc. (www.pasternack.com) Detoronics (www.Detoronics.com) Fischer Connectors SA (fischerconnectors.com/en)

18 Diverse Techniques in Cryogenic Engineering

18.1 STABILIZE TEMPERATURE OF "WET" HE-3 CRYOSTAT

The core components of a He-3 cryostat, charcoal sorption pump, 1K pot, and He-3 pot, are housed within the inner vacuum chamber— the IVC can. The sample is mounted in the IVC tail, which typically extends into the bore of a superconducting magnet, as illustrated in Figure 18.1(a).

The IVC is typically constructed from stainless steel rather than copper to reduces eddy current heating during experiments performed in magnetic fields. When the IVC can is fully submerged in liquid helium, it provides a stable thermal environment for the He-3 core. However, because stainless steel has relatively low thermal conductivity, the upper section of the IVC may rise above the liquid helium surface as the helium level drops, warm up, and consequently affect the sample temperature, as shown in Figure 18.1(b). To maintain temperature stability under these conditions, several approaches can be employed.

18.1.1 Copper Jacket

To help stabilize the IVC temperature, a copper jacket— typically comparable in length to the IVC belly section—can be mounted beneath the IVC top flange (Figure 18.2). As long as the lower end of the copper jacket remains submerged in liquid helium, the jacket can be maintained at a nearly uniform temperature of approximately 5 K. This configuration not only cools the IVC top flange to about 5 K but also serves as an effective radiation shield for the IVC belly, thereby helping to maintain a stable thermal environment for the He-3 insert core.

To minimize eddy current generation in varying magnetic fields, the copper jacket is typically fabricated from two half-cylindrical sections rather than a continuous cylinder.

18.1.2 Copper Coil–Foil

An alternative to the copper jacket is the copper coil–foil. The copper coil–foil consists of numerous insulated fine copper wires, typically 0.003–0.005 inch in diameter, bonded to a thin Kapton or polyethylene sheet with GE varnish (Figure 18.3). This structure combines good thermal conductivity with a high degree of mechanical flexibility.

The copper coil–foil can be bonded directly to the IVC can using Stycast 2850FT conductive epoxy, as shown in Figure 18.3. Similar to the copper jacket, it helps maintain a stable thermal environment for the cryostat core when the liquid helium level fluctuates. Although its fabrication is more labor-intensive, the final assembly is more compact, versatile, and adaptable to a wide range of cryostat geometries and experimental requirements.

DOI: 10.1201/9781042000050-18

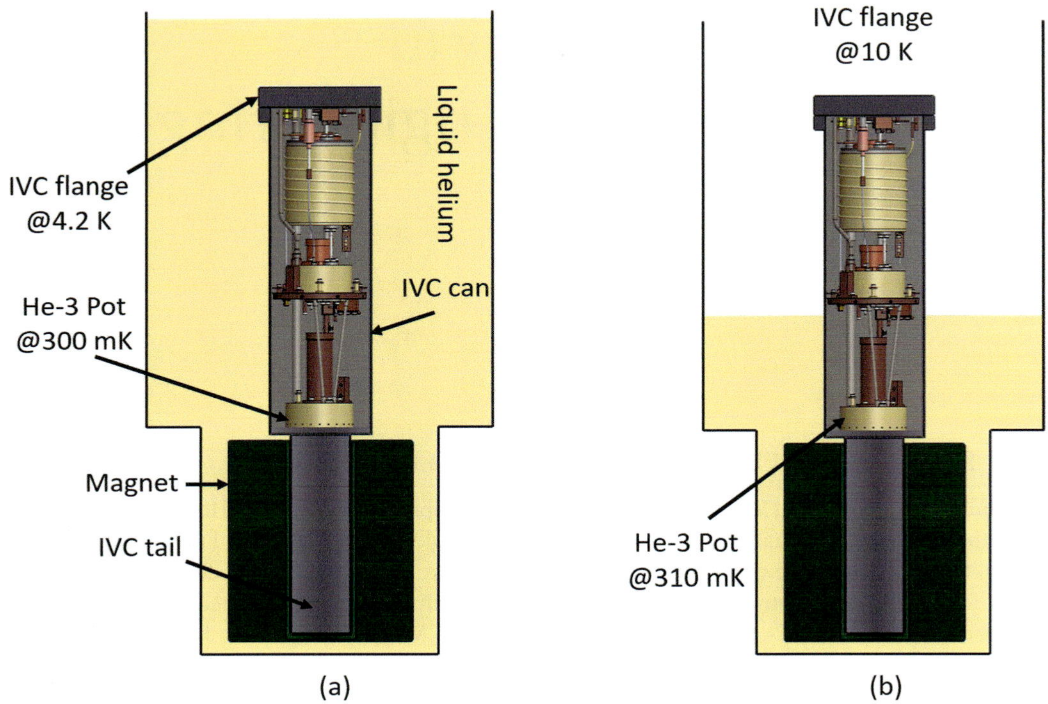

FIGURE 18.1 (a) Wet He-3 cryostat with superconducting magnet and (b) Lower liquid helium level will cause increase of He-3 pot temperature.

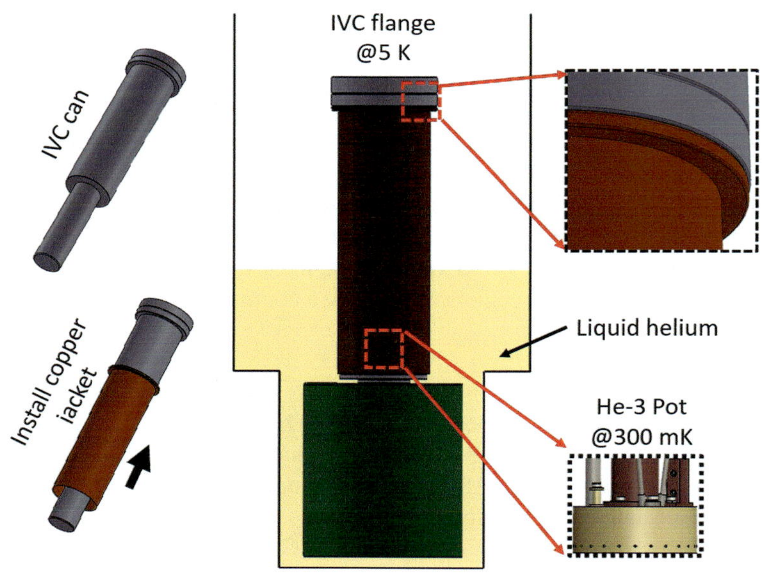

FIGURE 18.2 Copper jacket installed around IVC.

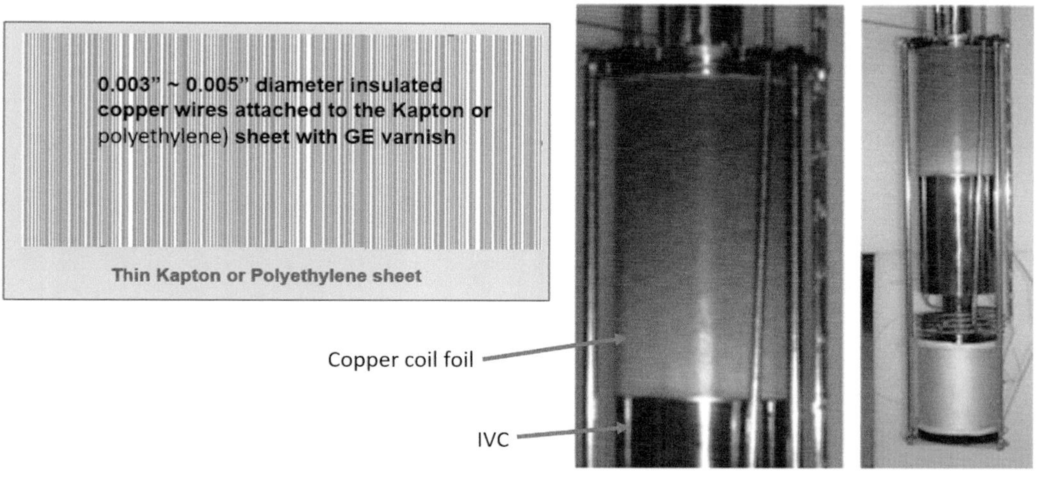

FIGURE 18.3 Copper coil Foil. (Projects accomplished by authors at Janis Research Company. Image courtesy of FormFactor, Inc.)

18.1.3 Copper Plating

The two solutions described above require sufficient radial clearance and therefore cannot be applied in space-limited configurations. Figure 18.4 shows a wet top-loading He-3 stainless steel He-3 pot with an outer diameter of 0.68 in housed within a tightly fitting IVC (0.805" ID; refer to Figure 3.2). Because of the extremely limited clearance, an alternative approach must be adopted.

In this approach, the cylindrical He-3 pot is fabricated from thin-walled (0.012 in) stainless steel tubing. The outer wall of the tubing is first machined to reduce its thickness by approximately 0.006 in. The outer surface of the thinned section is then copper-plated until the overall outer diameter slightly exceeds 0.68 in, and then the copper layer is machined to achieve a precise final outer diameter of 0.68 in. Finally, a narrow axial gap is introduced into the copper layer, as shown in Figure 18.4(b), to suppress eddy current formation during operation in magnetic fields.

18.2 CRYOSTAT SLIDING SEAL

A sliding seal provides a hermetic seal against a moving component that transitions between ambient and cryogenic conditions. Its primary application is to prevent cryogen loss or contamination when the component is inserted into or withdrawn from a cryogenic environment. A sliding seal assembly consists of two major sub-assemblies: an Astra O-ring, typically installed on the cryogen dewar, and a sliding seal tube, which is mounted on the moving component.

18.2.1 Astra O-Ring

An Astra O-ring consists of a stainless steel wound ribbon spring encapsulated in a Teflon (FEP or PFA) jacket (Figure 18.5) [1]. This sub-assembly provides excellent resistance to compression across a wide temperature range, from –250 °C to 260 °C. Unlike conventional elastomeric sealing materials such as Buna-N and Viton, which become brittle and prone to failure at cryogenic temperatures, the Astra O-ring maintains its resilience even under prolonged exposure to low-temperatures. In addition, the Teflon or PFA jacket minimizes adhesion to the sliding tube at cryogenic temperatures. This allows the tube to move smoothly through the stationary O-ring while maintaining seal integrity during dynamic motion.

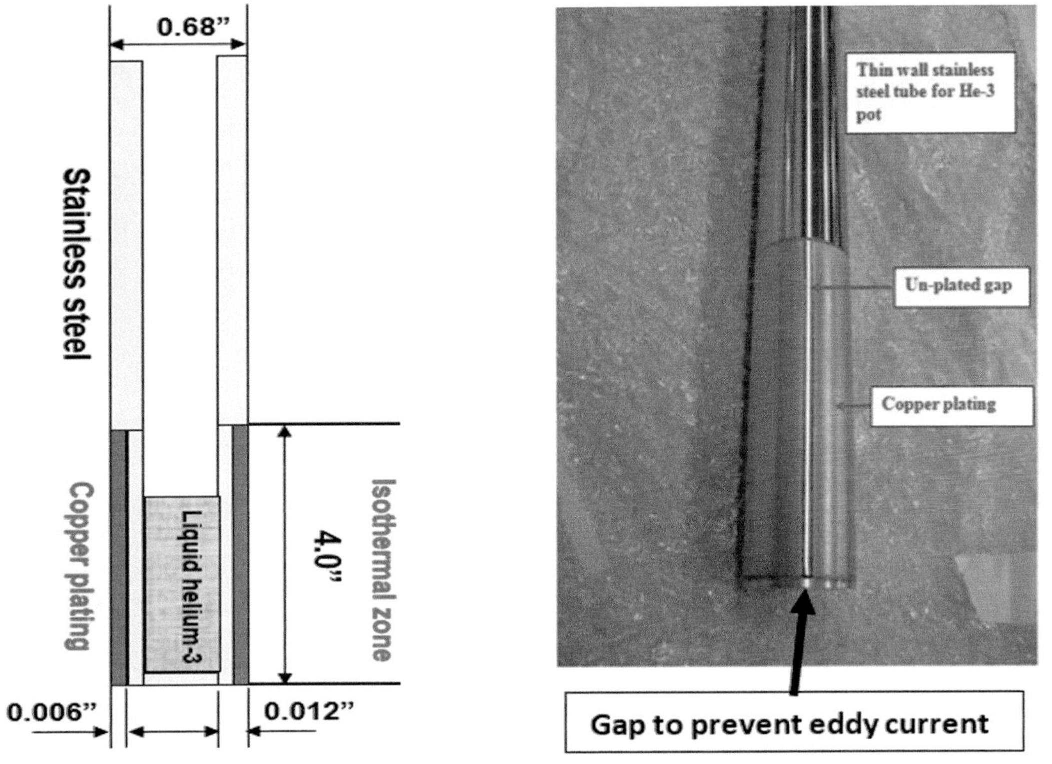

FIGURE 18.4 Copper plated isothermal zone. (Projects accomplished by authors at Janis Research Company. Image courtesy of FormFactor, Inc.)

Table 18.1 summarizes key parameters of commercially available Astra O-rings, including cross-sectional dimensions, inner diameters, tolerances, and compression force that is typically specified in pounds per linear inch. Among these parameters, the inner diameter and the tolerance are particularly critical, as they determine the outer diameter of the mating sliding seal tube. A reliable axial hermetic seal is achieved only when the sliding tube matches the specified diameter, allowing the O-ring to generate the proper radial compressive force.

18.2.2 Sliding Seal Tube

The sliding seal tube is typically made from a low-thermal-conductivity material, most commonly G-10, a fiberglass-reinforced epoxy laminate. The upper end of the tube is bonded to a mounting flange using Armstrong-12 epoxy [2, 3], as shown in Figure 18.6. The lower end features a chamfer to facilitate smooth insertion into the Astra O-ring.

The tube is mounted on a moving component, usually a cryostat insert. The outer diameter of the sliding seal tube is slightly larger than the inner diameter of the Astra O-ring. During installation, the O-ring is gently stretched to accommodate the interference fit. Once in place, the O-ring applies a firm radial compression against the tube's outer surface, providing an effective hermetic seal even during motion at cryogenic temperatures.

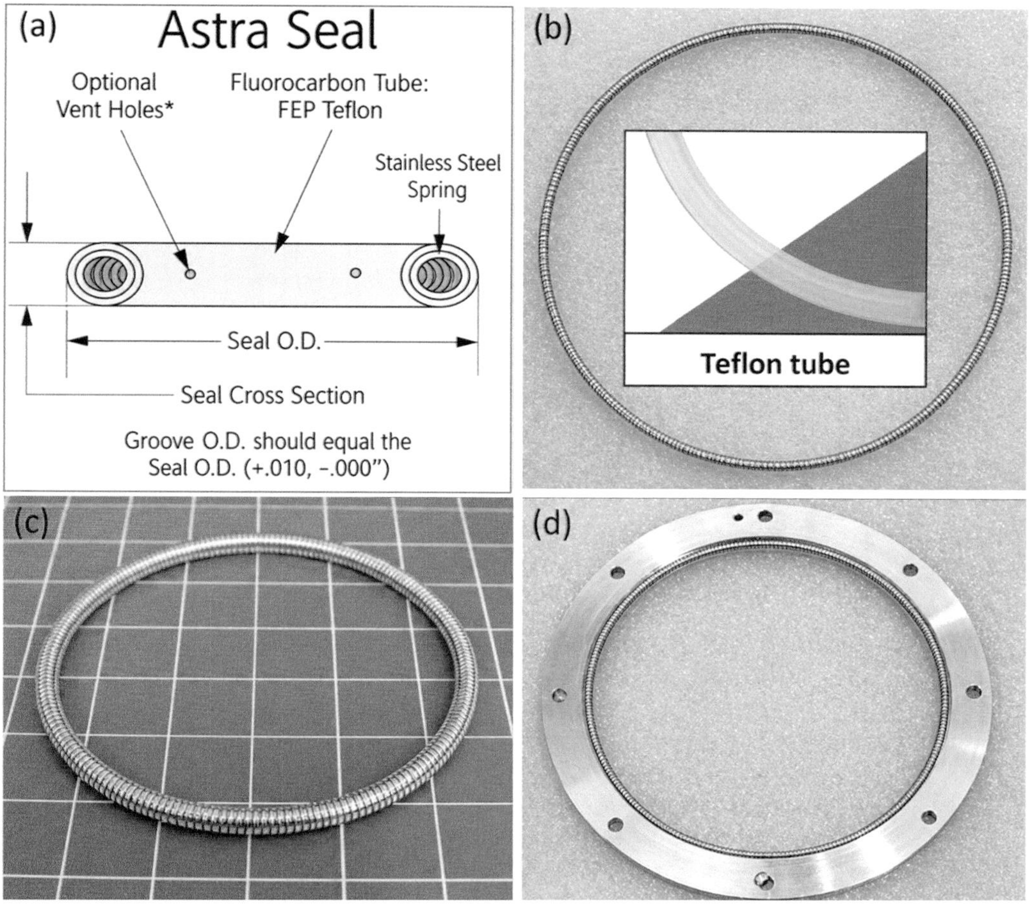

FIGURE 18.5 (a) Concept of Astra O-ring, Source: Information from Northern Engineering (Sheffield) Ltd. www.nes-ips.com/astra-seals/; (b) Stainless steel wound ribbon spring; (c) Astra O-ring (stainless steel wound ribbon spring encapsulated in Teflon tube) Source: Information from Ashton Seals; www.ashtonseals.co.uk/astra; (d) Astra O-ring seated inside the groove of an aluminum flange. (Projects accomplished by authors at Janis Research Company. Image courtesy of FormFactor, Inc.)

18.2.3 Cryostat with Complete Sliding Seal Assembly

In operation, the sliding seal separates the cryogen bath into two regions: one inside and one outside the sliding seal tube. This separation can introduce two major practical issues. First, differential helium vapor pressures between the inner and outer regions may lead to unequal cryogen levels. These discrepancies can complicate level measurements and create operational ambiguity. Second, if all helium vapor vents exclusively from inside the sliding seal tube, there is no vapor cooling available to intercept conductive heat load through the dewar neck. This condition can significantly increase cryogen boil-off rates, especially for liquid helium dewar.

Both issues can be mitigated through the incorporation of a vapor vent chamber (hereafter referred to as "the chamber") that including multiple flanges, as shown in Figure 18.7. The assembly consists of the following elements:

TABLE 18.1
(a) Commercially Available Astra O-ring (Astra O-Rings with Larger Inner Diameters up to 10 inches are Available), (b) Astra Seal® Tolerances

Table A

Cross Section (mm)	Squeeze (mm)	Groove Depth (mm)	Groove Width (mm)	Minimum ID (mm)	Force (lbs.)
1.91	0.56	1.35	2.67	12.70	40
2.40	0.61	1.78	3.18	19.05	40
2.62	0.71	1.90	3.53	19.05	45
2.92	0.76	2.16	3.94	31.75	40
3.00	0.76	2.24	3.96	31.75	45
3.10	0.76	2.34	4.07	31.75	50
3.53	0.76	2.79	4.95	31.75	65
3.80	0.84	2.92	4.95	38.10	85
4.00	0.89	3.10	5.08	38.10	70
4.32	1.02	3.30	5.84	38.10	90
4.50	1.02	3.48	6.02	38.10	100
4.88	1.12	3.76	6.48	44.45	110
5.34	1.14	4.19	7.11	50.80	140
6.35	1.27	5.08	8.25	76.20	110
6.99	1.27	5.71	8.89	88.90	95
8.00	1.52	6.48	9.90	101.60	80

Table B

Inside Diameter (mm)	Tolerance (±)
10.60–16.09	0.2
16.10–25.09	0.25
25.10–40.09	0.35
40.10–63.09	0.4
63.10–100.09	0.5
100.10–160.0	0.7
Over 160	0.50

Source: Information from Northern Engineering (Sheffield) Ltd; www.nes-ips.com/astra-seals/

- Flange #1: Insert top flange.
- Flange #2: Sliding seal tube mounting flange, bolted to Flange #1 using a standard O-ring seal to maintain vacuum integrity.
- Flange #3: Compressing flange that applies axial compression to the Astra O-ring against Flange #4.
- Flanges #4: Upper and lower flanges of the vapor vent sleeve, located between Flange #3 and Flange #5. The top surface of the sleeve features a chamfered groove where the Astra O-ring is seated (see Figure 18.8). When Flange #3 is tightened, the O-ring is slightly deformed, producing a radial compression force on the sliding seal tube.
- Flange #5: Dewar top flange. A standard O-ring seal is installed between the lower Flanges #4 and Flange #5 to secure vacuum integrity.

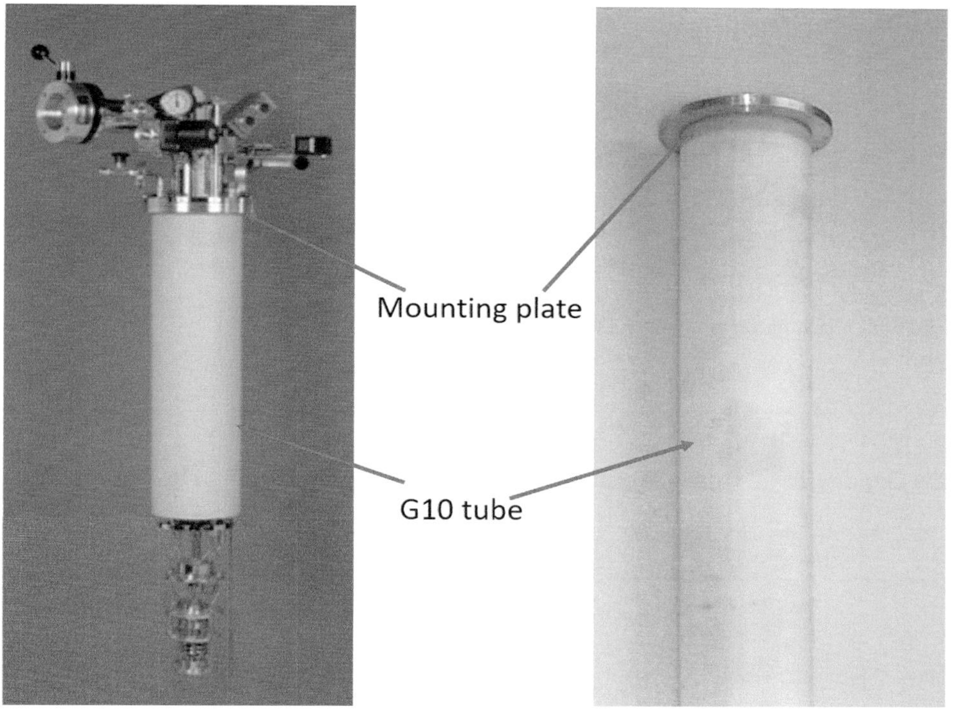

FIGURE 18.6 Sliding seal tube. (Projects accomplished by authors at Janis Research Company. Image courtesy of FormFactor, Inc.)

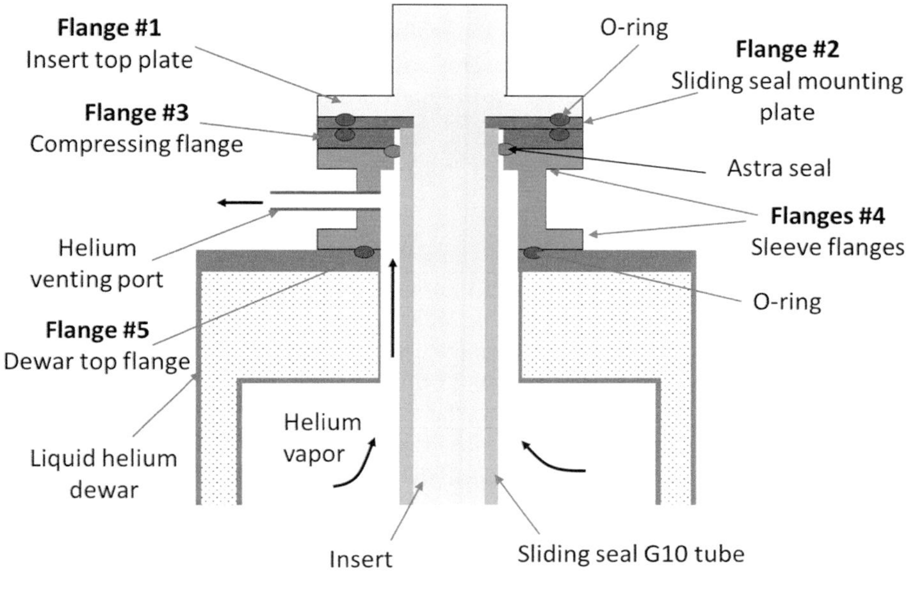

FIGURE 18.7 Cryostat with complete sliding seal assembly including a vapor vent chamber.

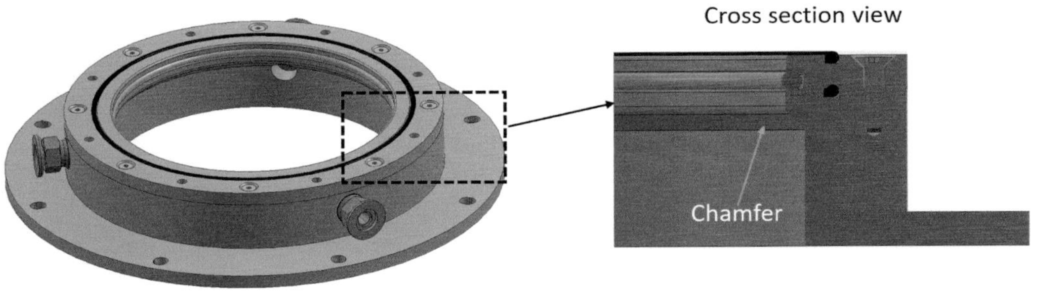

FIGURE 18.8 Chamfer on top flange of the vapor vent chamber.

TABLE 18.2
Thermal Conductivities of G-10 and Stainless Steel

	Temperature Range (K) 300 – 77	Temperature Range (K) 300 – 20	Temperature Range (K) 300 – 4
G-10	4.5×10^{-3} (W·cm^{-1}·K^{-1})	4.5×10^{-3} (W·cm^{-1}·K^{-1})	4.0×10^{-3} (W·cm^{-1}·K^{-1})
Stainless steel	0.123 (W·cm^{-1}·K^{-1})	0.109 (W·cm^{-1}·K^{-1})	0.103 (W·cm^{-1}·K^{-1})

Source: Data from author's note.

The small annular gap between the chamber and the sliding seal tube allows cold helium vapor to vent upward through the sleeve, providing convective cooling for both the dewar neck and the sliding seal tube.

To maintain pressure equilibrium between the regions inside and outside the sliding seal tube, pressure relief valves with identical burst pressures are installed on both the sleeve and Flange #1 (the insert's top flange). In practice, these two regions are often interconnected, allowing more effective pressure equalization.

18.2.4 Discussion

18.2.4.1 Stainless Steel Sliding Seal Tube

The sliding seal tube can also be fabricated from thin-walled stainless steel, which also exhibits relatively low thermal conductivity, as listed in Table 18.2.

Over the temperature range from room temperature to liquid helium temperature, the thermal conductivity of stainless steel is approximately 25 times higher than that of G-10. However, the effective thermal conductance of stainless steel tube can be significantly reduced by decreasing its wall thickness. Commercially available G-10 tubes typically have wall thicknesses between 0.063 and 0.093 in, whereas stainless steel tubes are available with much thinner walls, typically ranging from 0.020 to 0.028 in. It has been demonstrated that, with effective vapor cooling, the actual conductive heat loads of G-10 and stainless steel tubes from 300 K to 4 K are comparable and remain within the same order of magnitude. Figure 18.9(a) and (b) illustrates examples of sliding seal tubes made from G-10 and stainless steel, respectively.

In addition, cryostats typically impose strict spatial constraints, and thin-walled stainless steel sliding seal tubes offer several advantages:

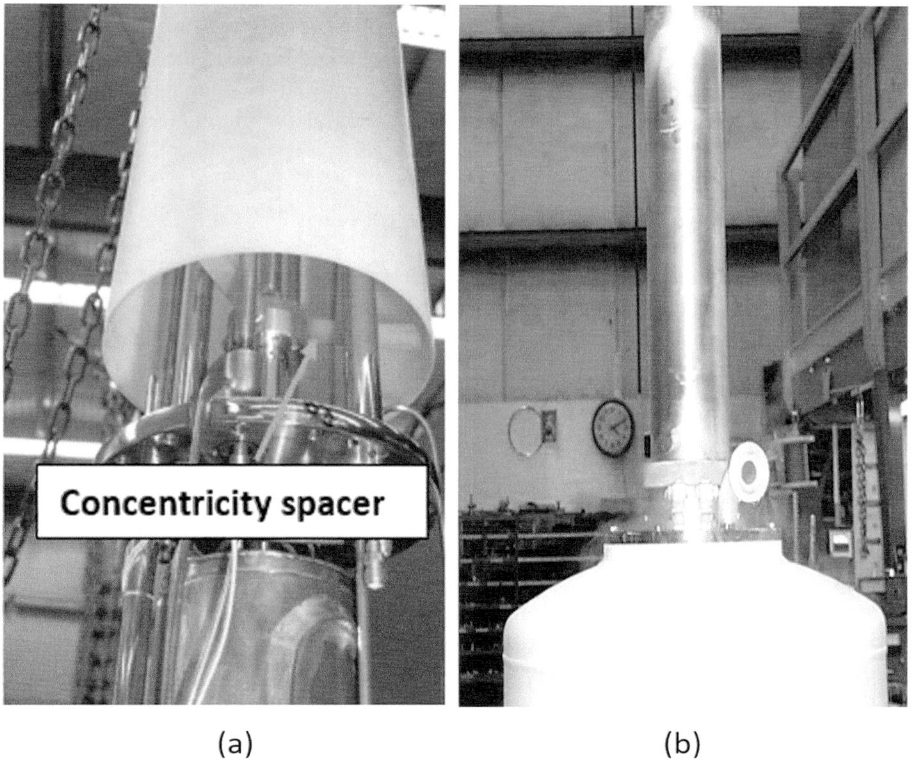

FIGURE 18.9 (a) G-10 sliding seal, (b) Stainless steel sliding seal. (Projects accomplished by authors at Janis Research Company. Image courtesy of FormFactor, Inc.)

- Space efficiency: The reduced walls of stainless steel tubes allow for more compact assemblies, making them ideal for cryostats with limited clearance.
- Dimensional uniformity: Stainless steel tubes generally exhibit excellent diameter consistency, supporting precise assembly and reliable sealing performance.
- Cost-effectiveness: Stainless steel is typically cheaper than G-10, providing a cost-effective solution without compromising structural integrity.

18.2.4.2 Ice Adhesion at Cryogenic Temperatures

Although Astra O-rings do not inherently adhere to G-10 or stainless steel at cryogenic temperatures, the formation of water ice can cause the O-ring to stick to these surfaces. To mitigate this risk, the sliding seal should be moved slowly, and the tube should be continuously defrosted during removal from a cold cryostat. If ice formation occurs, a heat gun may be used to assist with defrosting. Applying excessive force to break the ice must be avoided, as it can damage the Astra O-ring and potentially compromise the seal's integrity.

18.2.4.3 Lubrication-Free Operation

Lubrication of the Astra O-ring is not recommended. Common lubricants, including Apiezon or Dow Corning grease, solidify at cryogenic temperatures. Once solidified, these materials can cause the O-ring to adhere to the sliding seal tube, impeding motion and degrading sealing performance.

18.2.4.4 Concentric Spacer

A concentric spacer, as shown in Figure 18.9(a), may be installed between the insert and the sliding seal tube to improve alignment, particularly for long tubes with larger diameters. Proper alignment ensures uniform compression of the O-ring, which is essential for maintaining reliable hermetic sealing.

18.3 VACUUM SEAL AT CRYOGENIC TEMPERATURES

Hermeticity is a critical requirement in cryogenic engineering and represents a fundamental design consideration. Achieving a hermetic seal at cryogenic temperatures typically relies on compressing an interstitial material at the sealing interface. Common examples include ConFlat (CF) flange seals, VCR seals, taper seals, and indium wire seals.

18.3.1 Indium Wire Seal

Indium is particularly well-suited as an interstitial material due to its unique mechanical properties. As a very soft metal, indium readily conforms to surface imperfections, ensuring a reliable hermetic seal. Furthermore, indium retains its malleability at cryogenic temperatures rather than becoming brittle, making it especially attractive for low-temperature applications.

Commercial indium wires are available in diameters of 1.0 mm (0.040"), 1.5 mm (0.060"), and 2.32 mm (0.093"). The diameter of 1.0 mm is commonly used for most laboratory-scale applications. While the larger diameter of 1.5 mm, are typically used for large flanges subjected to heavy mechanical loads, the largest diameter of 2.32 mm is used only for cases when the sealing groove is unusually wide and deep.

18.3.1.1 Groove Seal Design

Most indium seals employ a groove-seal design in which an indium wire is placed in a machined groove while the mating surface remains flat. The groove typically begins with a shallow straight section and then transitions into a semicircular profile, as illustrated in Figure 18.10. The groove depth is generally designed to be about 50% or slightly more of the indium wire diameter to ensure proper compression (see Appendix A4).

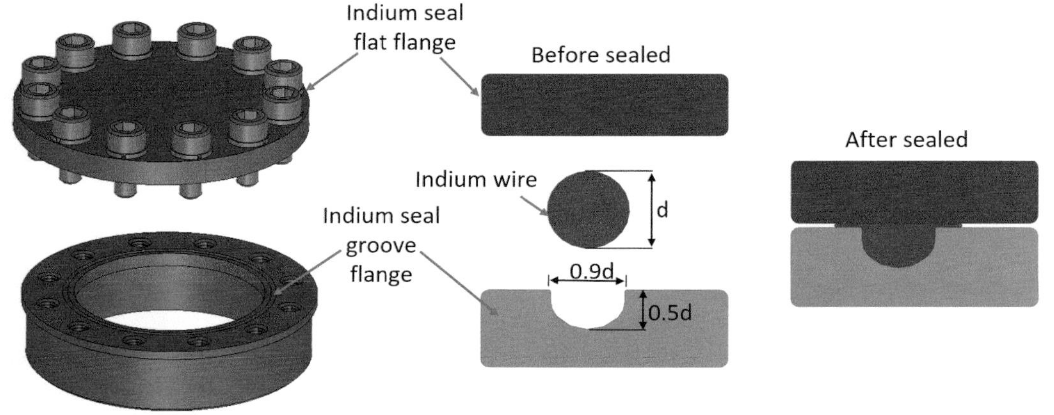

FIGURE 18.10 Groove seal design. (Projects accomplished by authors at Janis Research Company. Model courtesy of FormFactor, Inc.)

18.3.1.2 Corner Seal Design

The corner-seal design is formed by compressing an indium wire between the sharp edges of a pair of flanges, as illustrated in Figure 18.11. In this configuration, the female flange incorporates a square-step groove that is slightly deeper than the corresponding square-step protrusion on the male flange (see Appendix A4). The groove diameter is also made marginally larger than that of the protrusion.

This design is commonly employed in the *mini-indium seals*, which typically have flange diameters of approximately 0.5" and do not provide sufficient space for a dedicated indium groove. For these seals, an indium wire with a maximum diameter of 1.0 mm is recommended. Proper tolerances between the two flanges are critical to ensuring a reliable seal at cryogenic temperatures. For example, when using a 1.0 mm (0.040") diameter indium wire, the recommended tolerances are about 0.002" in groove depth and 0.003" in diameter between the male and female flanges, as illustrated in Figure 18.11.

18.3.1.3 General Guidelines for Indium Wire Seal

- Both brass and stainless steel are suitable materials for flanges used with indium wire seals. Regardless of material choice, the sealing surfaces of the mating flanges must be flat and free of scratches. Scratches that cross the sealing groove are particularly detrimental, as they can create leak paths and compromise hermeticity.
- Only clean, fresh indium wire should be used when making an indium O-ring. Indium wires should be thoroughly cleaned with isopropyl alcohol or acetone prior to installation to remove surface contaminants that could degrade sealing performance. A common practice is to cut both ends at a shallow angle and press them together so that the angled faces overlap slightly. This method improves continuity at the joint and enhances sealing reliability.
- Bolts or screws should be tightened gradually in a crisscross pattern to ensure uniform compression of the indium wire. Tightening should continue until slight deformation of the indium is felt. After the initial tightening, the assembly should be allowed to rest for approximately 30 minutes to permit stress relaxation in indium, after which the bolts should be re-tightened. This process should be repeated two or three times until no further relaxation is observed, while avoiding excessive tightening.
- Applying the correct torque is essential to achieving a reliable indium seal and preventing damage to the flanges. Recommended torque values depend on factors such as indium wire diameter, flange size, material, and surface finish. As an example, for indium wire diameters of 1.0 mm and 1.5 mm, typical torque values are approximately 25 kg·cm and 56 kg·cm per

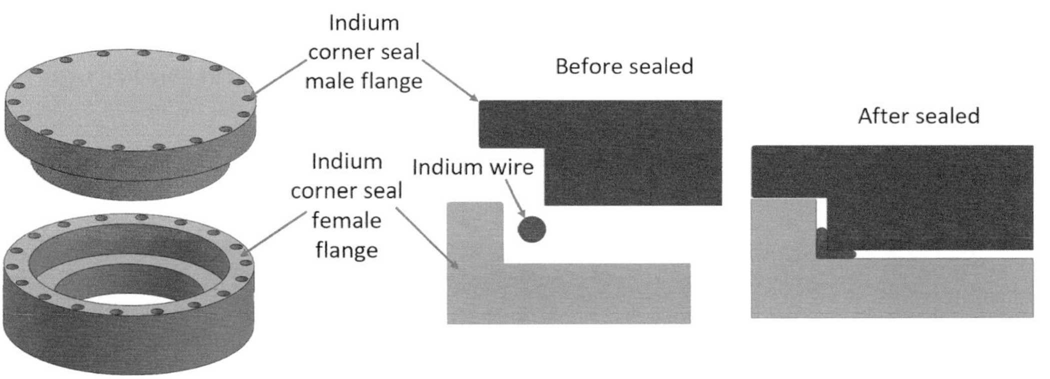

FIGURE 18.11 Corner seal design. (Projects accomplished by authors at Janis Research Company. Model courtesy of FormFactor, Inc.)

bolt, respectively [4]. The use of a torque wrench is strongly recommended rather than relying on a standard wrench.

- Although bolts are often made from the same material as the flanges, selecting bolts with a different coefficient of thermal contraction can increase clamping force at cryogenic temperatures. For instance, brass bolts can increase the effective clamping force when used with stainless steel flanges because brass contracts more upon cooling.
- After compression and repeated cryogenic cycling, it is common for the two flanges to become bonded together by the indium wire. To enable safe and controlled separation, threaded holes for jack screws should be incorporated as a standard design feature. The jack screws enable the components to be pushed apart uniformly, preventing damage to the sealing surfaces and eliminating the need for excessive prying forces.
- After a thermal cycle, residual indium often remains adhered to the flange and must be removed before installing a new seal. However, hard tools that could scratch the sealing surface should be avoided. A wooden stick with a tapered or pointed end is effective for removing residual indium without damaging the flange.
- In general, applying vacuum grease to indium wire is not recommended because it can interfere with proper metal-to-metal contact and reduce sealing effectiveness. In some cases, however, a very light coating may be used to reduce adhesion between the indium and flange during thermal cycles. The decision to use vacuum grease should be based on the specific application and user experience.
- For large indium-sealed flanges, bolt spacing plays a critical role in maintaining hermeticity. In applications such as a demountable, indium-sealed bottom flange of a liquid helium reservoir, the seal must withstand large forces generated by the pressure differential across the flange. For example, an 18 in-diameter flange may experience an downward force of approximately 3,690 lb., requiring a sufficient number of bolts to provide adequate clamping force and maintain a reliable hermetic seal through thermal cycles. To ensure adequate and uniform compression of the indium seal, the chord length—the straight-line distance between adjacent bolts—should not exceed approximately 1.0", as shown in Figure 18.12. This requirement

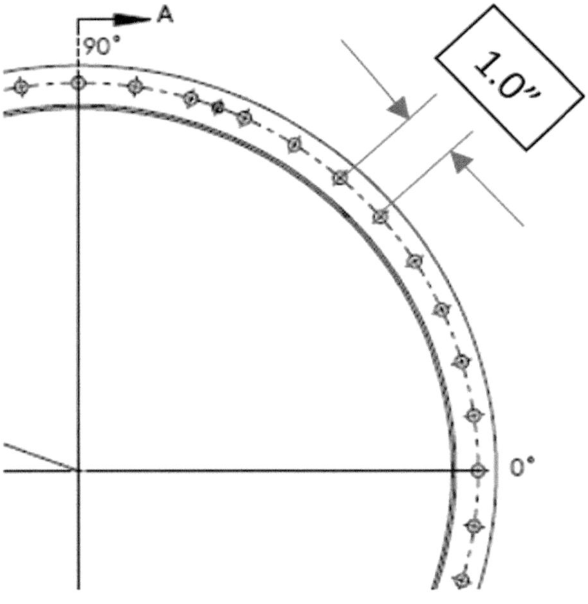

FIGURE 18.12 Cord length between adjacent bolts for larger indium-sealed flanges.

determines the total number of bolts and ensures sufficient clamping force during thermal cycling [5].

- Indium is a type I superconductor and may generate unwanted flux-avalanche signals when it transitions into the superconducting state. In applications sensitive to magnetic noise, indium can, in principle, be replaced by plastic materials such as PTFE (Teflon) or Kapton. Pure gold wire is another viable alternative due to its excellent ductility and chemical stability, but the high cost makes it impractical for most users. Copper, while well suited for CF gaskets, is generally unsuitable for indium-style groove seals because of its comparatively limited ductility.

18.3.2 Taper Seal

18.3.2.1 Brass Taper Seal Design

Brass taper sealing is a widely used technique for achieving reliable hermetic seals in IVC cans with diameters up to 2.5″ at cryogenic temperatures, including super-leak-tight joints at 2.17 K or lower [6]. Compared to conventional vacuum-sealed cryostats, cryostats with brass taper seals allow for a more compact overall design, making them particularly suitable for space-limited environments. Furthermore, taper-sealed cryostats offer faster and more convenient sample changes.

The brass taper assembly consists of two primary components: the male taper assembly mounted on the cryostat and the female taper assembly mounted on the IVC can, as illustrated in Figure 18.13. The male assembly is constructed from a brass bushing and a brass male taper. The bushing is silver-soldered to the stainless steel cryostat flange, and the male taper is then soft-soldered to the bushing using 60/40 solder. Similarly for the female assembly, a brass bushing is silver-soldered to the stainless steel IVC can, and a brass female taper is soft-soldered to the bushing. If the IVC can itself is made of copper or brass, the female taper can instead be soldered directly to the can using 60/40 solder.

To achieve a hermetic seal at cryogenic temperatures, the male and female brass tapers must mate precisely. For this reason, it is highly recommended that both components be machined in the same batch to ensure dimensional compatibility. Even minor deformations, scratches, or dents can cause vacuum seal failure. If a brass taper is damaged, it can be removed by heating the soft solder joint. However, in this case, both the male and female tapers should be replaced to ensure reliable sealing.

18.3.2.2 Brass Taper Seal Operation

Closing a Taper-Sealed IVC Can

- Clean both tapered sealing surfaces with lint-free wipes and alcohol or acetone.
- Carefully inspect the surfaces to ensure they are free of deformations, scratches, dents, debris, or dirt. If minor damages are present, the tapers may need to be reworked or replaced.
- Apply a thin, uniform layer of clean DuPont Molykote or Dow Corning high-vacuum grease to both taper surfaces.
- Mate the IVC can to the cryostat by engaging the male and female tapers.
- Rotate the IVC can back and forth three or four times against the cryostat to evenly distribute the grease. Proper engagement is typically indicated by a slight "oozing" of grease at the joint.
- Pump down the IVC, allowing atmospheric pressure to hold the can securely against the cryostat.
- Perform a helium leak check of the taper joint. The joint is considered hermetically sealed if the leak rate is on the order of 10^{-8} mbar·L/s or lower.

Opening a Taper-Sealed IVC Can

- Support the IVC can to prevent it from falling abruptly when the seal is released.

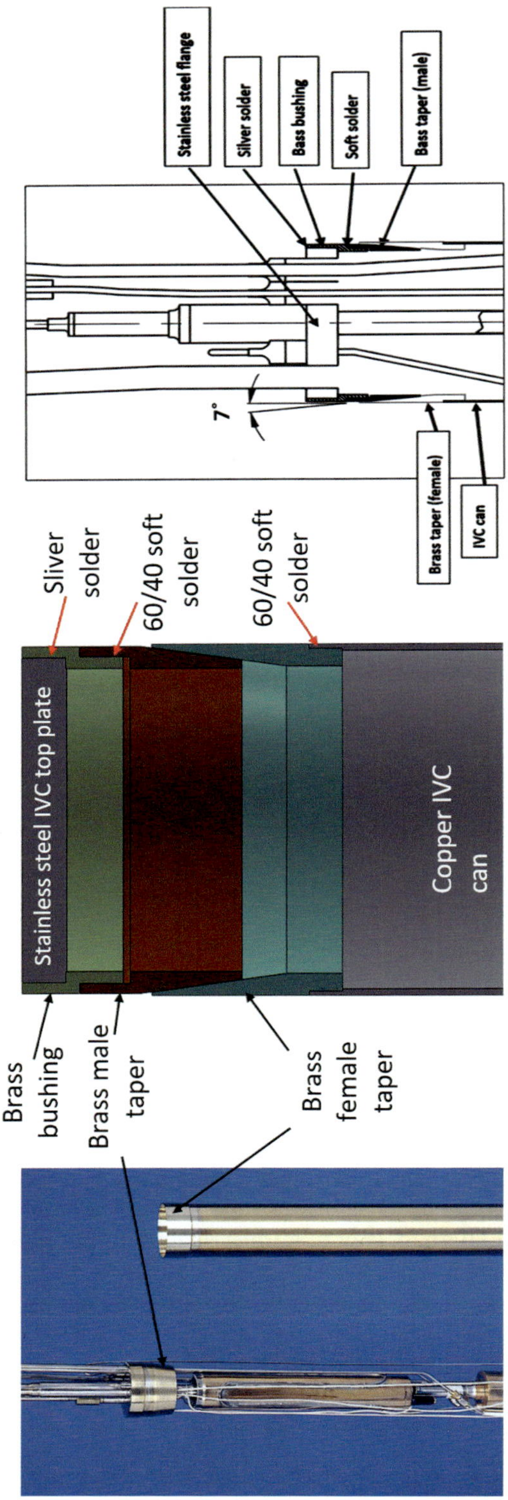

FIGURE 18.13 Brass taper design. (Projects accomplished by authors at Janis Research Company. Image courtesy of FormFactor, Inc.)

- Vent the IVC to ambient pressure only after the entire cryostat has warmed to room temperature.
- Remove the support and detach the IVC can from the cryostat at the taper joint.
- If the can is stuck, gently tap and wiggle the bottom of the IVC from the side until it comes free.
- Handle the brass tapers with care at all times to avoid scratches, dents, or other damage.

18.3.2.3 Supporting Fixture

Under normal operating conditions, a taper-sealed IVC is held in place by the pressure difference between the inside and outside of the can. In some situations, however, this pressure difference may be small, and the taper seal alone may not provide adequate mechanical support. To prevent accidental dropping of the IVC can, a gravity support fixture can be installed at the bottom of the IVC can (Figure 18.14).

18.3.2.4 Nonmetal Taper Seal

In applications where the presence of metal near the sample must be avoided, a non-metallic taper-sealed cell can be constructed using Stycast Epoxy 1266 in place of a brass taper seal. The overall design and operating procedures for the nonmetallic taper seal are essentially the same as those for the brass version, although greater mechanical force is typically required to press the two taper pieces together to ensure a reliable seal. A conceptual design of the nonmetallic taper seal is shown in Figure 18.15.

The experimental cell assembly consists of three main components: an upper section with a male taper and external threads, a lower section with a female taper, and a compression ring with internal threads. The compression ring screws onto the upper part and applies axial force to press the female taper against the male taper to form a seal. At room temperature, helium gas can slowly permeate through Stycast Epoxy 1266 at a rate of approximately 10^{-6} mbar·L/s. When cooled to liquid-nitrogen temperatures or below, however, the leak rate decreases significantly to the order of 10^{-8} mbar·L/s.

18.3.3 Conflat Flange Seal

The CF flange is a widely used commercial vacuum sealing standard that is also well suited for cryogenic applications. In a CF joint, both mating surfaces are machined with knife edges that bite into

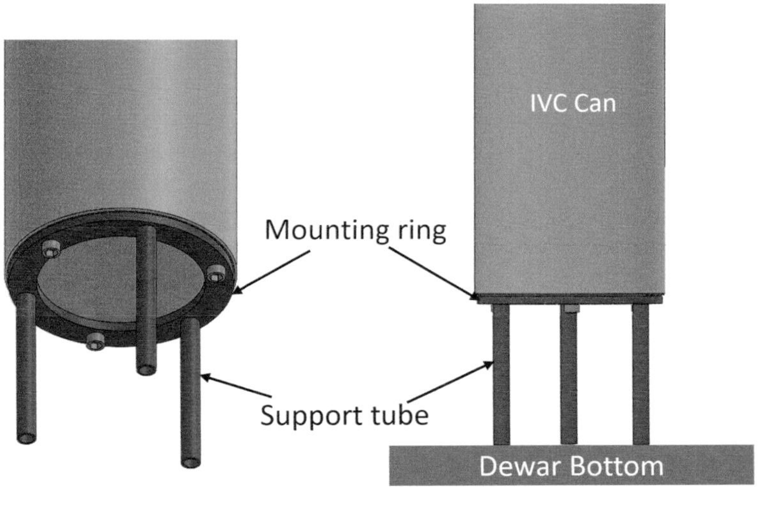

FIGURE 18.14 Fixture to secure the female part.

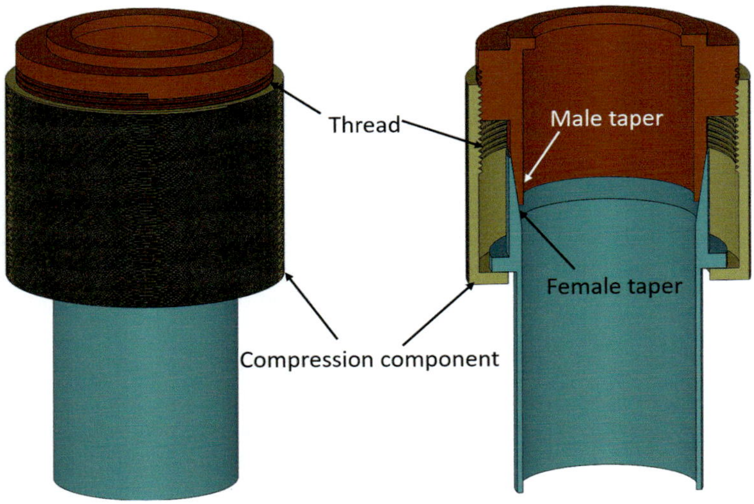

FIGURE 18.15 Nonmetallic taper seal made of Stycast 1266 epoxy.

a soft copper gasket to create a reliable vacuum seal, as shown in Figure 18.16(a). Copper remains ductile at low temperatures and can accommodate differential thermal contraction, making CF seals highly reliable under repeated thermal cycling.

Unlike indium seals, which can be fabricated in a wide range of sizes, CF flanges are available only in discrete, standardized diameters. For the same vacuum inner diameter, a CF flange has a larger overall footprint and is therefore less compact than an indium seal, which can limit its use in space-constrained cryogenic applications. In addition, commercial copper gaskets are generally not available for CF flanges larger than 27 in. in diameter, and therefore for such large flanges, custom copper-wire O-rings are typically used to seal the vacuum.

Although CF flanges are a standard commercial component, the knife-edge design can vary slightly among different manufacturers. Two mating CF flanges with different knife-edge profiles will not form a hermetic seal, even at room temperature. Therefore, it is essential to ensure that both flanges are fabricated by the same manufacturer to ensure compatibility and reliable sealing performance.

18.3.4 VCR Seal

A Swagelok VCR seal, commonly referred to simply as a VCR seal, is a type of high-purity, metal-to-metal face seal fitting widely used in vacuum and cryogenic systems. A standard VCR assembly consists of bare glands, a threaded male nut, a threaded female cap, and a gasket. The glands, cap, and nut are typically made of stainless steel, while gaskets are most commonly fabricated from stainless steel or copper. In specialized applications, nickel and aluminum gaskets may be used: aluminum gaskets are easier to seal, whereas nickel gaskets are preferred for harsh chemical environments and high-temperature conditions.

As shown in Figure 18.17, the gasket is positioned between the two mating glands. Tightening the nut into the cap compresses the gasket, creating the seal. The thin, flat geometry of the gasket results in only slight plastic deformation, producing a robust, leak-tight seal even at millikelvin temperatures (Figure 18.18). Additional assembly details can be found in Reference [8].

The VCR seal offers multiple advantages, including a compact design, ease of installation, the ability to disassemble and reassemble the fitting multiple times by simply replacing the gasket. Furthermore, because they rely on metal-to-metal sealing, VCR fittings eliminate the need

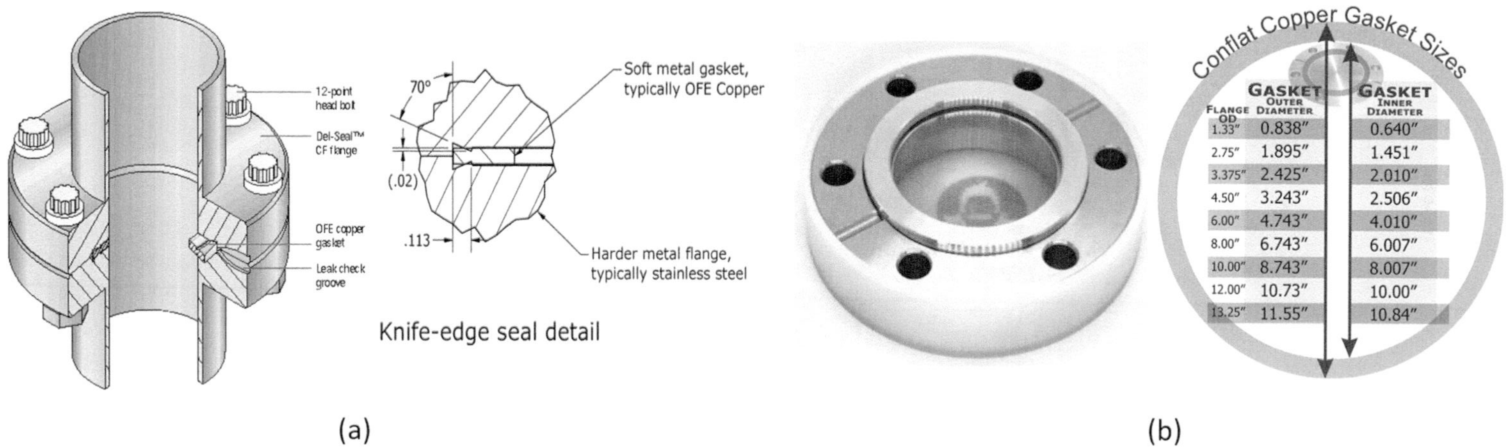

FIGURE 18.16 (a) Seal with Conflat flange, (b) Standard commercial copper gaskets compared to the outer diameters of matching Conflat flanges. Source: Information from Kurt J. Lesker Company, www.lesker.com/) (public Domain) [7].

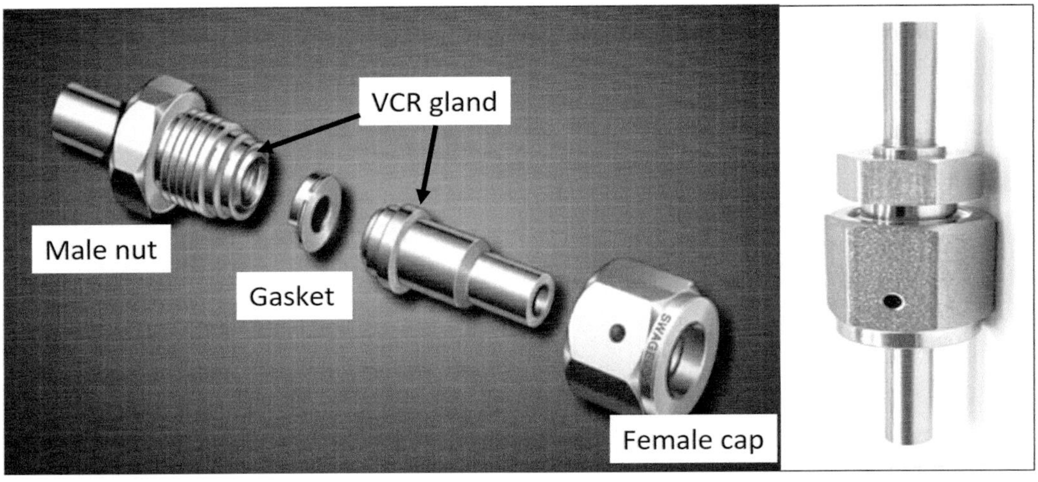

FIGURE 18.17 VCR assembly. Source (a): Information from Kurt J. Lesker Company, www.lesker.com/) (public Domain), Source (b): Information from Swagelok Pittsburgh, https://pittsburgh.swagelok.com (public Domain).

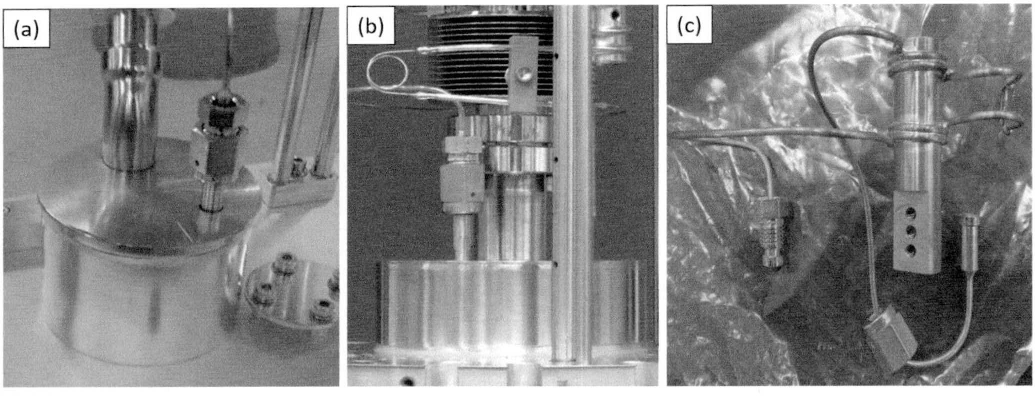

FIGURE 18.18 (a) VCR seal installed on He-3 cryostat, (b) VCR seal installed on dilution refrigerator, (c) VCR seal installed on the heating block. (Projects accomplished by authors at Janis Research Company. Image courtesy of FormFactor, Inc.)

for elastomeric O-rings and are therefore well suited for ultra-high-vacuum and the cryogenic applications.

18.3.5 Cryogenic Hermetic Seal Summary

The advantages and limitations of commonly used cryogenic hermetic seals are listed in Table 18.3.

18.4 HERMETIC PLASTIC SAMPLE TUBE AT LIQUID HELIUM TEMPERATURES

To minimize eddy-current heating in low-temperature, high-magnetic-field experiments, nonconductive materials are preferred for constructing cryostat components, such as the He-3 pot in

TABLE 18.3
Summary of Often Used Cryogenic Hermetic Seal

Type of seal	Primary Advantage	Primary Limitation
Indium seal	Flexible in design; fits flanges of any size	Requires thorough cleaning of Indium residue when re-making the seal
Taper seal	Compact; fast sample change; no elastomeric O-rings needed	Delicate; requires replacing the whole taper seal assembly due to minor damage; limited size options
Conflat seal	UHV compatible	Limited size options
VCR seal	Compact, easy to assemble and disassemble; no elastomeric O-rings needed, UHV compatible	Fit tubes with smaller diameters only

Source: Data from author's note.

a He-3 cryostat or the mixing chamber in a dilution refrigerator [9]. In this section, we present the design and fabrication of hermetically sealed plastic sample tubes capable of operation at liquid-helium temperatures.

18.4.1 Design and Fabrication

Three functional prototype designs were developed using different material combinations, including G-10 and black Resdel epoxy tubes. All assemblies were bonded and sealed using Stycast 1266 epoxy.

Design #1: G-10 Tube (Figure 18.19)

1. Base tube preparation
   A G-10 tube with an outer diameter of 0.602" and a wall thickness of 0.020" is used as the base tube.
2. Initial epoxy coating
   Two thin layers of Stycast 1266 epoxy are applied to the outer surface of the G-10 tube.
3. Top-end assembly
   A stainless steel, indium-sealed flange is bonded to a G-10 collar. The flange–collar assembly is then attached to the top of the G-10 tube using Stycast 1266 epoxy.
   Note: The G-10 collar is designed to shrink more than the stainless steel flange at low temperatures to enhance hermeticity.
4. Collar sealing
   Two thin layers of Stycast 1266 epoxy are applied to the outer surface of the G-10 collar.
5. Bottom sealing
   A G-10 cap is bonded to the bottom of the tube, and its outer surface is coated with Stycast 1266 epoxy.

Design #2: G-10 Tube with Kapton Protection Layers (Figure 18.20)

1. Base tube preparation
   A G-10 tube with an outer diameter of 0.602" and a wall thickness of 0.020" is used as the base tube.
2. Initial epoxy coating
   Two thin layers of Stycast 1266 epoxy are applied to the outer surface of the G-10 tube, the same procedure as in Design #1.

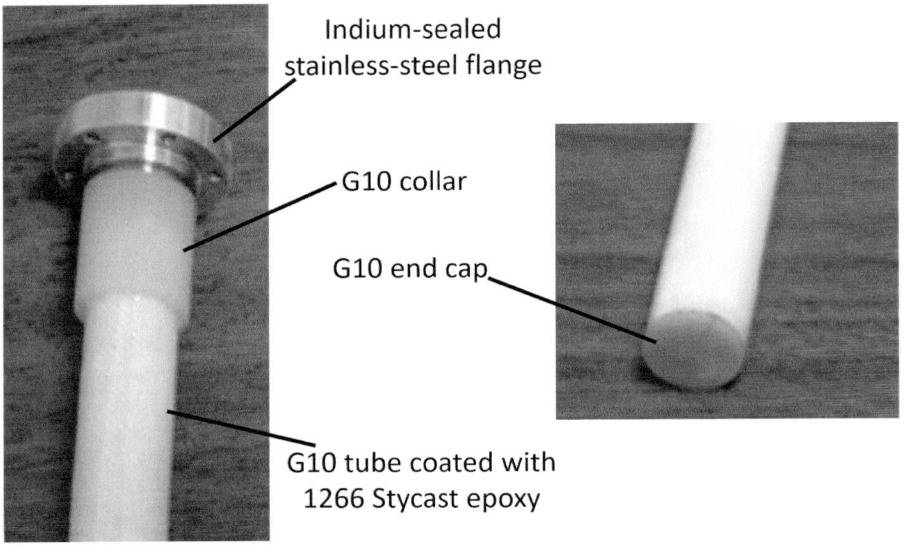

FIGURE 18.19 Design #1: G-10 tube coated with Stycast 1266. (Projects accomplished by authors at Janis Research Company. Image courtesy of FormFactor, Inc.)

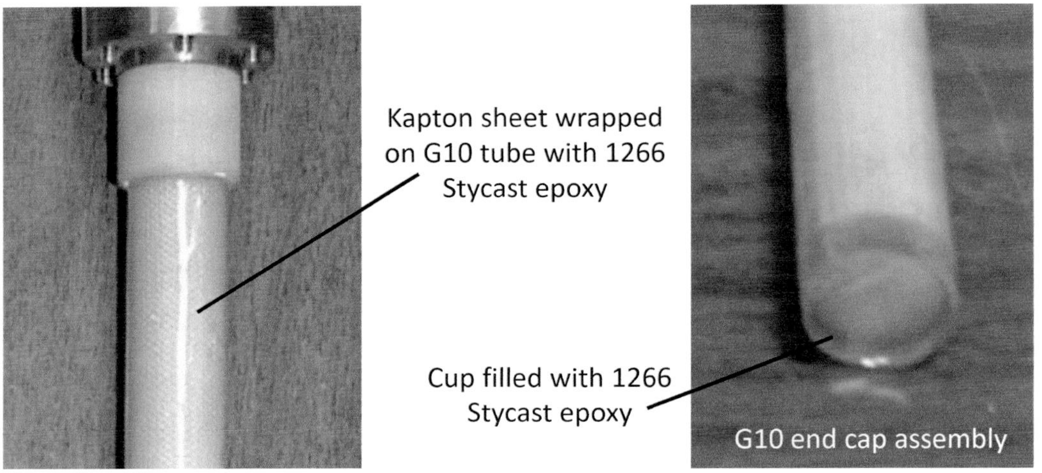

FIGURE 18.20 Design #2: G-10 Tube with Kapton Protection Layers. (Projects accomplished by authors at Janis Research Company. Image courtesy of FormFactor, Inc.)

3. Kapton wrapping
   A single layer of 0.002″-thick Kapton tape is wrapped around the tube with an overlap of approximately 0.25″. The overlap region is precoated with Stycast 1266 epoxy to ensure proper bonding.
4. Compression wrapping
   Nylon tape is tightly wrapped around the Kapton layer to press it firmly against the tube during curing. After curing, the nylon tape is removed.

5. Top-end assembly
   A stainless steel, indium-sealed flange is bonded to a G-10 collar, and the flange–collar assembly is attached to the top of the G-10 tube using Stycast 1266 epoxy.
6. Collar sealing
   A layer of Stycast 1266 epoxy is applied to the outer surface of the G-10 collar.
7. Bottom sealing
   A G-10 end cap is bonded to the bottom of the tube and coated with Stycast 1266 epoxy.
8. Kapton cup formation
   The Kapton layer is extended beyond the bottom edge of the G-10 cap to form a small Kapton cup.
9. Final bottom seal
   The Kapton cup is filled with Stycast 1266 epoxy to complete the bottom seal.

Design #3: Resdel (Black) Epoxy Tube [10] (Figure 18.21)

1. Base tube preparation
   A Resdel (black) epoxy tube with an outer diameter of 0.615″ and a wall thickness of 0.028″ is used [10].
2. Top-end assembly
   A stainless steel, indium-sealed flange is bonded to a G-10 collar, and the flange–collar assembly is attached to the top of the Resdel tube using Stycast 1266 epoxy.
3. Outer surface treatment
   No epoxy is applied to the outer surface of the Resdel epoxy tube.
4. Bottom sealing
   An aluminum end cap is bonded to the bottom of the Resdel tube using Stycast 1266 epoxy.

18.4.2 Test and Conclusions

For hermetic testing, the stainless steel, indium-sealed flange at the top of each sample tube was connected to a helium leak detector, as shown in Figure 18.22. All three sample tubes were tested at both room temperature and cryogenic temperatures.

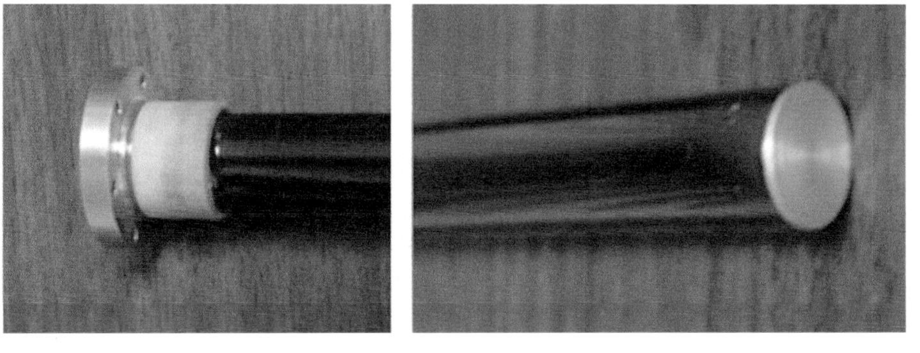

FIGURE 18.21 Design #3: Using Resdel (black) epoxy tube. (Projects accomplished by authors at Janis Research Company. Image courtesy of FormFactor, Inc.)

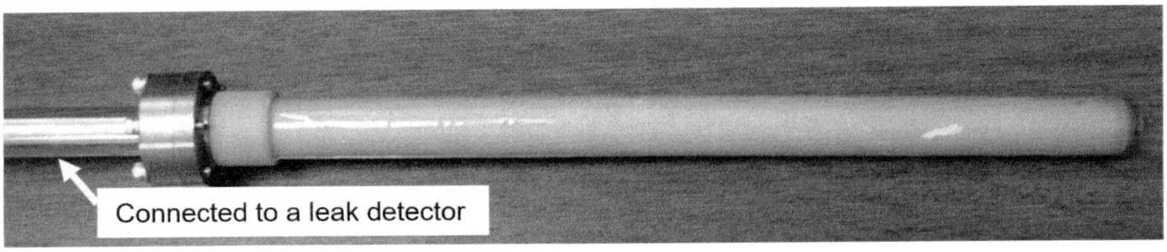

FIGURE 18.22 Sample tube installed on the test fixture. (Projects accomplished by authors at Janis Research Company. Image courtesy of FormFactor, Inc.)

- Room Temperature Results
  Design #1 and Design #2, which employed G-10 tubes, exhibited a measurable helium diffusion rate of less than 10^{-6} mbar·L/s through the G-10 material. Design #3, based on a Resdel epoxy tube, showed a significantly lower diffusion rate of approximately 10^{-8} mbar·L/s, indicating superior hermetic performance at room temperature.
- Cryogenic Temperature Results
  When cooled to liquid nitrogen temperature (~77 K) or lower, no helium diffusion was detected in any of the three designs. Leak checks at liquid helium temperature (~4.2 K) were conducted by submerging the tubes in liquid helium. During these tests, the helium background remained stable at approximately 10^{-8} mbar·L/s for all three designs, confirming the absence of detectable leaks.
- Thermal Durability
  All sealed assemblies withstood multiple thermal cycles and thermal shocks without mechanical failure or degradation in hermeticity.

These test results demonstrate that all three plastic sample tube designs provide reliable hermetic sealing under extreme cryogenic conditions. While the Resdel-based tube (Design #3) exhibited the lowest helium permeability at room temperature, all designs performed equivalently well at cryogenic temperatures, confirming their suitability for use in low-temperature experimental environments.

18.5 RACE-TRACK CONSTRUCTION FOR RESEARCH LABS [11]

18.5.1 Introduction

The hyperfine splitting of atomic hydrogen in a magnetic field is shown in Figure 18.23. In this diagram, two of the spin-up states are classified as *low-field seekers*—quantum states in which atomic hydrogen has lower energy in weaker magnetic fields. While these states are not the absolute ground state of hydrogen, they become energetically favorable in regions of reduced magnetic field strength.

A quadrupole magnet, built from four identical racetrack-shaped coils and featuring a zero magnetic field at its center, provides an effective configuration for trapping low-field-seeking atomic hydrogen. Figures 18.24 and 18.25 presents a scaled schematic of such a magnet used in a magnetostatic trap for spin-polarized atomic hydrogen.

When operated at a current of 30 A, the quadrupole magnet generates a nearly linear radial magnetic field gradient of about 1.0 tesla per centimeter, with zero field at the center. This field configuration is well suited for trapping spin-up atomic hydrogen. The following sections outline the key techniques used in fabricating the racetrack coils.

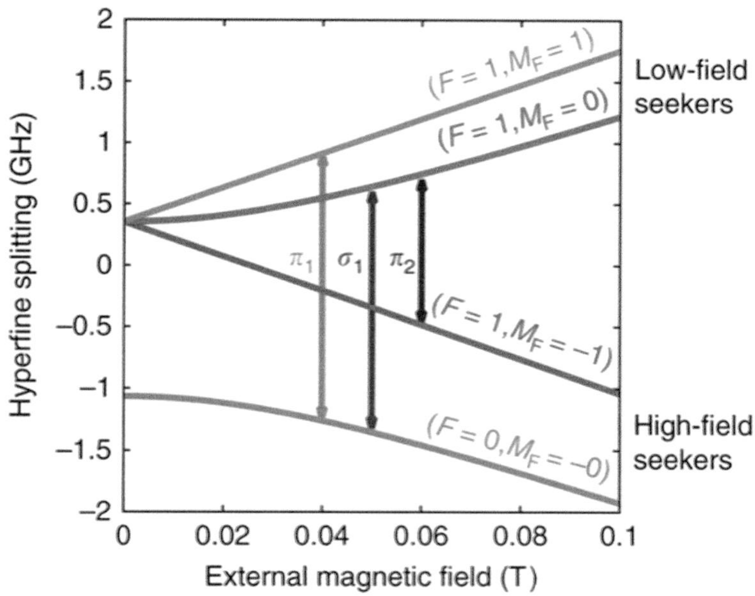

FIGURE 18.23 Quantum states of atomic hydrogen in magnetic field.

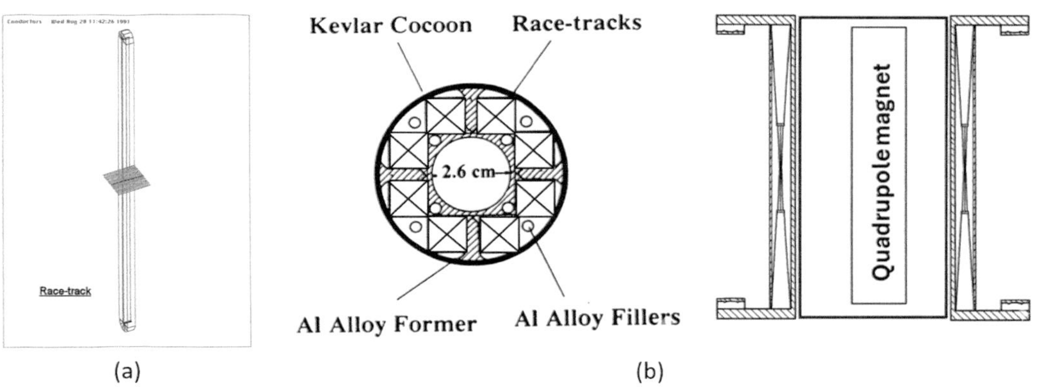

FIGURE 18.24 (a) Race track, (b) Quadrupole magnet. (Reprint with permission from the author for his project.) [11].

18.5.2 Challenges

Winding racetrack coils for quadrupole magnet fabrication is significantly more challenging than winding a solenoid. In a solenoid, the circular geometry generates centripetal tension toward the axis, which naturally holds the wire against the winding form. In contrast, the racetrack coil contains a straight section about 50.2 cm long (Figure 18.24), where the wire must rest on a flat surface. This region lacks the self-securing tension of solenoidal winding, making precise wire placement and mechanical stability more difficult to achieve. As a result, fabricating a stable, highly homogeneous quadrupole magnet is challenging.

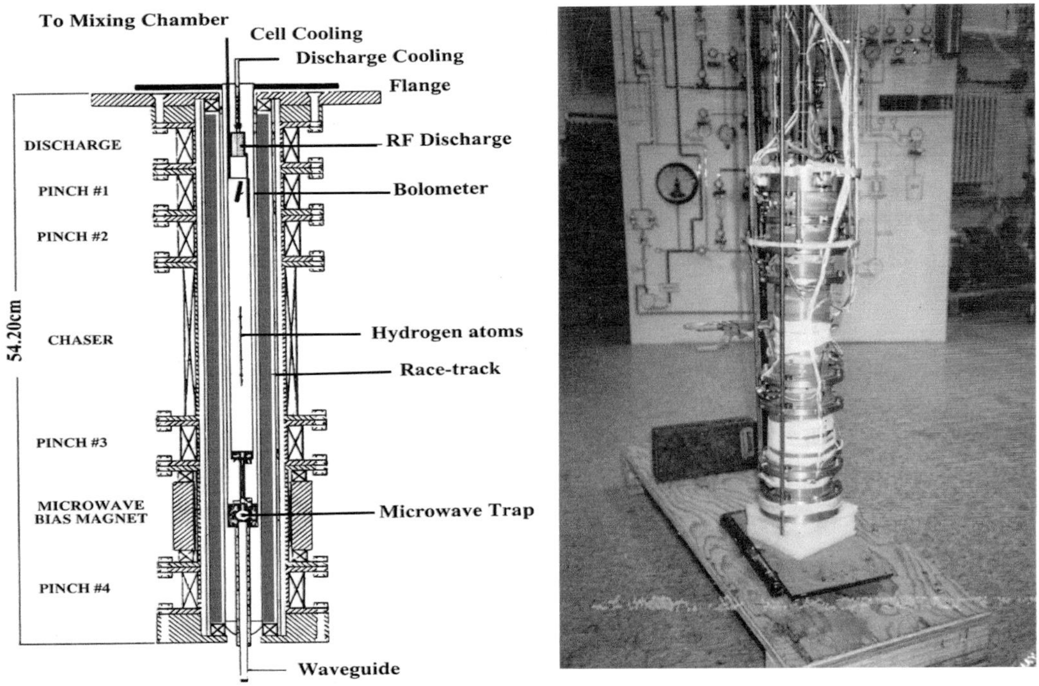

FIGURE 18.25 Magnetostatic trap for atomic hydrogen. (Reprint with permission from the author for his project.) [11].

18.5.3 Solution

18.5.3.1 Material selection and wet winding with epoxy

The racetrack coils were wound with multifilament copper-clad NbTi superconducting wire [12], with the following specifications:

- Copper-to-NbTi ratio: 3:1
- Bare wire diameter: 0.25 mm
- Insulation: Formvar coating

As illustrated in the schematic diagrams in Figure 18.26, the winding direction is perpendicular to the plane of the paper. The coils were stacked in a racetrack configuration, with each layer containing 49 or 50 turns. The gray background in the figures denotes a 1:1 mass ratio mixture of Epon Epoxy 816 and Epon V40. In Figure 18.26(b), no tension is applied in the direction perpendicular to the winding, highlighting the inherent mechanical instability of the straight sections during fabrication. To address this, a *wet-winding* technique was employed in the wires were stacked layer by layer and, after completion of each layer, coated with the epoxy mixture shown Figure 18.26(a) [13]. Once cured, the epoxy bonded the wires firmly together, providing the mechanical stability required for the racetrack winding.

To compensate for the absence of natural tension in the racetrack geometry, the winding was clamped from both sides using Teflon jigs, as illustrated in Figure 18.27. Teflon was chosen because it does not adhere to Epoxy 816, allowing easy removal after semi-curing.

A critical step in this process is the timing of jig removal: the epoxy must be allowed to partially cure—sufficiently firm to bond securely to the previous layer without peeling, yet still soft enough

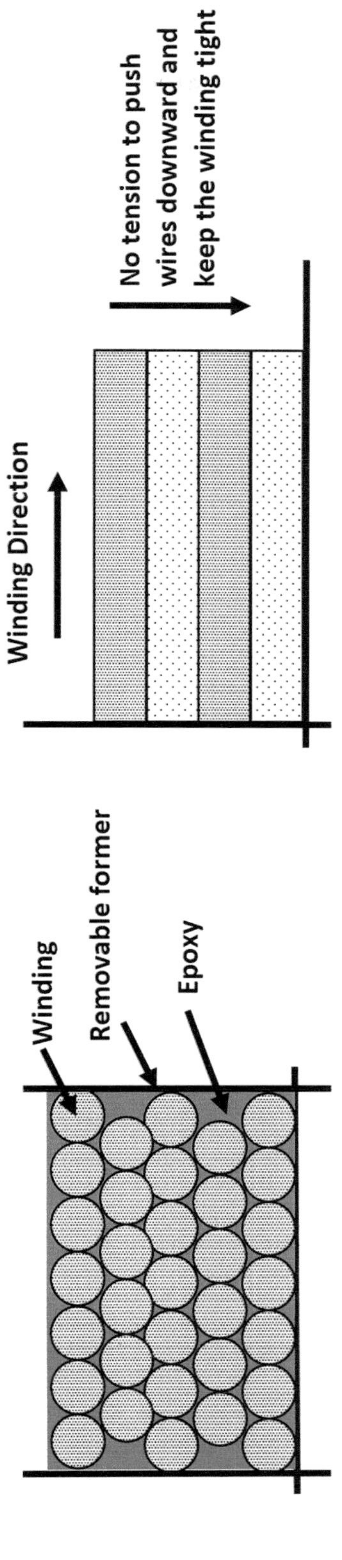

FIGURE 18.26 (a) The winding direction is normal to the paper. Winding layers were stacked in race-track winding; each layer has 49/50 wires in the race-track shown in figure. The gray background is Epon Epoxy 816 and Epon V40. (b) No tension normal to the winding direction.

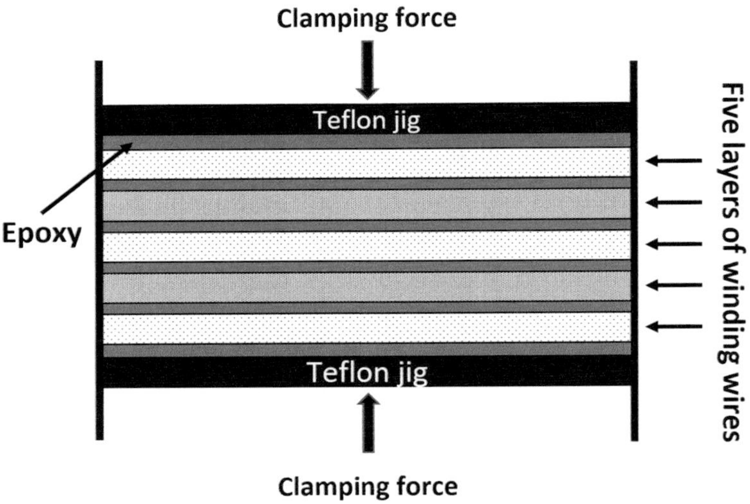

FIGURE 18.27 Schematic diagram of clamp race-track winding.

to permit surface cleaning and provide mechanical support for the next winding. This partially cured state also stabilizes the new layer during the next winding step.

Each racetrack coil consists of 25 winding layers, with clamping applied after every five layers to press each group firmly against the previous one and maintain tight packing before continuing with the next set of windings.

18.5.3.2 Final Assembly

The four identical racetrack coils were mounted onto a quadrupole magnet former machined from 6061 aluminum. Curved aluminum fillers were placed between the former and the coils to ensure proper alignment and mechanical support. The completed quadrupole magnet measured approximately 20 inches in length and featured an 1.0-inch clear bore.

For the final reinforcement, the entire unit was tightly wet-wrapped with 10 layers of Kevlar string, totaling about 8,000 turns [14]. This Kevlar overwrap prevents deformation caused by radial repulsive forces when the magnet is energized.

REFERENCES

1. Northern Engineering Sheffield Ltd (NES) Brooklands Park, Haigh Moor Dr, Dinnington, Sheffield S25 2JY, United Kingdom Email: sales@nes-ips.com Website: www.nes-ips.com/astra-seals
2. ResinLab, LLC (Armstrong Adhesives A-12) N109 W13300 Ellsworth Drive, Germantown, WI 53022, USA Website: www.ellsworth.com
3. A-12 epoxy is a 1:1 two-part resin–hardener system mixed and cured at room temperature or under mild heat. It is used for structural and vacuum-compatible bonding applications.
4. A. Roth, *Vacuum Technology*, Third Edition, North-Holland, (Elsevier) Amsterdam, Netherlands, (1990).
5. E. Oberg, F.D. Jones, H.L. Horton, and H.H. Ryffel, *Machinery's Handbook*, 29th Edition, Industrial Press, (2008).
6. E.T. Swartz, Charcoal-Pumped 3He Cryostats for Storage Dewars, *Review of Scientific Instruments*, 58(5), 881–889, (1987).
7. Kurt J. Lesker Company (public domain) 1925 Route 51, Jefferson Hills, PA 15025, USA Website: www.lesker.com

8. Swagelok Company (public domain) 29500 Solon Road, Solon, OH 44139, USA Website: www.swagelok.com.
9. Ewout ter Haar, Raymond Wagner, Clemens M. C. M. van Woerkens, Stephen C. Steel, Giorgio Frossati, and Ladislav Skrbek, Plastic Dilution Refrigerators, *Journal of Low Temperature Physics*, 99(1/2), 151–166, (1995).
10. Resdel Corporation 174 Route 50, Mays Landing, NJ 08330, USA Website: www.resdel.com
11. Zuyu Zhao, Isaac F. Silvera, and Meritt Reynolds, A Novel Magnetic Trap for Spin-Polarized Hydrogen, *Journal of Low Temperature Physics*, 89(3/4), 703–706, (1992).
12. Supercon, Inc 830 Boston Turnpike, Shrewsbury, MA 01545 Website: www.supercon-wire.com
13. Miller-Stephenson Chemical Company, Inc. 55 Backus Ave, Danbury, CT 06810, USA Website: www.miller-stephenson.com
14. DuPont de Nemours, Inc. 974 Centre Road, Wilmington, DE 19805, USA Website: www.dupont.com

Appendix

A1 IMPEDANCE FABRICATION AND MEASUREMENT

A1.1 Overview

Impedance for continuous flow He-3 cryostat is made of CuNi capillary with the following typical parameters:

- Outer diameter: 0.020”
- Inner diameter: 0.010” ~ 0.012”
- Length: approximately 10 cm, depending on the project requirements
- Impedance value: 10^9 ~ 10^{10} cm^{-3}
- Length reduction option: a non-annealed manganin wire with a diameter of approximately 0.006” ~ 0.008” can be inserted into the CuNi capillary to increase the flow resistance and thereby reduce the required capillary length.

A1.2 Fabrication

Prepare a clean copper tube with the following typical dimensions:

- Outer diameter: 0.062”
- Wall thickness: 0.014”
- Inner diameter: ~0.034”
- Length: depends on the application

Assembly Procedures

1. Insert the CuNi capillary into the copper tube to a minimum depth of 2 cm. *Ensure that the overlap between the CuNi capillary and the copper tube is sufficient to prevent flux from being drawn into the capillary due to capillary action, which could partially block the channel and alter the flow impedance.*
2. Solder the CuNi capillary to the copper tube using a soldering iron and 430 solder, along with Superior 30 flux (blue flux).
3. Clean off the flux residue using acetone, followed by alcohol.
4. Silver-solder the VCR fittings onto both ends of the impedance assembly.

A1.3 Measurement

The impedance can be measured with a BET set-up as illustrated in Figure A1 [1, 2].

a) Impedance calculation worksheet

$$Z = \frac{P_1 - P_2}{\eta \cdot V} \qquad \text{(A1.1)}$$

P: Pressures (torr)
η: He-4 gas viscosity at room temperature (torr·s)

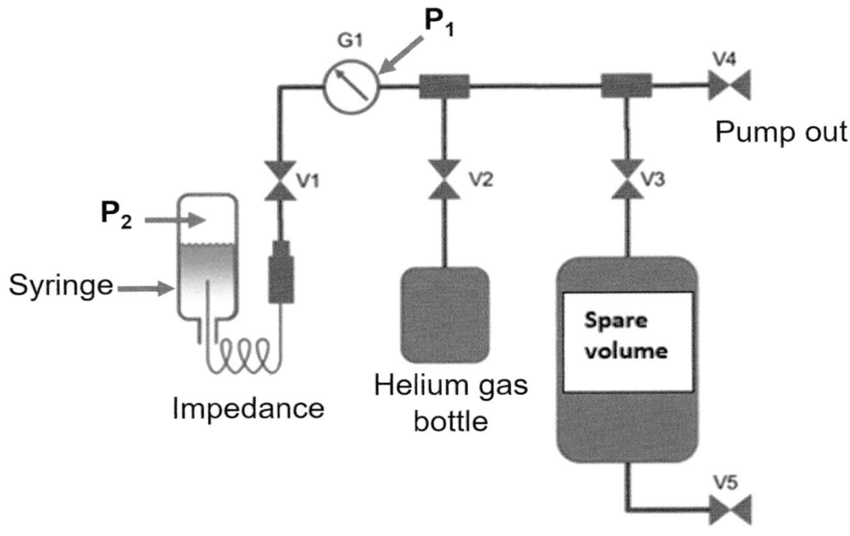

FIGURE A1 Impedance measurement set-up. (Reprint with permission of Taylor & Francis Group, *Cryogenic Engineering and Technologies: Principles and Applications of Cryogen-Free Systems*, Taylor & Francis Group, LLC, (2020), edited by Zuyu Zhao and Chao Wang.)

V: Volume of the He-4 gas collected inside the syringe per unit time (cm^3/s)

b) Measurement Procedures
- Fill a syringe with water.
- Connect the inlet of the impedance device to a helium gas bottle, and regulate the inlet pressure to be higher than atmospheric pressure.
  For example, if the ambient pressure is 763 torr and the inlet pressure is 848 torr, the pressure difference across the capillary is 85 torr.
- Submerge the exhaust end of the impedance device into the water-filled syringe.
- Helium gas from the pressurized bottle flow through the capillary and bubble into the syringe. Measure the volume of gas collected in the syringe over a fixed time interval to calculate the flow rate (e.g., 0.095 cm^3/s).
- The viscosity of helium at room temperature is approximately 1.463×10^{-7} torr·s.
- Using these parameters, the impedance of the device can be calculated using Equation A1.1. For the example above, the calculated impedance is 6.1×10^9 cm^{-3}.

A1.4 Impedance Heating Block

Refer to Chapter 5, Section 5.3.3

A1.5 Examples of Impedance Parameters

Table A1 lists the parameters of several impedance devices used in author's projects.

A2 CMN THERMOMETER MEASUREMENT

A2.1 Working Principle

As illustrated in Figure A2, the CMN device functions as a transformer. It consists of a primary coil and a secondary coil, which wound concentrically around the CMN thermometer. The relationship

TABLE A1
Parameters of Several Impedances Used in Author's Projects

Cryostat Type	O.D. (in)	I.D. (in)	Length (cm)	Impedance (cm^{-3})	Heating Block	Manganin Wire (in)
Wet	0.020	0.008	7.5	7.5×10^9	No	0.006
Wet	0.020	0.008	7.5	1.19×10^{10}	Yes	0.006
Wet	0.020	0.008	14.0	6.5×10^9	Yes	No
Wet	0.020	0.008	13.0	5.65×10^{10}	Yes	0.006
Wet	0.020	0.008	9.2	6.3×10^9	Yes	No
Dry	0.020	0.010	n/a	3.3×10^{10}	Yes	No
Dry	0.020	0.010	n/a	3.2×10^{10}	Yes	0.008

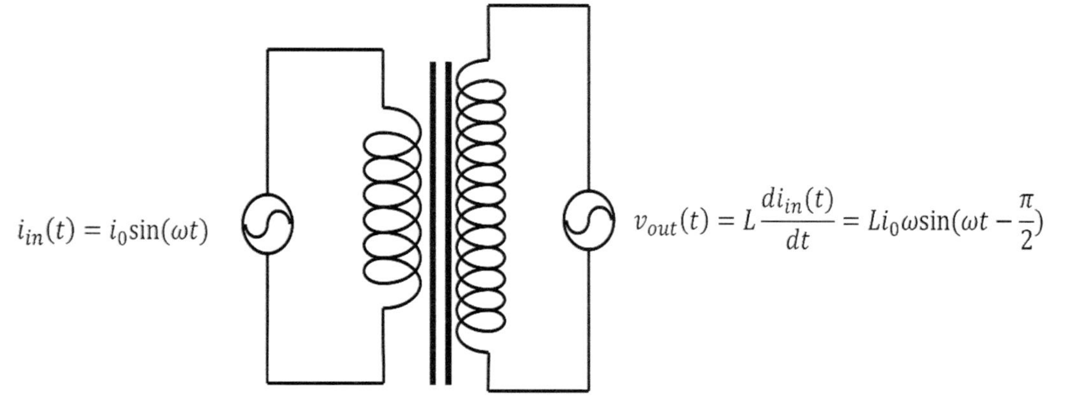

FIGURE A2 Principle of the measurement on CMN.

between its temperature (T) and susceptibility (χ) follows Curie-Weiss Law, and the susceptibility has a linear relationship with the mutual inductance (L_{12}) of the device. Therefore, mutual inductance is a linear function of the temperature and can be expressed as,

$$L_{12} = A + B\chi(T) \tag{A2.1}$$

where *A* and *B* are calibration constants determined for CMN thermometer.

A2.2 Mutual Inductance Measurement

Lakeshore model LS372 resistance bridge [3] is used for the mutual inductance measurement. An AC current is applied to the primary coil, for the inductance measurement, generating a temperature-dependent AC voltage in the secondary coil that is proportional to the mutual inductance L_{12}. By measuring the temperature dependence of the AC voltage, the temperature is determined (refer to Equation 9.2).

It is highly recommended that users considers the following points during the measurement:

1. The LS372 bridge converts the signal into an equivalent resistance proportional to the measured voltage. The temperature dependence of this resistance is then used to determine the temperature.

2. Since the output signal is 90° out of phase with the input signal (see Figure A2), the LS372 reads the imaginary component of the resistance.
3. Due to the arbitrary assignment of coil polarity during fabrication or installation, the measured imaginary resistance may be either positive or negative.
4. The LS372 resistance bridge offers two excitation modes: current mode and voltage mode. However, the voltage mode is simulated, and, internally, the instrument still applies current excitation. Users should consult the LS372 manual for further details.
5. To minimize the self-heating at ultra-low temperatures, the auto-range function for the CMN channels is typically disabled in the measurements. Instead, a fixed low-level excitation is used across the entire temperature range. Typical excitation settings include:
   - Voltage mode: 200 mV across 200 mΩ, or 63.2 μV across 200 mΩ
   - Current mode: 1 mA through 200 mΩ, or 316 μA through 200 mΩ
6. If self-heating is observed with excitation currents of 1 mA or 316 μA, the current should be reduced. However, the highest excitation current that does not introduce detectable self-heating should be used to maximize signal quality.
   Self-heating can be identified by continuously monitoring the CMN channel: a monotonic decrease in the measured resistance over time indicates self-heating.
7. Since the output voltage of the secondary coil is proportional to the excitation angular frequency (ω, as shown in Figure A2), the highest available frequency on the LS372 (18.2 Hz) is used in our measurements to optimize signal quality.

A3 EMISSIVITY OF COMMONLY USED MATERIALS

Table A2 lists the emissivity values of commonly used laboratory materials.

TABLE A2
Emissivity Values of Commonly Used Laboratory Materials, Compiled from Multiple Sources (The Emissivity Data are for Reference Only).

	Temperature-Dependent Emissivity		
Material	**300 K**	**77 K~ 80 K**	**4.2 K**
Aluminum [4]	0.03	0.018	0.011
Aluminum (unpolished) [5]	0.25	0.12	0.07
Aluminum (polished mechanically) [5]	0.20	0.10	0.06
Aluminum (polished Electronically) [5]	0.15	0.08	0.04
Aluminum (polished mechanically) [6]	0.03	0.018	0.011
Aluminum (polished mechanically) [7]	0.04	0.048	0.011 (@2 K)
Aluminum [7]		0.03	
Aluminum (oxidized) [7]	0.31		
Aluminum (as found) [8]	0.12	0.07	0.05
Aluminum (polished mechanically) [8]	0.1	0.04	0.02
Brass [4]	0.035		0.018
Brass (polished) [5]	0.03	0.029	0.018
Brass (polished) [6]	0.03	0.029	0.018 (@2 K)
Brass (oxidized badly) [6]	0.60		
Brass (polished) [7]	0.03	0.029	0.018 (@2 K)
Brass (oxidized) [7]	0.60		
Copper [4]	0.018	0.008	0.0050
Copper (polished) [5]	0.03	0.019	0.015

TABLE A2 (Continued)

Material	Temperature-Dependent Emissivity		
	300 K	**77 K~ 80 K**	**4.2 K**
Copper (polished) [6]	0.03	0.015 ~ 0.018	0.0062 ~ 0.015
Copper (oxidized badly) [6]	0.78		
Copper (polished) [7]	0.02	0.015 ~ 0.019	0.0062 ~ 0.015
Copper (oxidized) [7]	0.6		
Copper (as found) [8]	0.12	0.06	0.04
Copper (polished mechanically) [8]	0.07	0.04	0.02
Copper (gold plated) [8]	0.018	0.015	0.015
Gold [4]	0.02	0.01	
Gold [5]	0.20	0.10	
Gold [6]		0.026	
Gold plate on copper [6]		0.025 ~ 0.027	
Gold [7]		0.026	
Nickel [4]	0.04	0.022	
Nickel [5]	0.045	0.022	
Nickel (polished) [6]	0.04	0.022	
Nickel plate on copper [4]		0.033	0.027 (@20 K)
Nickel plate on steel [6]	0.045		
Silver [4]	0.02	0.008	0.0044
Silver [5]	0.022	0.008	0.005
Silver [6]		0.008	0.0044
Silver [7]	0.02 ~ 0.08	0.008	
Silver plate on copper [4]		0.017	0.013 (@20 K)
Silver plate [6]		0.0083	
Stainless steel [4]	0.08	0.048	
Stainless steel [5]	0.20	0.12	0.10
Stainless steel [6]	0.08	0.048	
Stainless steel [7]	0.074	0.048	
Stainless steel (as found) [8]	0.3	0.09	0.08
Stainless steel (polished mechanically) [8]	0.12	0.048	0.02
Most non-metal [8]	> 0.8		
G-10 [8]	0.8 ~ 0.9		

A4 ENGINEERING DRAWINGS OF MINI-INDIUM SEALS

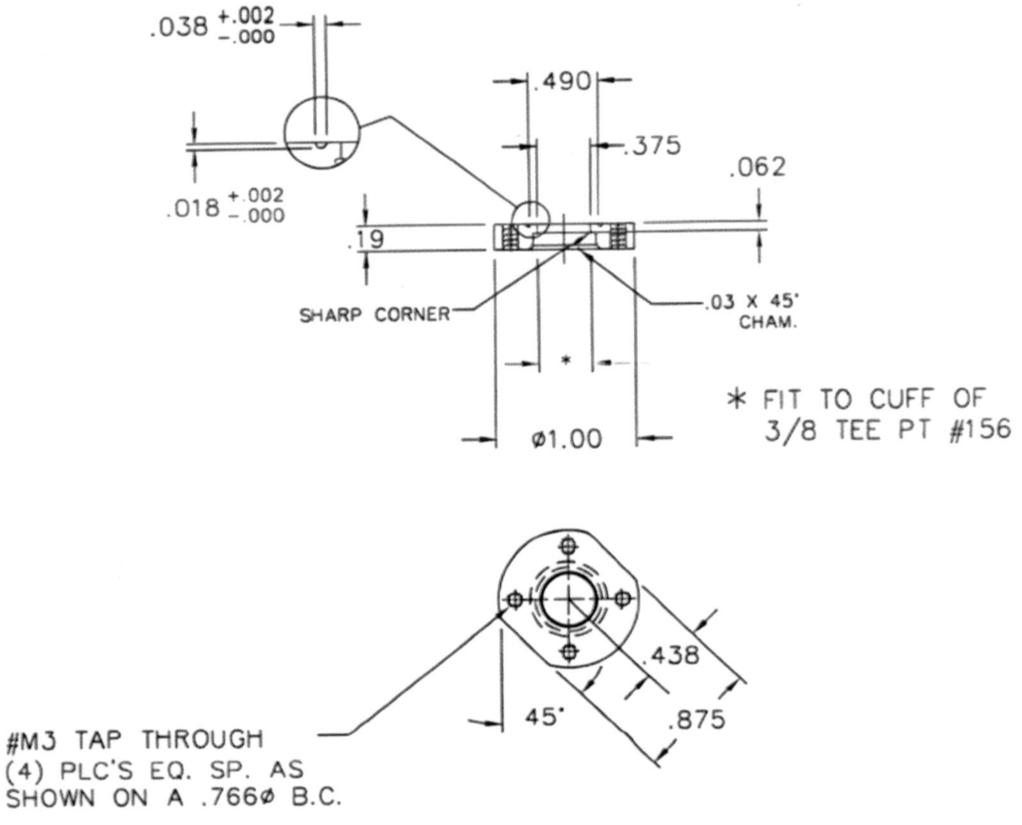

FIGURE A3 Engineering drawing of groove seal design.

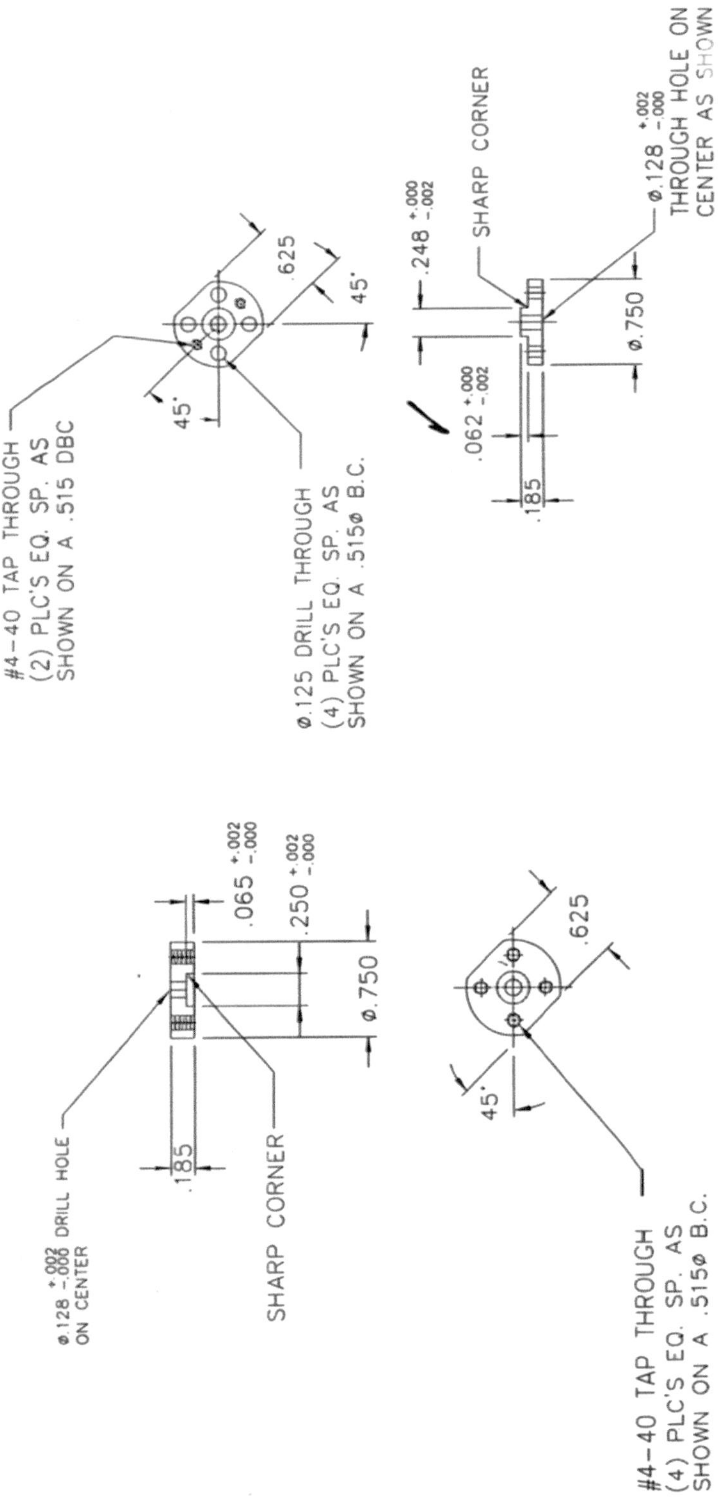

FIGURE A4 Engineering drawing of corner seal design.

REFERENCES

1. Zuyu Zhao and Chao Wang (Eds.), *Cryogenic Engineering and Technologies: Principles and Applications of Cryogen-Free Systems*, CRC Press (Taylor & Francis Group) Boca Raton, FL, USA., (2020).
2. Stephen Brunauer, Paul H. Emmett, and Edward Teller, Adsorption of Gases in Multimolecular Layers, *Journal of the American Chemical Society*, 60, 2309–2319, (1938).
3. Lake Shore Cryotronics, Inc. Model 372 AC resistance bridge operation manual. www.manualslib.com/manual/1888991/Lake-Shore-372.html
4. T. Flynn, *Cryogenic Engineering*, Second Edition, Values are the absorptivities for a radiation from a source at 300K. Normal and hemispherical values are included indiscriminately, Marcel Dekker, New York, NY, USA, (2005).
5. Giorgio Frossati, *Low Temperature Techniques*, Leiden Cryogenics B.V. (not published).
6. S.S. Yan and G. Lu, *Principles and Methods of Low Temperature Physics Experiments*, (China): Science Press, (1985).
7. R. Scott, *Cryogenic Engineering*, Van Nostrand, Princeton, NJ, USA, (1979).
8. These values are based on internal data from Janis Research Company, LLC. and were used by the authors in their projects.

Index

For Product Safety Concerns and Information please contact our EU representative GPSR@taylorandfrancis.com Taylor & Francis Verlag GmbH, Kaufingerstraße 24, 80331 München, Germany